P9-AQB-807

# VLSI SYSTEMS DESIGN FOR DIGITAL SIGNAL PROCESSING

# VLSI SYSTEMS DESIGN FOR DIGITAL SIGNAL PROCESSING

## Volume I: Signal Processing and Signal Processors

**B. A. Bowen**

*Department of Systems and Computer Engineering*
*Carleton University*
*Ottawa*

and

**W. R. Brown**

*Department of National Defense*
*Ottawa*

PRENTICE-HALL, INC., ***Englewood Cliffs, New Jersey 07632***

*Library of Congress Cataloging in Publication Data*

Bowen, B. A.
VLSI systems design for digital signal processing.

Bibliography: v. 1
Includes index.
Contents: v. 1. Signal processing and signal processors.
1. Signal processing—Digital techniques. 2. Integrated circuits—Very large scale integration. I. Brown, W. R. (William Roy) II. Title. III. Title: V.L.S.I. systems design for digital signal processing.
TK7868.D5B68 621.3819'592 81-13854
ISBN 0-13-942706-6 (v. 1) AACR2



**Editorial/production supervision and interior design: Nancy Milnamow**
**Cover design: Dawn Stanley**
**Manufacturing buyer: Gordon Osbourne**

Printed in the United States of America

10 9 8 7 6 5 4 3 2 1

ISBN 0-13-942706-6

Prentice-Hall International, Inc., *London*
Prentice-Hall of Australia Pty. Limited, *Sydney*
Prentice-Hall of Canada, Ltd., *Toronto*
Prentice-Hall of India Private Limited, *New Delhi*
Prentice-Hall of Japan, Inc., *Tokyo*
Prentice-Hall of Southeast Asia Pte. Ltd., *Singapore*
Whitehall Books Limited, *Wellington, New Zealand*

# Global Preface

## PREAMBLE

The rapid advances being made in the field of digital component technology are having profound effects on all aspects of digital systems design. Nowhere are these effects being felt more strongly than in the design of high performance systems for such applications as digital signal processing.

The classical architectural alternatives for realizing the high performance requirements of signal processing systems must be re-examined in the light of the exponential increases occurring in component capabilities. This two volume set brings together a wide variety of logical concepts that impact the design of such systems which acknowledge and take advantage of modern component technology.

Classically, the implementation of digital signal processing techniques has been constrained by the limitations of component technology. System realizations have tended toward one of two general approaches depending on performance requirements. Where high performance real time processing requirements were stringent, the trend was toward inflexible dedicated hardware systems. Where a lower performance was tolerable, the more flexible approach of a software implementation on a general purpose computer was often taken.

Initial attempts to provide both flexibility and high performance resulted in the evolution of supercomputers. These systems tended to be expensive and difficult to program and often achieved their full processing potential for only a limited class of problems. Over the past decade, the use of specialized high-speed array processors attached to a conventional general-purpose host computer has become the standard approach to the implementation of general purpose digital signal processors.

Modern component technology has recently opened a floodgate of high performance components that have dramatically altered the cost/performance

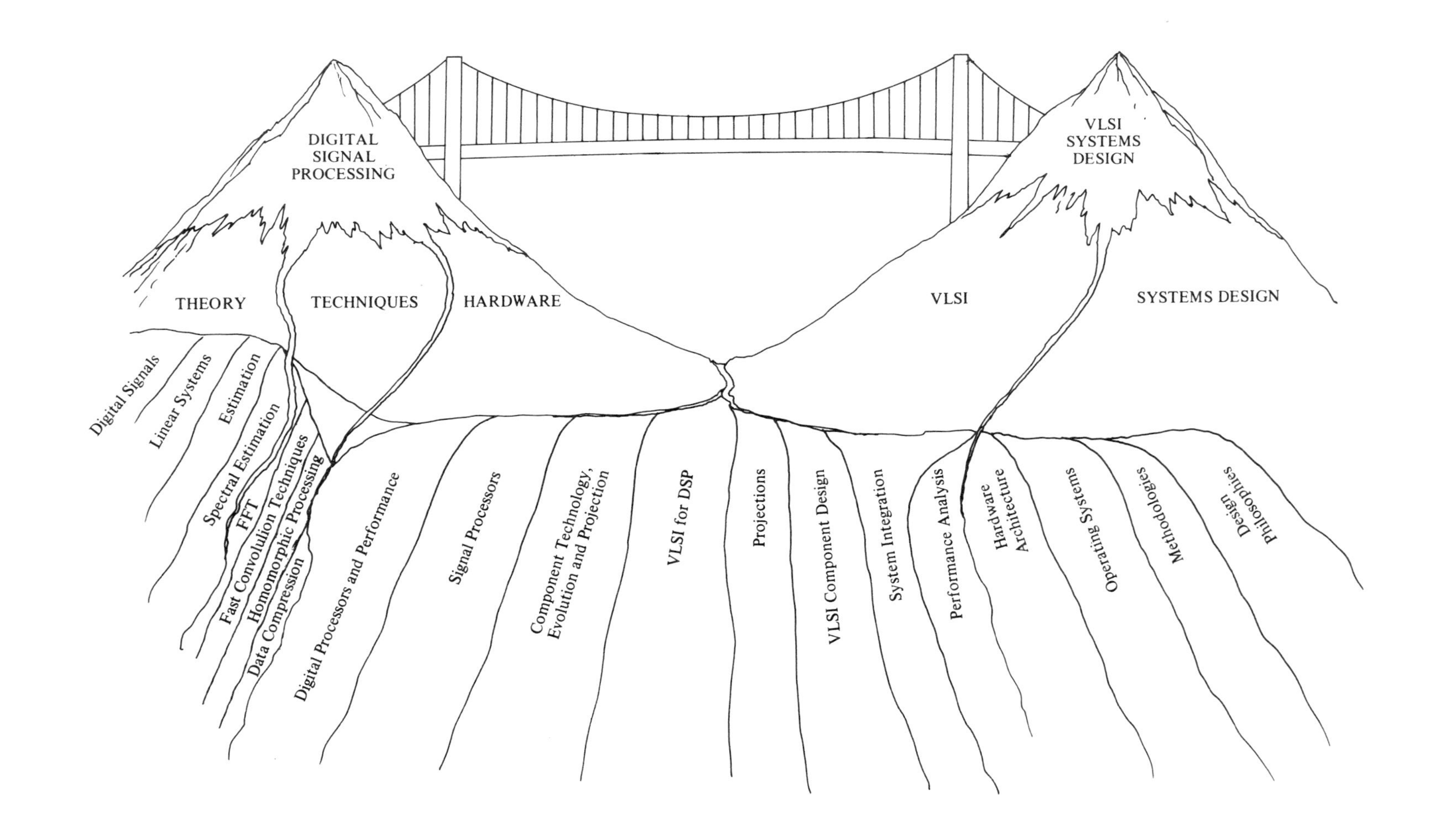
DIGITAL SIGNAL PROCESSING
VLSI SYSTEMS DESIGN
THEORY
TECHNIQUES
HARDWARE
VLSI
SYSTEMS DESIGN
Digital Signals
Linear Systems
Estimation
Spectral Estimation
FFT
Fast Convolution Techniques
Homomorphic Processing
Data Compression
Digital Processors and Performance
Signal Processors
Component Technology, Evolution and Projection
VLSI for DSP
Projections
VLSI Component Design
System Integration
Performance Analysis
Hardware Architecture
Operating Systems
Methodologies
Design Philosophies

criteria for all digital systems. In particular, the feasibility of systems consisting of programmable and configurable chips is exerting its influence on a wide range of new application areas.

In order to exploit the capabilities of these new components for the solution of signal processing problems, it is necessary for those involved in signal processing and for digital systems architects to find a bridge between their respective fields. It is the overall purpose of these three volumes to explore both worlds and establish such a bridge in the form of a systems design philosophy and methodology. For this bridge to support traffic it was found necessary to establish firm foundations on both sides and to create examples of safe passage.

A fundamental concept of the philosophy presented in Volume II is that design should be driven by systems requirements while accommodating implementation constraints. To present a design methodology and illustrate its application to digital signal processor design, it was found necessary to establish firmly the computational requirements and implementation constraints of signal processors. It is to this end that Volume I is dedicated. From a pedagogical point of view, Volume I is designed to be a senior course in digital signal processing with a strong engineering emphasis on the architecture of real processors. The content of the three volumes reflects the two major topics indicated in the title: systems design and signal processing. The material on signal processing is presented first since the first step in any design process must be the identification of the item to be designed. We wish to emphasize, however, that the goal is to unite these two major areas under the influence of VLSI (Very Large Scale Integration) component technology as a single coherent topic. To discuss "VLSI Systems Design for Digital Signal Processing" as a single topic, we must define that topic and what it encompasses.

There appear to be two possible interpretations of the title:

1. the design of VLSI components (themselves complex systems) intended for use in digital signal processing applications,
2. the design of digital signal processing systems that utilize VLSI components.

It is the second interpretation that we address in these volumes.

The subject matter encompassed tends to become almost cosmological in nature, seeming to immerse the reader (and at times, the authors) in an endless interlocked array of concepts which appear to expand without limit. Indeed, during the course of structuring the material there were times when it appeared that organizing a course of instruction to teach a centipede to walk would be an easier task. However, a concerted effort has been made to weave the required concepts together to form a coherent pattern.

In general, Volume I concerns digital signal processing and the basic concepts of digital signal processor architectures, while Volume II addresses the concepts of systems design and presents the bridge between the two fields in light of VLSI technology. It is in the areas of processor architectures and component technology that the underlying concepts of both signal processing and systems design merge.

It is recommended that the reader examine the introductions to Volumes I and II in sequence. In this way, an overall picture of the scope of the material to be presented, its organization and interrelation should emerge. This overall picture will be enhanced by a further reading of the introductions and summaries to each chapter prior to attempting the main body of the text. The bibliographies have been organized to suggest two levels of additional reading: first, tutorial expositions, usually in text books; second, more advanced or, perhaps, specialized reading, most often in papers. By these means, a complete personal road map can be constructed.

All this material is intended for use by electrical engineers at the senior undergraduate or introductory graduate level. While the material in Volume I is concerned specifically with digital signal processing, the design philosophies and methodology of Volume II are applicable to the design of digital systems in general. Indeed, the central theme of Volume II may be considered as systems design with digital signal processing representing a generalized case study framework within which the concepts of systems design are illustrated.

The choice of digital signal processing as the subject of the case study seems justifiable (aside from being of interest to the authors). Current trends and projections regarding digital component technology indicate that the capability to fabricate high performance digital signal processors at a reasonable cost is increasing dramatically. With increased performance and reduced cost the demand for signal processing systems can also be expected to increase. Thus, engineers entering the field of digital systems design need a broad familiarity with the basic concepts of signal processor design.

## ACKNOWLEDGEMENTS

It is not unusual for professors to be supported by their students. This is never more true than in the preparation of books. Many students received early courses of instruction from incomplete manuscripts (and perhaps thought). Their reactions and comments formed powerful stimulus for modification and enhanced clarity. We hope we learned enough. John Bird read an early version of Chapters 1-4 of Volume I and his detailed knowledge of the mathematics proved invaluable.

Several of our colleagues read earlier versions of Volumes I and II. Their comments are particularly appreciated. Professors George and Coll made a host of suggestions which strongly influenced our choice of order and presentation. Professor Coll also provided valuable insight into the pragmatics of adaptive processing from his long experience in this field.

Grateful acknowledgement must also go to Karl Karlstrom of Prentice-Hall. As other P-H authors will agree, Karl provides that hearty laugh and subtle encouragement to press over the rough parts of a project. And of course, a phone call always meant a chance to talk to Rhoda Haas, and that added a further sparkle.

Finally, we are grateful to Elaine Carlyle who, with her innumerable skills and infinite patience, forms a super system in her own right, and without whom the pragmatics of preparing a manuscript could easily have become a serious matter.

# Introduction to Volume I

In order to design digital signal processors that take advantage of advanced component technology, we must first understand the computational requirements and, as well, have an appreciation for the constraints imposed by technology. It is the purpose of this volume to present an overview of the important elements of background theory, processing techniques, and hardware evolution in a manner that will:

provide for the reader who is unfamiliar with digital signal processing a concise presentation of the major concepts;
lead to the identification of the basic computational requirements associated with digital signal processing theory and techniques;
set the stage for the systems design discussions of Volume II by establishing an appropriate perspective on how the dynamic nature of component technology effects the implementation of signal processors.

The material of this volume is organized basically in two parts. The first five chapters form Part A which reviews the underlying theory and basic techniques of digital signal processing. Chapter 5 provides the bridge from processing to processors. Chapters 6 and 7 make up Part B and concern the architecture and performance limitations of processors and hardware evolution.

It is not our intention to present a comprehensive treatment of the theoretical aspects of digital signal processing in Part A. The aim here is to place in perspective the wide range of theoretical material that forms the foundation of digital signal processing.

While a great deal of mathematical terminology and symbolism is presented in Part A (often in a somewhat encyclopaedic terseness), the major goal is to give the reader an overall grasp of signal processing theory, the link between this theory and some common processing techniques and, most importantly, the implied computational requirements for the implementation of these techniques in terms of digital hardware and/or software systems.

Chapter 5 provides a focal point for the material of Part A and establishes the context for moving from processing theory to processors. Since the move from theory to implementation must always take place within some context, Chapter 5 examines the general characteristics of a range of digital signal processing application areas with a focus on their influences on processing systems. The fundamental computational operations are summarized within an overall perspective of the interaction of theory, applications, and implementation issues.

Having established our foundations in digital signal processing theory, we proceed in Part B to establish the initial foundations of digital systems architecture. As in Part A, our approach here is to review the underlying concepts and to identify those major issues which influence the design of signal processors. We begin with an examination of the basic von Neumann machine architecture which embodies most of the fundamental concepts of digital computing architecture. The performance limitations and principal approaches to performance enhancement are explained through an examination of two early super computers. We then examine a selection of digital signal processors which illustrates the evolution of architectural concepts. The goal here is to expose answers to such questions as:

What implementation problems and performance limitations were faced?
How were these problems overcome or circumvented?
What problems have not been overcome?
Has anything changed that might allow unsolved problems of the past to be overcome now?
How does the specific system application affect the implementation, or vice versa?

In considering such questions, we note that the evolution of component technology has had a profound impact on signal processor performance (perhaps even more so than the implementation of classic architectural concepts for performance enhancement). Thus, Part B (and Volume I) concludes with an overview of integrated circuit technology evolution, some projections for future advances in VLSI components, and their impact on signal processors.

B. A. Bowen
W. R. Brown

# Contents

# Contents

## THE BRIDGE



## Part B: SIGNAL PROCESSORS

# VLSI SYSTEMS DESIGN
# FOR
# DIGITAL SIGNAL PROCESSING

PART A

# SIGNAL PROCESSING

## INTRODUCTION

The topic of digital signal processing (DSP) may be approached in one of two ways. The first follows the historical development of the theory. This approach considers signals and signal processing in the analog domain and then maps the results to the digital domain for implementation. The second approach is to consider signal processing in terms of operations on the numbers which characterize a signal. In this approach the first concern is with the numerical representation of digital signals. The concepts of processing can then be developed in a purely digital context. This second approach leads to a broader range of issues and exposes implementation mechanisms which are more global than those available in the analog world. Indeed, there seems to be a major conceptual disadvantage to basing our concepts in the analog world; digital processing deals with numbers and operations on them, so practitioners must build a familiarity with such representations. It is therefore the second approach that we have chosen for our overview of digital signal processing theory.

Several excellent texts are currently available on the general theory of digital signal processing, the majority of which were published since the mid-1970s (see annotated bibliography). The general structure of the presentation in virtually all of these texts follows the same basic pattern.

- A discussion of deterministic discrete-time signals, and discrete-time linear systems based on $Z$-transform techniques;
- A discussion of finite impulse response (FIR) and infinite impulse response (IIR) digital filters focusing on design techniques and quantization effects;
- A discussion of Hilbert transforms;

- A discussion of discrete Fourier transform (DFT) coefficient computation focused on the fast Fourier transform (FFT) algorithms followed by applications in the computation of correlation, convolutions and power spectra and the chirp-$z$ transform algorithm;
- A discussion of generalized linear filtering (homomorphic processing);
- A discussion on digital hardware and processor implementation examples, either special purpose, general purpose or both;
- A discussion of one or two application areas.

This presentation pattern is not exact for any of the texts but represents a composite view. Gold and Rader [3], who published the first comprehensive treatment of digital signal processing theory, seem to have set this presentation pattern with the exception of the last two points. The texts by Oppenheim and Schaffer [6] and by Rabiner and Gold [10], both published in 1975, are probably the most well known general treatments of digital signal processing theory. These texts both follow the same general structure with Rabiner and Gold including chapters on two-dimensional processing, speech, and radar applications but omitting generalized linearity. Oppenheim and Schaffer treat generalized linearity in detail and also include chapters on random signal modeling and introduce some of the problems of estimation theory associated with power spectrum estimation. Peled and Liu [9] give a brief introduction to the basic concepts of the theory, and concentrate on practical implementations and example processors. Tretter [16], on the other hand, tends toward the theoretical approach, specifically excluding the last three points of the general pattern. Tretter, however, emphasizes the time domain approach to linear systems through state space representations and the importance of estimation theory in signal processing, including chapters on linear parameter estimation and recursive estimation techniques.

These books form a cornerstone of DSP which have provided the basis for the wide dissemination of both the theory and the practice of the discipline. It is difficult to imagine anyone in the field who has not devoted many hours of study to these books (including the present authors).

No single book can carry every aspect of this field to its ultimate conclusion. It would be naive to think of or propose such a proposition. Further reading is necessary and a guided tour is presented in the bibliography. Rabiner and Gold suggested in their introduction that their book should follow Oppenhiem and Schaffer. Perhaps this book fits before both. It is an attempt to bring this important field to the undergraduate classroom.

Our presentation differs from the standard format chiefly in terms of emphasis. It is an overview that is aimed at identifying the computational requirements of digital signal processing so that we can get on with the task of systems design. Several areas—such as filter design techniques—are mentioned only briefly. While these topics are certainly of importance, a more detailed discussion contributes little to determining computational requirements. On the other hand, we do spend some time on the concepts in linear estimation theory since these concepts form the basis for the formulation of a wide range of signal processing problems in the time domain including the increasingly popular topic of adaptive signal processing.

Chapter 1 presents the fundamental concepts of digital signals. This presentation consists of three major sections. The first discusses the basic concepts and relationships of the digital representation of a signal to the real world in which signals appear as inherently analog entities. The second section deals with transform analysis of deterministic discrete-time signals. The third section develops a statistical model for random signals.

Chapter 2 reviews the elementary concepts of linear systems, followed by the introduction of infinite impulse (IIR) and finite impulse response (FIR) digital filters.

In Chapter 3 we present an introduction to detection and estimation theory. This material is presented in four sections. The first gives an overview of the detection problem. The second dicusses some of the basic concepts of estimation theory. In the third section we consider the general problem of linear mean-square estimation as the basis for time domain filter design. In the fourth section we discuss the estimation of power spectral densities in particular.

Chapter 4 deals specifically with digital signal processing techniques. We begin with a discussion of the fast Fourier transform (FFT) and its application to various computations such as correlation and convolution. The second section deals with the basic concepts of generalized linear filtering. In the third section we discuss the basic mechanisms for data compression. The fourth section introduces adaptive signal processing.

Chapter 5 forms a bridge from the theory of signal processing to the pragmatics of signal processors. In this chapter an examination of a variety of applications leads to the formulation of a set of basic computational operations which form the basis of all processing algorithms. These functions coalesce the theory and provide a broad view which transcends individual applications. In Part B a variety of processors will be examined and in each case the ability to execute algorithms based on sequences of these basic operations forms a methodology for both their evaluation and comparison.

CHAPTER 1

# Digital Signals

## 1.0 INTRODUCTION

This chapter contains three major sections which deal first with digital signal acquisition and representation, then in turn with the modeling and analysis of deterministic and of random signals.

Digital signals do not occur naturally in the same sense that analog signals do. They must be created for the purposes of storage and manipulation. Our first concern is thus with the conversion of analog signals to a numerical representation, and the re-creation of analog signals from these numbers. In Section 1.1 signal acquisition, representation and reconstruction are explored.

In the world of continuous signals a useful mathematical model based on Laplace and Fourier transforms has emerged and is well known. This model is so powerful both as a description and as a mechanism for the design of linear systems that it is common to forget that it is only a model. In the digital world an equally useful mathematical structure exists based on an extension of Laplace and Fourier transform concepts. In Section 1.2 the formalism for representing and manipulating deterministic digital signals by $Z$-transform techniques is summarized.

The application of the $Z$-transform to the analysis of digital signals is limited by the necessity for obtaining a deterministic mathematical representation of the signal. Since many signals of interest cannot easily be represented in a deterministic manner, it is necessary to have a model for nondeterministic signals. In Section 1.3 the concepts of statistical signal modeling are introduced. This model has been developed and refined over the last several decades and in retrospect is both appealing and useful.

The presentation is self contained; however, it rests on a background of Laplace and Fourier transform theory and assumes some background in basic probability and statistics. No attempt is made to provide proofs in a formal sense since this is not our goal; many excellent texts exist which fulfill this function and suitable references are suggested in the bibliography.

## 1.1 BASIC CONCEPTS

### 1.1.0 Introduction

Signal Processing using a digital computer implies that analog signals must be adequately represented in a digital format (i.e., as sequences of numbers). To obtain these numbers, it is necessary first to convert analog signals into a suitable digital format. The concept of an analog signal is somewhat intuitive; any physical phenomenon which varies in time, or in any other appropriate independent variable, could be thought of as a signal. For our purposes a signal is usually the output of a transducer which converts the physical phenomenon into a voltage (or current), and the independent variable is most often time.

We define digital and analog signals as follows:

Digital Signal: A digital signal is an ordered sequence of numbers, each of which is represented by a finite sequence of bits (i.e., a finite length digital word). A digital signal is therefore defined only at discrete points in time and takes on only one of a finite set of discrete values at each of these points.

Analog Signal: The adjective "analog" appears to have originated in the field of analog computing where a voltage (or current) is used as an analog of a physical variable in a differential equation. In a broad sense an analog signal is any signal that is not a digital signal. There are three subclasses:

i) continuous in both time and amplitude;
ii) continuous in time but taking on only discrete amplitude values (e.g., the type of analog signal used to approximate a digital bit stream in serial communication);
iii) continuous in amplitude but defined only at discrete points in time (e.g., sampled analog signals such as in charge transfer devices, also called sampled data signals).

A digital signal can represent samples of an analog signal at discrete points in time. The fidelity of this representation will be discussed later; first we shall discuss the mechanisms for obtaining the samples. There are two distinct operations involved in the transition from an analog to a digital signal:

1. Sampling: The transition from continuous-time to discrete-time.
2. Sample Quantization: The transition from a continuous to a discrete amplitude representation.

After transforming an analog signal to a digital representation and performing the necessary processing it is often necessary to transform the digital result back into an analog form. In some cases a signal may be initially generated in a digital processor and be required to be ouput in analog form. The transition from digital to analog involves a similar two-step procedure as analog to digital conversion only in the reverse sense.

In this section we shall review the conceptual basis for these conversion processes and for the representation of signals in digital format.

We consider first the sampling of an analog signal which generates a discrete time signal. In Section 1.1.2 we explore the conditions on the analog signal and the sampling interval which maintains a valid representation. In Section 1.1.3 the quantization of the amplitude sample is considered. This is an appropriate section to discuss the whole problem of analog to digital converters. Finally in Section 1.1.4 the various techniques for representing the sample numbers in a processor are discussed.

### 1.1.1 Continuous-time and Discrete-time Signals

Consider the sinusoidal analog signal $v(t)$ represented by Figure 1.1(a). A set of samples of the amplitude of this signal is illustrated in Figure 1.1(b). The sampled values give the exact value of $v(t)$ only at the sample times and thus represent a discrete-time version of $v(t)$. If the duration of the time interval between samples is constant this approximation can be expressed as $v(nT)$, where $T$ is a constant called the sample period and $n$ is an integer index. This is referred to as "uniform sampling," and is widely used in practice due to the ease of implementation and modeling.

Intuitively the sample period $T$ must have some effect on how accurately the original signal is represented by the sampled approximation. As an example of the ambiguities that can arise, the signal $v'(t)$ represented in Figure 1.1(c) has exactly the same samples as the signal $v(t)$ from part (a) when both signals are sampled at the same rate. Thus, the discrete-time signal $v(nT)$ does not necessarily define the original continuous-time signal from which it was derived. Conditions under which the representation is valid and in what sense the samples represent the original are of importance and will be discussed in the next section.

The mechanisms for obtaining a sample of an analog signal can be modeled in two ways. Ideally an instantaneous value of the signal is obtained at the sample time. This corresponds to a delta function as shown in Figure 1.2(a). Such a model is attractive for analysis but in practice the delta function is impossible to generate. A more realistic model is to consider the modulation of the original analog signal by a uniform train of pulses of width $P$ spaced at intervals of the sample period $T$, as illustrated in Figure 1.2(b). Such a sampling signal and its implications will be discussed in the next section. The result of either model is to transform a continuous-time signal into

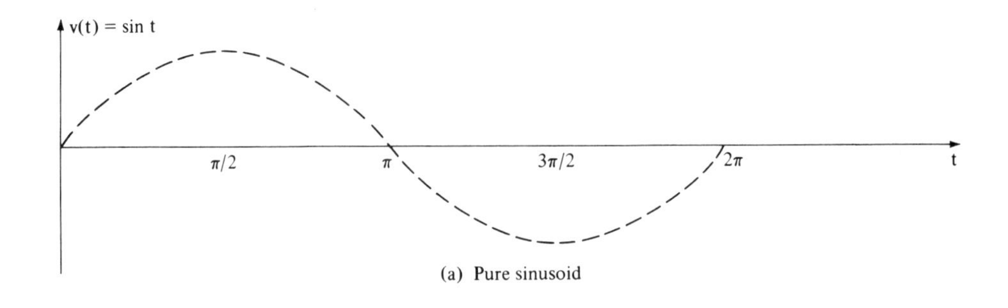


(a) Pure sinusoid

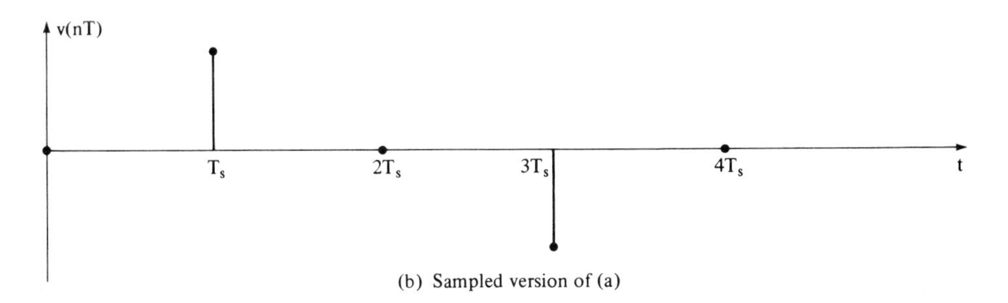


(b) Sampled version of (a)

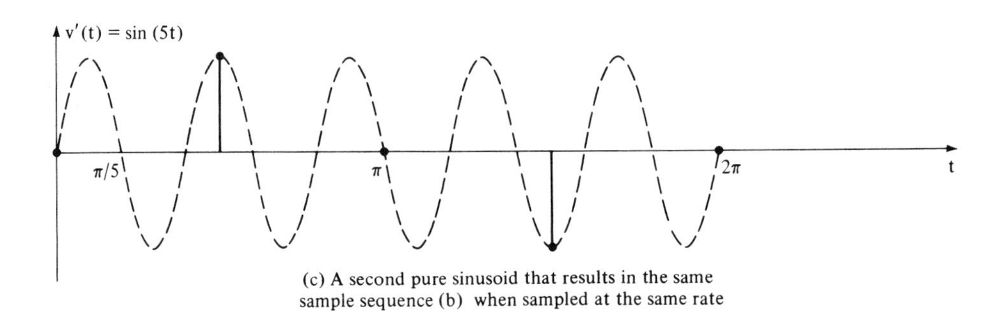


(c) A second pure sinusoid that results in the same sample sequence (b) when sampled at the same rate

**Figure 1.1**: Sampling of pure sine waves.

a discrete-time signal, which is the first step of the conversion process. The requirements on the sample period $T$ for a valid representation will be discussed next.

### 1.1.2 The Sampling Theorem

The restrictions on the period $T$ of the pulse train $p(t)$ such that the ambiguity in the sampled version of a signal can be avoided are given by the Nyquist Sampling Theorem which states:

If the highest frequency component in a band-limited signal is $\omega_c$ radians per second, then for a sample period $T$ such that

$$2\pi/T \geqslant 2\omega_c$$

the original signal can be completely recovered from the sampled version by ideal lowpass filtering.

This theorem and its implications are of central importance in digital signal processing.

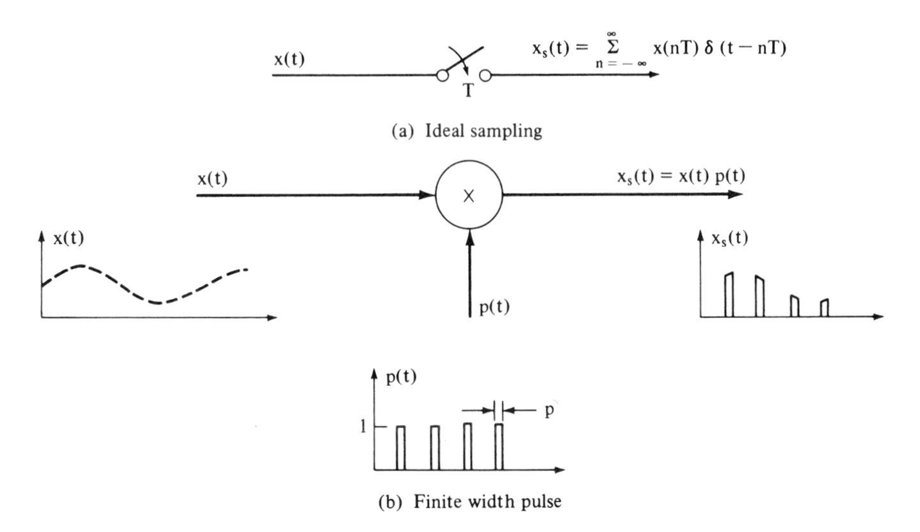


**Figure 1.2**: Sampling modeled as a modulation process.

To illustrate the implications, consider an analog signal $x(t)$ which is band-limited at $\omega_c$ as shown in Figure 1.3(a). Its Fourier transform is given by:

$$\mathcal{F}\,\{x(t)\} = X(\omega) = \int_{-\infty}^{\infty} x(t)e^{-j\omega t}dt \tag{1.1}$$

The spectrum of this signal is shown in Figure 1.3(a). When $x(t)$ is sampled with sample period $T$ the resulting sampled signal is

$$x_s(t) = x(t)p(t) \tag{1.2}$$

The sampling function $p(t)$ is periodic with period $T$ and can thus be expressed as a Fourier series i.e.,

$$p(t) = \sum_{n=-\infty}^{\infty} c_n e^{j\omega_s nt} \tag{1.3}$$

where $\omega_s = 2\pi/T$ is the sampling rate in radians per second and $c_n$ are the complex Fourier coefficients given by

$$c_n = 1/T \int_{-T/2}^{T/2} p(t) e^{-j\omega nt} dt = \frac{P}{T} \frac{\sin(n\omega_s P/2)}{n\omega_s P/2} \tag{1.4}$$

where $P$ is the pulse width as shown in Figure 1.2(b).

Therefore $x_s(t)$ as expressed in equation (1.2) can be written as:

$$x_s(t) = x(t)p(t) = \sum_{n=-\infty}^{\infty} c_n x(t) y e^{j\omega_s nt} \tag{1.5}$$

which has the Fourier transform

$$\mathcal{F}\{x_s(t)\} = \int_{-\infty}^{\infty} \left\{ \sum_{n=-\infty}^{\infty} c_n x(t) e^{j\omega_s nt} \right\} e^{-j\omega t} dt$$

$$= \sum_{n=-\infty}^{\infty} c_n X(\omega - n\omega_s) \tag{1.6}$$

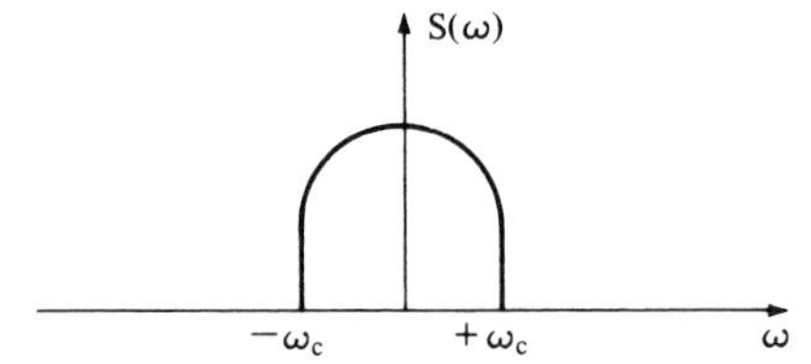


(a) Spectrum of original baseband signal

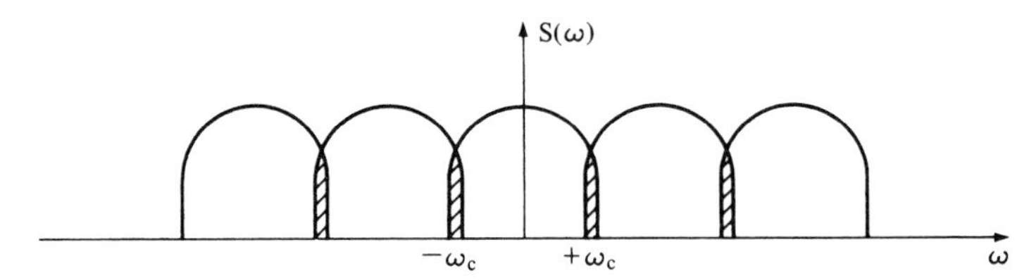


(b) Aliasing in the spectrum of the sampled signal when $\omega_s < 2\omega_c$

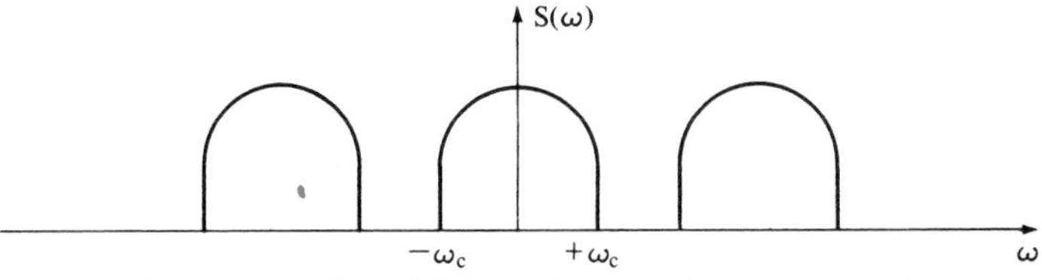


(c) Spectrum of sampled signal when $\omega_s > 2\omega_c$; no overlap of added spectral components

**Figure 1.3**: Sampled spectrums.

For ideal sampling, as shown in Figure 1.2(a), the pulse train $p(t)$ is a sequence of impulses or Dirac delta functions: thus, $c_n = 1/T$ and

$$\mathcal{F}\{x_s(t)\} = \frac{1}{T}\sum_{n=-\infty}^{\infty} X(\omega - n\omega_s) \tag{1.7}$$

In the case of ideal sampling the sampled signal $x_s(t)$ is a periodic sequence of impulses whose areas represent the individual sample values which correspond to the values of the signal $x(t)$ at the points in time $t = nT$. In practice ideal sampling can only be approximated and the sample values obtained for $x(nT)$ are averaged over the pulse duration $P$ as shown in equation (1.4).

Figures 1.3(b) and 1.3(c) illustrate the spectrum of the ideally sampled signal $x_s(t)$ for the cases of $\omega_s < 2\omega_c$ and $\omega_s > 2\omega_c$, respectively. It can be seen that the sampling operation adds new spectral components to the sampled signal that are translated replicas of the original spectrum. When the sample rate is less than twice the highest frequency contained in the original signal these added spectral components overlap. This overlapping is known as "aliasing". When aliasing occurs due to sampling at a rate less than $2\omega_c$, known as the "Nyquist rate," the structure of the baseband spectrum is altered and the original signal cannot be accurately recovered. If, however, the sample rate is equal to or greater than the Nyquist rate, there is no overlap and the original signal can be completely recovered from the sampled version by lowpass filtering.

The Nyquist rate of sampling forms the key which opens the door to the digital processing of analog signals. Note, however, the critical assumption that the original signal is band limited. This condition is seldom met exactly in practice and aliasing is always present to some extent. The function of "anti-aliasing" filters preceding the conversion process is to reduce a signal spectrum to an acceptable band limited approximation. A sampling rate higher than the Nyquist rate will also increase the separation of the spectra and therefore reduce the aliasing.

At this point the analog signal has been represented at discrete points properly spaced in time. In the next section, the conversion to a digital number will be discussed.

### 1.1.3 Analog-to-Digital Conversion (ADC)

As previously noted, the transition from analog to digital signal representation involves two steps: sampling and sample quantization. Since the ideally sampled signal $x_s(t)$ is exactly zero except at time $t = nT$, this signal can be represented as a sequence of numbers $x(n)$ representing sample values. In some sampled data systems, such as those involving electro-mechanical elements, or more recently in charge transfer devices, the representation of signal values may be continuous even though the signal is time sampled. However, in purely digital systems these sample values must be represented in some binary number code that can be stored in finite-length registers.

A binary sequence of length $n$ can represent at most $2^n$ different numbers. Thus if a signal is constrained in amplitude to be between $-v$ and $v$ volts, this interval can be divided at most into $2^n$ discrete numbers. Each number represents a range of signal values in the interval of size $E_0 = 2v/2^n$; thus the "quantization step size" or aperature inherently introduces an amplitude approximation which limits the potential accuracy of subsequent processing.

In Figure 1.5(a) the stair case transfer function illustrates the departure from the ideal caused by the finite bit length representation. A change in the least significant bit causes a step from one level to the next. Signal values between these steps must be approximated (and thereby we introduce the error).

There are three common types of sample quantizations that are used in practice:

1. Rounding: In rounding the quantization level nearest the actual sample value is used to approximate the sample. Figure 1.4(a) illustrates this type of quantization.
2. Truncation: In truncation the sample value is approximated by the highest quantization level that is not greater than the actual sample value. This type of quantization is equivalent to rounding the sample value less one half a quantization level. It is illustrated in Figure 1.4(b).
3. Signed Magnitude Truncation: This is a variation of truncation that is identical to truncation for positive sample values but for negative values the sample is approximated by the nearest quantization level that is not less than the actual sample value.

The digitized signal $x(nT)$ can be expressed as

$$x(nT) = x_0(nT) + e(nT),$$

where $x_0(nT)$ can be thought of as the actual signal value at $t = nT$, and $e(nT)$ is an added error signal called "quantization noise." The error signal $e(nT)$ is usually modeled as a uniformly distributed random sequence with the exact nature of the distribution dependent on the type of quantization involved. It is characterized by its variance which can be shown to be the same for both truncation and rounding, i.e.,

$$\sigma_e^2 = E_0^2/12 \tag{1.8}$$

where $E_0$ represents the quantization step size.* The statistical analysis of

*See Section 1.3 for a discussion of statistical averages and random sequence models.

quantization errors in analog-to-digital conversion is covered in considerable detail in several texts, a few of which are listed in the bibliography.

The "signal to noise ratio" introduced by quantization is usually defined in decibels as the ratio of peak-to-peak signal to rms noise, i.e.,

$$S/N = 20 \log(2^n\sqrt{12})\text{dB}$$
$$= 6.02n + 10$$

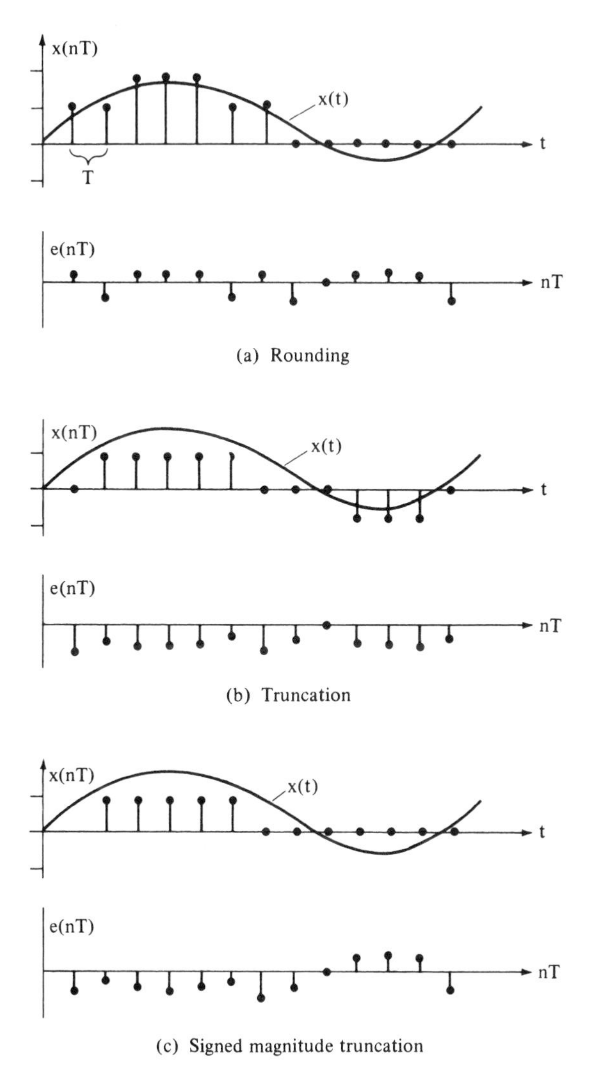


**Figure 1.4:** Quantization methods: in each case, $x(nT) = X_0(nT) + e(nT)$, where $x_{0(nT)} = x(t)|_{t=nT}$.

The $S/N$ ratio increases by 6 dB for each added bit in the converter.

The dynamic range of a converter is defined as the ratio of full scale voltage to quantization step size, i.e.,

$$\text{Dynamic Range} = 20 \log 2^n = 6.02n \text{ dB}$$

The required dynamic range of an ADC is an important attribute of a physical problem if the amplitude information of a signal is to be retained for processing.

It is obvious from the last three equations that the accuracy with which the sample is approximated is directly dependent on the quantization step size. Thus, it would appear that an arbitrary degree of accuracy could be obtained simply by decreasing quantization step size (i.e., increasing n). However, it must be remembered that the analog signal is ultimately limited in accuracy by the thermal noise of the analog components. Thus increasing digital sample accuracy beyond a certain point simply means a more accurate representation of the analog noise. It can be shown that a unity ratio of quantization step size to analog RMS noise level of the signal of interest is an optimal choice [7:C-5].

Real converters are analog devices and exhibit a further range of implementation errors, all of which may appear simultaneously, and which can change with time and temperature:

Offset Error: Analog value by which the transfer function fails to pass through zero: expressed either in milli-volts or as a percentage of full scale.

Gain Error: (also called scale factor) the difference between ideal and actual full scale value when the offset is zero.

Nonlinearity: A curved transfer function.

The first two can be dealt with by potentiometer adjustments: nonlinearity tends to be irreducible, although a digital error correction could be introduced to compensate.

The total error introduced by the converter must be accounted for in the design and choice of component. This problem is of a longstanding nature and excellent references are available on the details of error control.

The next problem is to code the numbers obtained from the ADC in a format suitable for further computation. This will be the subject of the next section.

### 1.1.4 Digital Number Representation

Digital numbers obtained from the ADC can be represented in a variety of formats. The most common are "fixed point," "floating point" and "block floating point."

### Fixed Point

There are $2^n$ different numbers that can be represented by an $n$-bit digital word. In a fixed point representation the position of the binary point is assumed fixed. The bits to the left of the point represent the integer part of the number and those to the right the fractional part.

There are three common formats for representing negative fixed point numbers: sign magnitude, one's complement and two's complement.

1. Sign-Magnitude: In an $n$-bit sign-magnitude number the leftmost bit represents the sign of the number (i.e., zero for positive and one for negative) and the remaining $(n - 1)$ bits represent the magnitude of the number. A problem with this format is the presence of two representations for zero (e.g., 1000 or 0000 for 4 bit sign-magnitude numbers).
2. One's Complement: In one's complement format, positive numbers are represented as in sign-magnitude. The negative of a number is obtained by complementing all the bits of the number (e.g., the negative of 0101 is 1010). This format also has the disadvantage of yielding two representations of zero (i.e., 0000 and 1111).
3. Two's Complement: In two's complement format, positive numbers are still represented as in sign-magnitude and one's complement. Negative numbers are represented as the two's complement of the positive number. To obtain the negative of a two's complement number, take the one's complement and then add 1 to the least significant (rightmost) bit. Thus, the two's complement of 0101 is 1010 + 1 = 1011. The largest positive number that can be represented in two's complement format is 0111 ($n = 4$) while the largest negative number that can be represented is 1000, which is one larger in magnitude than the largest positive number. However, there is only one representation for zero in two's complement format, which is an important advantage.

### Floating Point

The floating point representation of a number consists of two fixed point numbers representing a mantissa and an exponent. The floating point number $n$ is the product of the mantissa $m$ and the result of raising a specified base (often 2) to the power of the exponent a, i.e.,

$$n = m \cdot 2^a$$

Generally the mantissa is normalized to lie within a specific range (e.g., $\frac{1}{2} \leqslant m < 1$). The sign of a floating point number is represented by the sign bit of the mantissa, while the sign bit of the exponent indicates a negative exponent for numbers with magnitudes less than 0.5.

**Block Floating Point**

In block floating point format a single exponent is used for a block of fixed point numbers. The exponent is determined such that the largest fixed point number of the block is properly normalized. The main advantage of using a single exponent for a block of numbers is the saving in memory storage space or in transmission bandwidth.

Most analog-to-digital converters produce fixed point numbers. Some systems may contain analog prescaling amplifiers that allow sample values to be output in floating point or block floating point format.

Another technique of sample encoding common in digital communications applications is known as "differential encoding." In this technique, the output of the analog-to-digital converter is not the absolute value of the sample but rather the difference between the value of the current sample and the previous sample. The main advantage of differential encoding is that differences between successive sample values can generally be represented by fewer bits than the absolute sample. For this reason differential encoding techniques are of importance; for example, in data rate compression for digital communications. Differential encoding is not well suited however for arithmetic operations. Several of these issues are discussed further in Chapter 4.

The final overall problem is to convert digital numbers back to an analog signal. This will be discussed in the final two sections.

### 1.1.5 Signal Reconstruction

The process of reconstructing an analog signal from a digital signal is generally known as digital-to-analog conversion (DAC) and represents a two step procedure that is the reverse of analog-to-digital conversion.

The first step is the transformation from discrete sample amplitude representation to a continuous value. There is essentially no error inherent in this step since all that is involved is the generation of a signal value (voltage) specified by the digital sample value. A typical transfer function is shown in Figure 1.5(b).

The second step is to transform the discrete-time sample sequence to a continuous-time signal. The most straight-forward approach is simply to hold the sample value constant until the next sample value is received. This technique, known as "sample and hold," is illustrated in Figure 1.5(c). Another technique known as the "linear point connector" is shown in Figure 1.5(d). The sample and hold is, however, the most frequently used technique because of the ease of implementation.

The sharp amplitude changes resulting from the zero-order hold introduce high frequency components into the resulting analog signal. Therefore, the digital-to-analog converter is usually made up of a zero-order hold followed by a low pass filter as shown in Figure 1.5(e). This combination (called a recombination filter) has an effect on the accuracy of the analog representation of the digital signal, but it is relatively easily compensated. One method is to preprocess digitally the signal with the inverse transfer function of the zero-order hold and low pass filter combination so that the final analog output is an accurate representation of the digital signal.

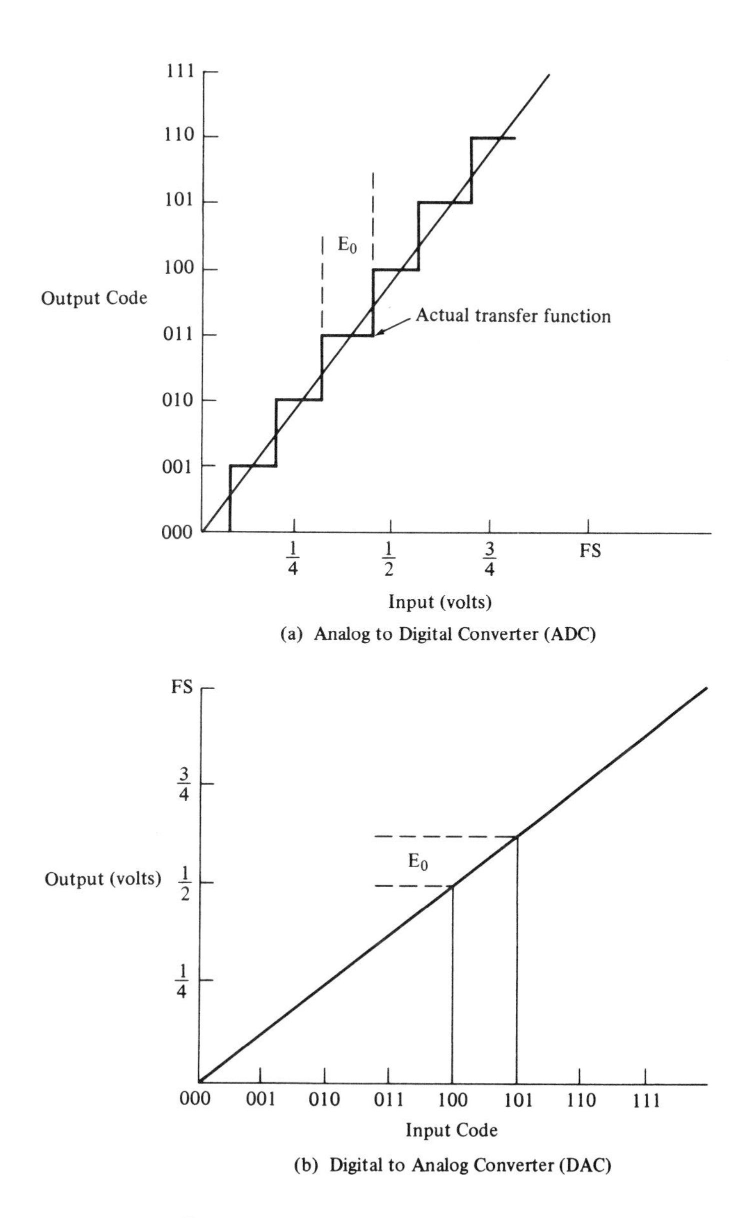


**Figure 1.5**: Converter transfer functions.

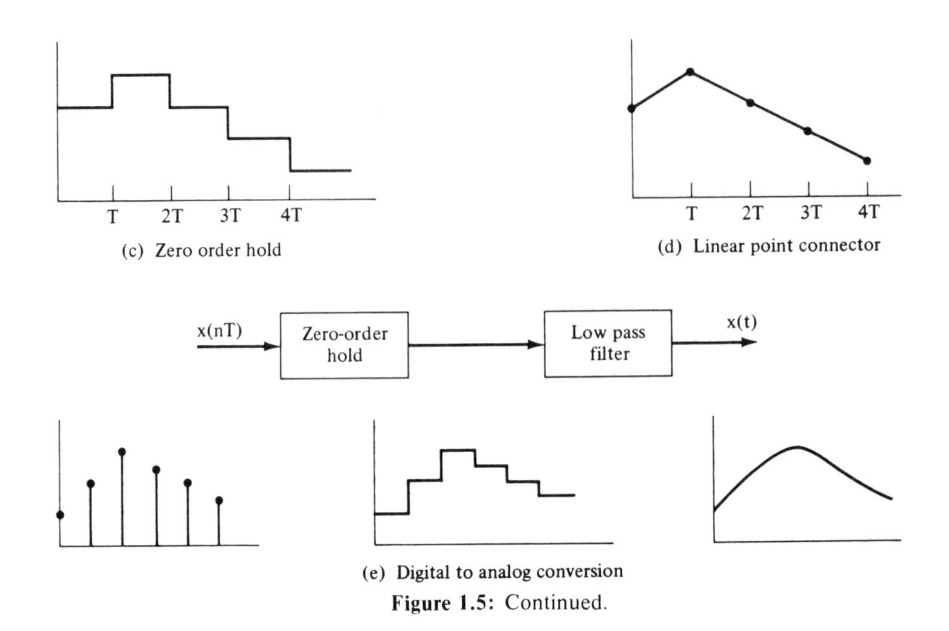


(e) Digital to analog conversion

**Figure 1.5:** Continued.

### 1.1.6 Signal Generation

In several applications of digital signal processing it is necessary to generate digital signals in the processor rather than converting analog signals to a digital form. For example, in the areas of speech synthesis, radar or sonar signal generation, or in simulation applications where there is no analog input signal, the digital signal must be generated either by special purpose hardware or software or both.

The requirement for the direct generation of digital signals can be considered in two categories: piecewise linear waveforms (i.e., ramp, sawtooth, triangle or square waves) and nonlinear waveforms (i.e., sinusoids and random noise sequences). Depending on the frequencies involved (or required sample rates) and the duty cycle for signal generation within a system, these signals may be generated either by dedicated hardware or by software. In either case the two common techniques are by the use of lookup tables or by implementing a suitable recursive or iterative algorithm.

Simple techniques for piecewise linear waveform generation tend to involve the iterative execution of simple linear relations and the utilization of overflows and sign bit manipulation. Nonlinear waveform generation is more generally implemented with table lookup and recursion techniques.

Consider the implementation of a simple digital frequency synthesizer for generating sinusoids. The two major functional requirements are for:

1. Angle generation in accordance with the required signal frequency and sample rate, and
2. Sine/cosine generation for the angles.

Figure 1.6 is a block diagram of a digital frequency synthesizer.

For generating random noise sequences a recursive technique is generally employed which, when supplied with a seed number, will generate random numbers within a specified range. Numerous algorithms for random number generation are available that can produce random sequences approximating a wide variety of distributions.

Another method of signal synthesis is the periodic excitation of a digital filter with an impulse. The waveform generated is the impulse response of the filter. This technique is often employed to generate complex waveforms, such as in the digital synthesis of speech. As we shall see when we examine digital filters in Chapter 2, this is essentially an implementation of a recursion technique for signal generation using a recursive filter.

### 1.1.7 Summary

This section has considered the input/output problem which is usually encountered in the processing of analog signals in a digital processor. It involves analog-to-digital conversion and the reverse, digital-to-analog conversion. These converters are available with a wide range of performance characteristics; both types induce a range of errors which must be considered in the specifications and selection phase of a project.

The input transition from an analog signal to a digital number is a two-step process involving sampling and sample quantization. There are several mechanisms for performing these two functions which have been the subject of our previous discussions. Our concern has been with the intrinsic errors introduced by the finite bit length used for quantizing the signal. The quantization step size was shown to be the full scale input signal range divided by $2^n$. This limitation introduces a quantization error which becomes an irreducible part of all future processing. In addition the ADC is an analog device

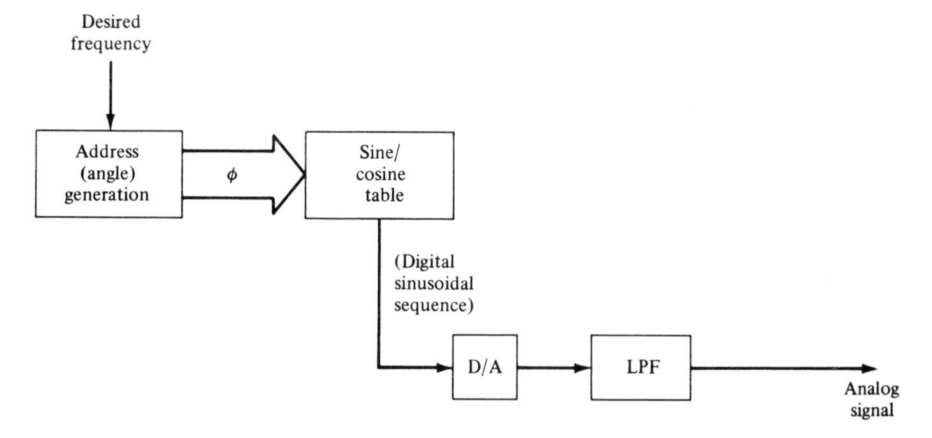


**Figure 1.6**: Digital frequency synthesizer.

(usually with several operational amplifiers and an analog switch); these circuits are subject to temperature effects as well as certain nonlinearities. The overriding conclusion is that digital processing must be preceeded by analog processing of a high quality. The designer of the whole system must be concerned with this problem, although complete treatment is beyond our scope.

The representation of an analog signal by digital numbers is based on the Nyquist sampling theorem. For a band-limited signal, a sampling rate of at least twice the highest frequency of the signal produces a spectrum in which no aliasing occurs. This becomes a second design issue, because most signals are not strictly band-limited and some compromise is necessary to preprocess with anti-aliasing filters and to adjust the sampling rate to insure that the distortion introduced by the spectral overlap is within error tolerances. This aspect of the design problem requires further study which is beyond the scope of this volume.

Amplitude variations in the incoming signal may contain important information. In order to capture this the ADC must have enough bits to insure a dynamic range consistent with this requirement. This was shown to be about 6 dB per bit. Good audio reproduction requires, for example, 60-90 dB which requires in turn 12 or more bits. The limiting problem is additive noise which ultimately limits the meaningful information as bits are increased. Once again good analog signal processing is required as these specifications become stringent.

In the processor, digital signals are represented by sequences of numbers. The ADC usually produces a fixed point code as its output. Depending on the processing requirement (and the processor) this is usually converted to two's complement or to a floating point format. The compromises here are based on the arithmetic capabilites of the processor and the requirement for storage and/or transmission. Block floating point is the most efficient from this point of view and many processors can process this format directly.

Digital to analog conversion is a somewhat simpler operation. Bits representing a signal value are added in a resistor chain to produce an output voltage. This voltage can be filtered to remove the high frequency components or interpolated. The DAC is also an analog device and is subject to the same class of problems as the ADC. Obviously the resistors used to add the bits must be accurate and stable.

This discussion of the analog interfaces to a digital processor has dealt only with the logical concepts involved. At the specifications phase, the concern is usually with the sampling rate, the dynamic range (i.e., the number of bits) and the error tolerances. These depend on the application area and more will be said about this in Chapter 5. At the implementation phase all of these specifications must come together in the choice of a converter and its interface to a processor. This is a field of expertise in its own right which is beyond our scope. Some good reading is indicated in the bibliography.

The discussions of representations and the various errors introduced by finite word lengths will be of further signifigance as we discuss the implementation of processing algorithms (e.g., a digital filter). Errors can introduce some interesting pathologies, which are the subject of intense concern in system design.

The next logical question is "How do we manipulate these numbers to achieve our processing requirements?" To answer this we must first develop a mathematical formalism for representing digital signals which can be used to specify processing algorithms. This is the goal of the next section.

## 1.2 TRANSFORM ANALYSIS OF DETERMINISTIC SIGNALS

### 1.2.0 Introduction

In this section a mathematical model is presented which provides the formal structure for analyzing and manipulating digital signals. Fortunately, the Laplace and Fourier transforms used for continuous signals have a direct counterpart in the digital domain. By simple extensions and suitable interpretation the Laplace transform becomes a $Z$-transform, and a discrete version of the Fourier transform is also found. The properties of these new mathmatical constructs are important and will now be explored and illustrated. To do this we first show how the $Z$-transform arises from the concepts of the Laplace transform of a discrete-time (sampled) signal. We then discuss the relationship of the $Z$-transform to the Laplace and Fourier transforms. In Section 1.2.2 several useful properties of the $Z$-transform are discussed. Section 1.2.3 introduces the important concepts and properties of the discrete Fourier transform (DFT) which are useful when dealing with finite duration and periodic signals.

#### The $Z$-Transform and its Relationship to the Laplace and Fourier Transforms

In continuous-time signal analysis the Laplace transform is defined as

$$X(s) = \mathcal{L}\{x(t)\} \triangleq \int_{-\infty}^{\infty} x(t)e^{-st}dt \tag{1.9}$$

where '$s$' is, in general, a complex number $(\sigma + j\omega)$. The values of $s$ for which the integral of equation (1.9) converges define the region of convergence (ROC) of the Laplace transform. The inverse relation is

$$x(t) = \mathcal{L}^{-1}\{X(s)\} = \frac{1}{2\pi j}\int_{\sigma-j\infty}^{\sigma+j\infty} X(s)e^{st}ds \tag{1.10}$$

where the limits indicate that the integration is carried out along a line parallel to the imaginary axis within the ROC.

The Fourier transform may be viewed as a special case of the Laplace transform where '$s$' is purely imaginary, i.e.,

$$X(\omega) = \mathcal{F}\{x(t)\} \triangleq \int_{-\infty}^{\infty} x(t)e^{-j\omega t}dt \tag{1.11}$$

and

$$x(t) = \mathcal{F}^{-1}\{X(\omega)\} = \frac{1}{2^{\pi}}\int_{-\infty}^{\infty} X(\omega)e^{j\omega t}d\omega \tag{1.12}$$

In Section 1.1 we discussed the concepts of ideal uniform sampling by modulating a continuous-time signal $x(t)$ with an impulse train to obtain a sequence of numbers $x(n)$ which represent the sampled signal

$$x_s(t) = x(t)p(t) = \sum_{n=-\infty}^{\infty} x(nT)\delta(t - nT) \tag{1.13}$$

If we take the Laplace transform of equation (1.13) we get

$$\begin{aligned} X(s) &= \int_{-\infty}^{+\infty} \sum_{n=-\infty} x(nT)\delta(t - nT)e^{-st}dt \\ &= \sum_{n=-\infty}^{\infty} x(nT)\int_{-\infty}^{+\infty} \delta(t - nT)e^{-st}dt \\ &= \sum_{n=-\infty}^{\infty} x(nT)e^{-snT} \end{aligned} \tag{1.14}$$

Similarly the Fourier transform of $x(n)$ is given by

$$X(\omega) = \mathcal{F}\{x(nT)\} = \sum_{n=-\infty}^{\infty} x(nT)e^{-j\omega nT} \tag{1.15}$$

If the complex exponential $e^{sT}$ is denoted by $z$ in equation (1.14) we get the expression

$$X(z) = \mathcal{Z}\{x(n)\} = \sum_{n=-\infty}^{\infty} x(n)z^{-n} \tag{1.16}$$

where we have also dropped the explicit inclusion of the sample period $T$ in the representation of the sample sequence $x(nT)$.

Equation (1.16) defines the $Z$-transform of a sequence $x(n)$ and is the discrete-time counterpart of the Laplace transform. It is of such fundamental importance that it is studied in its own right; although its relation to the Laplace transform remains useful when exploring its properties. If we set $z = e^{j\omega}$, we get

$$X(e^{j\omega}) = \sum_{n=-\infty}^{\infty} x(n)e^{-j\omega n} \tag{1.17}$$

which defines the Fourier transform of the discrete-time sequence $x(n)$ as a special case of the $Z$-transform.

The substitution of $z = e^{sT}$ made in moving from equation (1.14) to (1.16) represents a mapping (a bilinear transformation) between the $S$-plane and the $Z$-plane as illustrated in Figure 1.7. The relationship between the $S$-plane and the $Z$-plane is not a one-to-one mapping. Each horizontal strip in the $S$-plane of width $2\pi/T$ maps to the entire $Z$-plane. That portion of the strip in the left half $S$-plane maps to inside the unit circle on the $Z$-plane and the right half $S$-plane strip maps to outside the unit circle on the $Z$-plane. The $j\omega$ axis of the $S$-plane maps to the unit circle on the $Z$-plane with the $S$-plane origin and all points along the $j\omega$ axis that are multiples of $2\pi/T$

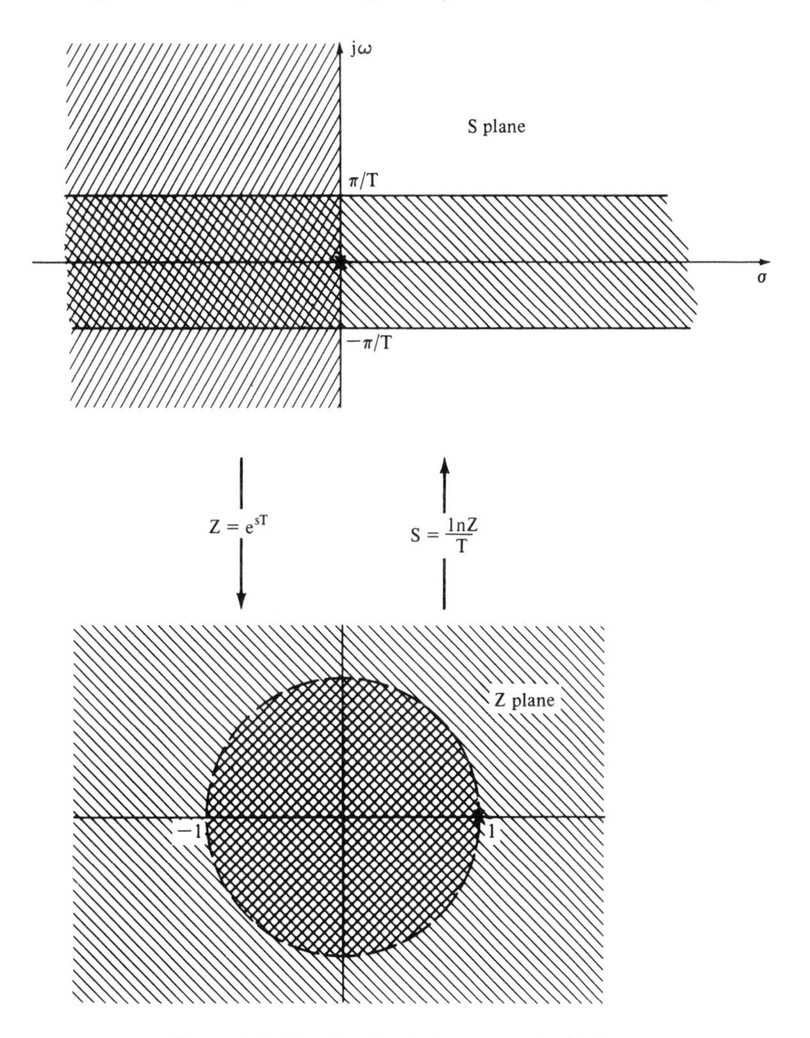


**Figure 1.7**: Mapping the $S$ plane onto the $Z$ plane.

mapping to the point $z = 1$. This mapping of the $S$-plane provides insight into the representation of frequency in digital signal processing. In the $S$-plane we relate the $j\omega$ axis to signal frequencies ranging from $-\infty$ to $+\infty$. When the $j\omega$ axis is mapped to the $Z$-plane for a sampled signal, it wraps around the unit circle on top of itself for every interval of length $2\pi/T$. When the signal being sampled contains frequency components extending beyond the $2\pi/T$ wide strip of the $S$-plane centered on the $\sigma$ axis, then these components overlap the baseband components when the $j\omega$ axis wraps around the unit circle of the $Z$-plane. This is another manner in which to view the phenomenon of aliasing discussed in Section 1.1.3. We see from this perspective the reason for requiring a band limited signal and a sample rate at least twice the highest frequency to avoid aliasing.

The result of the mapping $z = e^{sT}$ is that the Fourier transform of a sampled signal, $X(e^{j\omega})$, represents the frequency components of the signal in the range $-\pi/T \leqslant \omega \leqslant \pi/T$ summed with all other frequency components at multiples of $2\pi/T$ (both positive and negative). To translate to real positive frequencies we note the relationship that the interval of $2\pi$ radians represented by going once around the unit circle is equivalent to a range of signal frequencies of $-f_s/2$ to $f_s/2$, where $f_s = 1/T$ is the sample frequency in Hz. Since the interval from $\pi/T$ to $2\pi/T$ is the same as the interval $-\pi/T$ to 0 we can consider only the positive frequencies of the range $0 \leqslant f \leqslant f_s$ as the baseband region represented by the unit circle of the $Z$-plane. We note that this spectrum will be symmetric about the point $f_s/2$.

The $Z$-plane may appear as an unusual space in which to work; however consider that the entire frequency axis is now on the unit circle. This will prove useful when finding the frequency components of a digital signal.

Similar to the one-sided or unilateral Laplace transform defined as

$$\mathcal{L}_+\{x(t)\} \triangleq \int_0^{\infty} x(t)e^{-st}dt \tag{1.18}$$

the one sided $Z$-transform is defined as

$$\mathcal{Z}_+\{x(n)\} \triangleq \sum_{n=0}^{\infty} x(n)z^{-n} \tag{1.19}$$

The one-sided $Z$-transform is useful when dealing with causal sequences (i.e., sequences which are zero or undefined for $n < 0$). Since negative time is not a real world concept, all digital sequences resulting from the analog-to-digital conversion of an analog signal are causal.

The defining equation of the $Z$-transform, (1.16), has the general form of a power series known as a Laurent series and will generally converge when $z$

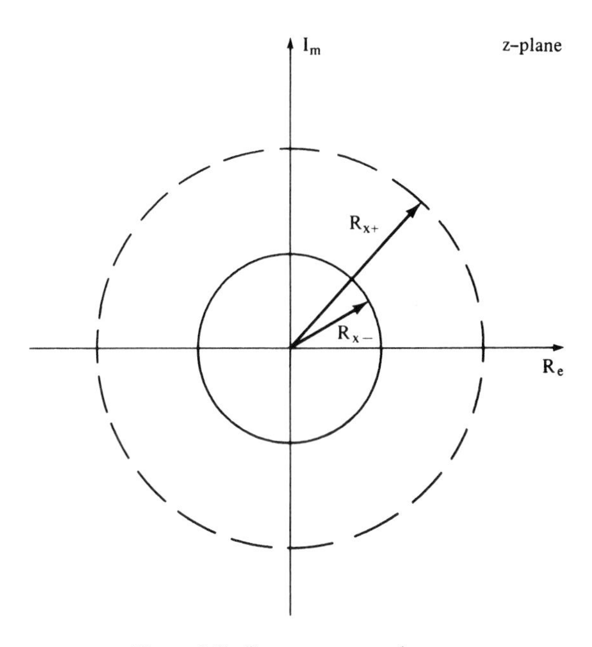


**Figure 1.8**: Convergence regions.

takes on certain values in an annular region of the $Z$-plane known as the region of convergence (ROC),

$$R_{x-} < |z| < R_{x+}$$

as illustrated in Figure 1.8.

When the $Z$-transform of a sequence exists and can be expressed as a ratio of polynomials in $z$ (i.e., as a rational fraction) the roots of the numerator polynomial represent those values of $z$ for which $X(z) = 0$ and are known as the zeros of $X(z)$. Similarly, the roots of the denominator polynomial represent those values of $z$ for which $X(z)$ is infinite and are known as the poles of $X(z)$. Clearly poles do not occur within the region of convergence. The $Z$-transform of a sequence is often illustrated graphically by a pole-zero plot of the complex plane as shown in the following examples for some common elementary discrete time signals.

***Example 1:***

The Unit Sample Sequence is defined as

$$x(n) = \delta(n) = \begin{cases} 1, & n=0 \\ 0, & n \neq 0 \end{cases}$$

Thus

$$X(z) = \sum_{n=-\infty}^{\infty} \delta(n) z^{-n} = 1$$

and the region of convergence is the entire $Z$-plane. This sequence is similiar to the delta function used for continuous time signals.

If the unit sample is shifted such that

$$x(n) = \delta(n-N) = \begin{cases} 1, & n=N \\ 0, & n \neq N \end{cases}$$

then

$$X(z) = z^{-N}$$

This is an impulse delayed by $N$ periods.

***Example 2:***

The Unit Step Sequence is defined as

$$x_u(n) = \begin{cases} 1, & n \geqslant 0 \\ 0, & n < 0 \end{cases}$$

Therefore

$$X_u(z) = \sum_{n=0}^{\infty} z^{-n}$$

which is a geometric series that converges for $|z| > 1$ to

$$X_u(z) = \frac{1}{(1 - z^{-1})}$$

The region of convergence is the entire $Z$-plane outside the unit circle. $X_u(z)$ has a zero at the origin and a pole at $z = 1$. The pole-zero plot is illustrated in Figure 1.9(a).

The unit step sequence is a special case of the general exponential sequence

$$x(n) = \begin{cases} a^n, & n \geqslant 0 \\ 0, & n < 0 \end{cases}$$

for which the $Z$-transform has the general form

$$X(z) = \sum_{n=0}^{\infty} (az^{-1})^n = \frac{1}{(1 - az^{-1})} \quad , |z| > a$$

and a pole-zero plot as shown in Figure 1.9(b). For a $<1$ this represents an exponential decay.

***Example 3:***

$$x(n) = \begin{cases} nc^n, & n \geqslant 0 \\ 0 & n < 0 \end{cases}$$

$$X(z) = \sum_{n=0}^{\infty} nc^n z^{-n}$$

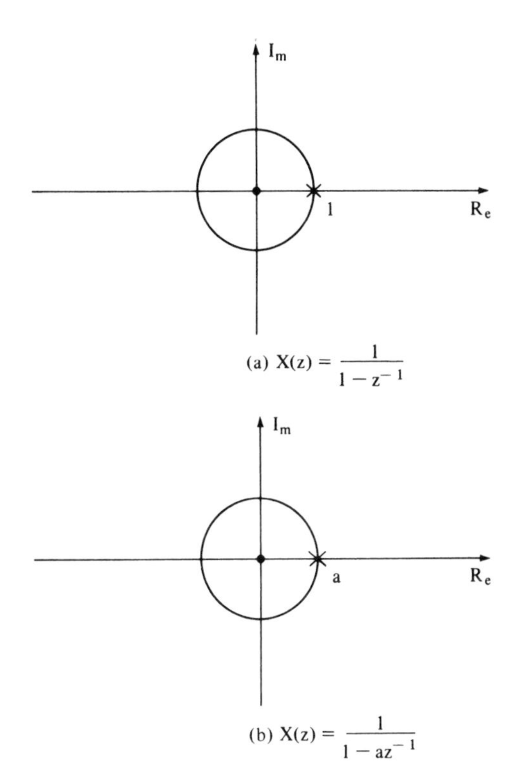


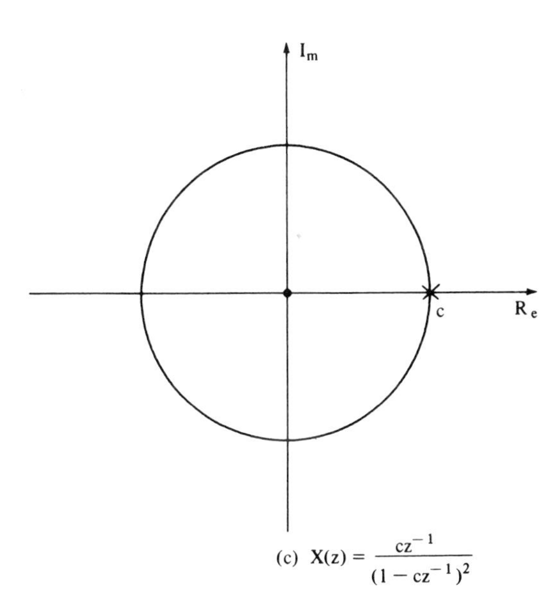


**Figure 1.9:** (a) Simple pole at $Z = 1$. (b) Simple pole at $Z$ at $a$. (c) Second order pole at $Z = c$.

By rewriting the definition of $X(z)$ in a slightly different form we get

$$X(z) = z^{-1}d/dz^{-1}\left[\sum_{n=0}^{\infty}(cz^{-1})^n\right]$$

$$= cz^{-1}/(1 - cz^{-1})^2, \ |z| > c$$

The pole-zero plot is shown in Figure 1.9(c): $X(z)$ has a second order pole at $z = c$.

The inverse relationship for the $Z$-transform can be derived from the inverse Laplace transform equation as

$$x(n) = 1/2\pi j \oint_C X(z)z^{n-1}dz \tag{1.20}$$

where $C$ is an appropriate contour within the region of convergence that encloses the origin. For rational $Z$-transforms, the above contour integral can often be evaluated using the residue theorem, i.e.,

$$x(n) = \sum \text{(residues of } X(z)z^{n-1} \text{ at the poles inside } C)$$

or by a procedure of partial fraction expansion and long division.

***Example 4*** (The Inverse $Z$-Transform):

Suppose

$$X(z) = \frac{(1 - e^{-aT})z}{(z - 1)(z - e^{-aT})}$$

Find the time function which this represents. Assume the sample interval is $T$. Expand $X(z)$ by partial fractions, i.e.,

$$X(z) = \frac{z}{z - 1} - \frac{z}{z - e^{-aT}}$$

From a transform table

$$x(nT) = 1 - e^{-anT}$$

describes a time function whose values at the sample points are

$$x(nT) = \sum(1 - e^{-anT})\delta(t - nT)$$

If $x(nT)$ represents samples above the Nyquist rate, $x(t)$ can be recovered by ideal low pass filtering.

### 1.2.2 Properties of the Z-Transform

A number of basic properties of the Z-transform are of importance. These properties are directly analogous to those of the Laplace transform and will be used freely in what follows.

1. Linearity: The Z-transform is a linear operation, i.e.,

$$\mathfrak{Z}\{ax_1(n) + bx_2(n)\} = aX_1(z) + bX_2(z)$$

where $a$ and $b$ are real constants and the region of convergence is:

$$\max\left[R_{x_1^-}, R_{x_2^-}\right] < |z| < \min\left[R_{x_1^+}, R_{x_2^+}\right]$$

where the ROC of $X_1(z)$ and $X_2(z)$ are

$$R_{x_1^-} < |z| < R_{x_1^+} \text{ and } R_{x_2^-} < |z| < R_{x_2^+}$$

respectively.

2. Delay Property:

$$\mathfrak{Z}\{x(n-k)\} = z^{-k}X(z), \; R_{x-} < |z| < R_{x+}$$

where $k$ is an integer. This property emphasizes the shift introduced by $z^{-1}$.

3. One-Sided Advance Property: For a causal sequence $x(n)$ with Z-transform $X(z)$ the one-sided Z-transform of $x(n+k)$, $k > 0$, is

$$\mathfrak{Z}_+\{x(n+k)\} = z^kX(z) - \sum_{n=0}^{k-1} x(n)z^{k-n}$$

4. Time Reversal: If the Z-transform of $x(n)$ is $X(z)$, $R_{x-} < |z| < R_{x+}$ then

$$\mathfrak{Z}\{x(-n)\} = X(z^{-1}), \; 1/R_{x+} < |z| < 1/R_{x-}$$

5. Multiplication by $a^n$: If the Z-transform of $x(n)$ is $X(z)$, $R_{x-} < |z| < R_{x+}$ then

$$\mathfrak{Z}\{a^nx(n)\} = X(a^{-1}z), \; |a|R_{x-} < |z| < |a|R_{x+}$$

6. Multiplication by $n$:

$$\mathfrak{Z}\{nx(n)\} = -z\frac{dX(z)}{dz}, \; R_{x-} \; |z| < R_{x+}$$

7. Complex Conjugate Property: If $x(n)$ is a complex sequence with Z-transform $X(z)$ which converges for $R_{x-} < |z| < R_{x+}$ then

$$\mathfrak{Z}\{x^*(n)\} = X^*(z^*), \; R_{x-} < |z| < R_{x+}$$

where * denotes complex conjugate.

8. Symmetry of $X(z)$ for Real Sequences: If $x(n)$ is a real sequence with $Z$-transform $X(z)$, and the ROC includes the unit circle then

$$X(-z)\,\big|_{z=e^{j\omega}} = X^*(z)\,\big|_{z=e^{j\omega}}$$

That is, the magnitude of $X(e^{j\omega})$ is even and the argument of $X(e^{j\omega})$ is odd, i.e.,

$$|X(e^{j\omega})| = |X(-e^{j\omega})|$$

and

$$arg\Big[X(e^{j\omega})\Big] = -arg\Big[X(-e^{j\omega})\Big]$$

9. Convolution of Sequences: The discrete-time convolution of two sequences $x(n)$ and $y(n)$ is defined as

$$x(n) * y(n) = \sum_{k=-\infty}^{\infty} x(k)y(n-k)$$

Given two sequences, $x(n)$ and $y(n)$ with $Z$-transforms $X(z)$, $R_{x-} < |z| < R_{x+}$ and $Y(z)$, $R_{y-} < |z| < R_{y+}$, the $Z$-transform of the convolution of $x(n)$ with $y(n)$ is given by the product of their respective $Z$-transforms, i.e.,

$$\begin{aligned}\mathcal{Z}\{x(n) * y(n)\} &= \sum_{n=-\infty}^{\infty}\left\{\sum_{k=-\infty}^{\infty} x(k)y(n-k)\right\}z^{-n} \\ &= \sum_{k=-\infty}^{\infty} x(k) \sum_{n=-\infty}^{\infty} y(n-k)z^{-n} \\ &= \sum_{k=-\infty}^{\infty} x(k)z^{-k}Y(z) \\ &= X(z)\,Y(z)\end{aligned}$$

The ROC is

$$\max[R_{x-},\ R_{y-}] < |z| < \min[R_{x+},\ R_{y+}]$$

10. Multiplication of Sequences—The Complex Convolution Theorem: Given two sequences $x(n)$ and $y(n)$ with $Z$-transforms $X(z)$, $R_{x-} < |z| < R_{x+}$, and $Y(z)$, $R_{y-} < |z| < R_{y+}$, the $Z$-transform of the product of these two sequences is given by

$$\mathcal{Z}\{x(n)y(n)\} = \frac{1}{2\pi j}\oint_C X(\nu)\,Y(z/\nu)\,d\nu/\nu$$

$$\max\Big[R_{x-},\ R_{y-}\Big] < |z| < \min\Big[R_{x+},\ R_{y+}\Big]$$

where $C$ is a closed curve encircling the origin that lies within the overlap of the individual regions of convergence of $X(z)$ and $Y(z)$.

11. Parseval's Theorem: Parseval's Theorem relates the time-domain sum of the product of two sequences (or the sum of the squares of a sequence of samples) to a $Z$-transform domain integral.

 Given two sequences $x(n)$ and $y(n)$ with $Z$-transforms $X(z)$, $R_{x-} < |z| < R_{x+}$, and $Y(z)$, $R_{y-} < |z| < R_{y+}$, such that $R_{x-} R_{y-} \leqslant |1| \leqslant R_{x+} R_{y+}$, then

$$\sum_{n=-\infty}^{\infty} x(n)y(n) = \frac{1}{2\pi j} \oint_C X(z)\, Y(z^{-1})\, dz/z$$

 where $C$ is a closed curve encircling the origin with

$$\max\left[R_{x-},\ 1/R_{y+}\right] < |z| \leqslant \min\left[R_{x+},\ 1/R_{y-}\right]$$

 For the special case of $z = e^{j\omega}$ (i.e., the Fourier transform) we have $|z| = 1$ giving

$$\sum_{n=-\infty}^{\infty} x(n)y(n) = \frac{1}{2\pi} \int_{-\pi}^{\pi} X(e^{j\omega})\, Y(e^{-j\omega})\, d\omega$$

 or

$$\sum_{n=-\infty}^{\infty} |x(n)|^2 = \frac{1}{2\pi} \int_{-\pi}^{\pi} |X(e^{j\omega})|^2 d\omega$$

 For the case of a finite duration sequence $x(n)$ of length $N$, Parseval's theorem can be expressed as

$$\sum_{n=0}^{N-1} |x(n)|^2 = 1/N \sum_{k=0}^{N-1} |X_k|^2$$

 where the quantities $X_k$ are the discrete Fourier transform coefficients of $x(n)$, as discussed in the next section.

12. Frequency Translation: If a sample sequence $x(n)$ whose $Z$-transform converges on the unit circle (i.e., its Fourier transform exists) is multiplied by a complex exponential sequence $z_0^n$, $z_0 = e^{j\omega_0}$, then the Fourier transform of the product is

$$\begin{aligned}
\mathcal{F}\{z_0^n\, x(n)\}\Big|_{z_0 = e^{j\omega_0}} &= \sum_{n=-\infty}^{\infty} e^{j\omega_0 n} x(n) e^{-j\omega n} \\
&= \sum_{n=-\infty}^{\infty} x(n) e^{-j(\omega-\omega_0)n} \\
&= X(e^{j(\omega-\omega_0)n})
\end{aligned}$$

 Thus, multiplication by $e^{j\omega_0 n}$ represents complex frequency translation. This property provides a convenient method of carrying out frequency translation on digital signals and represents one of the fundamental computational requirements for digital signal processing discussed in Chapter 5. This property leads also to the concepts of analytic signals, Hilbert transforms and band pass sampling. Further discussion of this is beyond the scope of this volume; however, the interested reader can examine their development in [10].

The properties stated in this section are presented without proofs. Some are easily derived directly from the definition of the $Z$-transform. For proofs of these and other properties of the $Z$-transform the reader is referred to the references indicated in the bibliography.

### 1.2.3 The Discrete Fourier Transform

We have shown that discrete-time sequences can be represented in terms of the $Z$-transform (or the Fourier transform as a special case of the $Z$-transform). These representations are, however, continuous functions defined by an infinite series. A somewhat more useful transform domain for implementation on digital hardware is the discrete Fourier transform (DFT) of an $N$-point finite duration sequence defined as

$$X_k = \sum_{n=0}^{N-1} x(n) e^{-j(2\pi/N)nk}, \quad k = 0, 1, \ldots, N-1 \tag{1.21}$$

The DFT of an $N$-point finite sequence results in another $N$-point finite sequence which represents harmonically related frequency components of $x(n)$.

The inverse discrete Fourier transform (IDFT) relates the original sequence values $x(n)$ to the discrete Fourier coefficients $X_k$.

$$X(n) = \frac{1}{N} \sum_{k=0}^{N-1} X_k e^{j(2\pi/N)nk}, \quad n = 0, 1, \ldots, N-1 \tag{1.22}$$

There are several derivations and interpretations of the DFT contained in various texts. We shall present here a discussion relating the DFT to the $Z$-transform of a discrete time sequence. In practice only a finite number of samples in the time domain are ever available. The spectral components of such a set are of importance. It is the DFT that provides a method for determining these spectral components.

Consider a sequence $x(n)$ that has nonzero values only within a finite range $N$ of index values, i.e.,

$$x(n) = \begin{cases} x(n), & 0 \leqslant n \leqslant N-1 \\ 0 & \text{otherwise} \end{cases}$$

The $Z$-transform of this sequence is

$$X(z) = \sum_{n=0}^{N-1} x(n) z^{-n}$$

If $X(z)$ is evaluated on the unit circle of the $Z$-plane (i.e., at $z = e^{j\omega}$), we obtain the Fourier transform relation

$$X(e^{j\omega}) = \sum_{n=0}^{N-1} x(n) e^{-j\omega n}$$

which is a continuous function of $\omega$ . If $X(e^{j\omega})$ is evaluated at discrete values of $\omega$ uniformly spaced at intervals of $2\pi/N$ (i.e., sample $X(e^{j\omega})$ in frequency) we obtain only $N$ distinct sample values due to the periodicity of the complex exponential $e^{j\omega}$ , i.e.,

$$e^{j(2\pi/N)k} = e^{j(2\pi/N)(k+nN)}, \quad 0 \leqslant k \leqslant N-1, \ -\infty < n < \infty$$

Therefore

$$X_k = X(z)\big|_{z = e^{jk2\pi/N}} = \sum_{n=0}^{N-1} x(n) e^{-j(2\pi/N)nk}, \quad k = 0, 1, \ldots, N-1 \tag{1.23}$$

and we have an important observation:

> The DFT of a finite duration sequence is equivalent to evaluating the $Z$-transform at $N$ uniformly spaced intervals on the unit circle.

It is important to note at this point that the $N$ DFT coefficient values represent specific values of $X(e^{j\omega})$ at both positive and negative values of $\omega$. If the sample sequence $x(n)$, $n = 0, \ldots, N-1$ is real, then the DFT coefficient sequence $X_k$ will exhibit a symmetry property as illustrated in Figure 1.10(a) and (b) for the case of $N$ even or odd respectively, [i.e., $X_k = X_{N-k}$]. This is an important symmetry property that will be drawn upon in Chapter 4 to describe a method of increasing the efficiency of an algorithm for computing DFT coefficients known as the Fast Fourier Transform (FFT).

We have seen that evaluating the $Z$-transform on the unit circle [i.e., $z = e^{j\omega}$] results in the Fourier transform. Hence the DFT gives us a set of Fourier coefficients for specific values of angular frequency $\omega$ over a range of $0 \leqslant \omega \leqslant 2\pi/T$ radians/sec. Assuming Nyquist sampling of a properly band-limited signal these DFT coefficients relate to the frequency content of the signal. In Chapter 3 we shall consider spectral estimation and make use of this DFT property.

If $x(n)$ is a periodic sequence with a fundamental period of $N$ samples, then all the information in the sequence is contained in one period and can thus be represented as a finite duration sequence of any $N$ consecutive samples of the original periodic sequence.

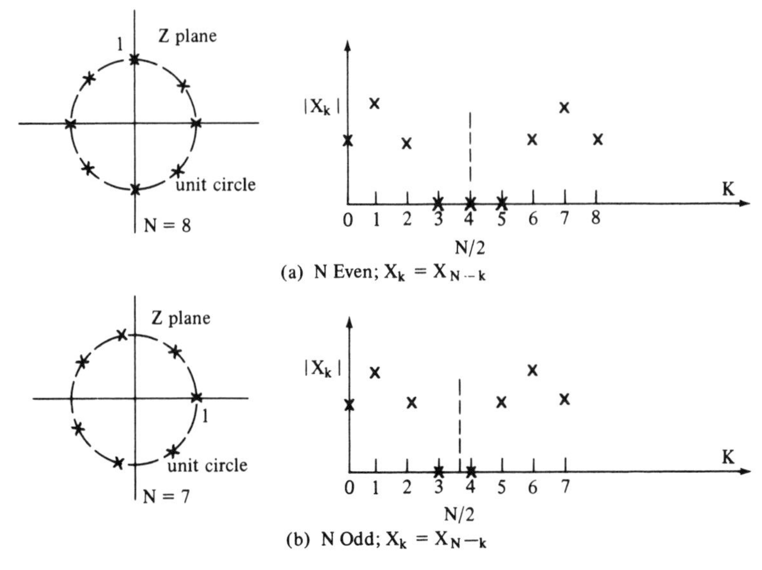


**Figure 1.10**: Symmetry properties of DFT coefficients.

To see that the original sample values can be recovered from the $N$ DFT values $X_k$ consider the sum

$$
\begin{aligned}
S_n &= \sum_{k=0}^{N-1} X_k e^{j(2\pi/N)nk} \qquad , \; 0 \leqslant n \leqslant N-1 \\
&= \sum_{k=0}^{N-1} \left[ \sum_{i=0}^{N-1} x(i) e^{-j(2\pi/N)ik} e^{j(2\pi/N)nk} \right] \\
&= \sum_{i=0}^{N-1} x(i) \left[ \sum_{k=0}^{N-1} e^{j(2\pi/N)k(n-i)} \right]
\end{aligned}
$$

By noting that the summation in brackets in the last equation is equal to $N$, for $(n-i) = mN$, $m$ an integer, and zero otherwise, we see (after some thought) that

$$S_n = Nx(n)$$

Therefore

$$x(n) = \frac{1}{N} \sum_{k=0}^{N-1} X_k e^{j(2\pi/N)nk} \qquad , \; 0 \leqslant n \leqslant N-1 \tag{1.24}$$

and we have a second important observation:

The DFT coefficients of a finite duration sequence uniquely determine the original sequence values.

These results for a finite sequence are of fundamental importance in signal processing. The fast Fourier transform (FFT) to be discussed later is an algorithm for quickly evaluating the discrete Fourier coefficients. Equation (1.24) is the cumulative product of two complex number sequences. This requires $(N-1)^2$ complex multiplications and $N(N-1)$ complex additions. This is a formidable computational load if real time constraints exist; e.g., consider a 1024 point transform and a multiply time of 700 n.s. with an add time of 500 n.s. (these numbers are typical of the ILLIAC IV, studied in Chapter 6).

We have presented the DFT as a sample sequence of values of the $Z$-transform of a finite duration sequence. We might expect the DFT to exhibit properties similar to those of the $Z$-transform. This is indeed the case. A few of these properties are summarized below:

1. Linearity

$$\text{DFT}\{c_1 x(n) + c_2 y(n)\} = c_1 X_k + c_2 Y_k$$

2. Time Reversal

$$\text{DFT}\{x(-n)\} = X_{-k} = X_{N-k}$$

3. Shift Properties

$$\text{DFT}\{x(n-r)\} = X_k e^{-j(2\pi/N)rk}$$
$$\text{IDFT}\{X_{k-r}\} = x(n) e^{-j(2\pi/N)nr}$$

4. Complex Conjugates

$$\text{DFT}\{x^*(n)\} = X^*_{-k} = X^*_{N-k}$$

5. Symmetry Properties
   (a) If $x(n)$ is even, that is $x(-n) = x(n)$, then its DFT is also even, $X_k = X_{-k} = X_{N-k}$.
   (b) If $x(n)$ is odd, that is $x(-n) = -x(n)$, then its DFT is also odd, $X_k = -X_{-k} = -X_{N-k}$.
   (c) If $x(n)$ is real, its DFT has an even real part and an odd imaginary part, i.e., $X_k = X^*_{-k} = X^*_{N-k}$.
   (d) If $x(n)$ is purely imaginary then its DFT has odd real part and even imaginary part, i.e., $X_k = -X^*_{-k} = -X^*_{N-k}$.

6. Circular Convolution: The circular convolution of two periodic sequences $x(n)$ and $y(n)$, each of period $N$, is defined as

$$x(n) \circledast y(n) = \sum_{k=0}^{N-1} x(k) y(n-k) = \sum_{k=0}^{N-1} y(k) x(n-k)$$

The product of the DFTs of a single period of the individual sequences is equal to the DFT of their circular convolution, i.e.,

$$\text{DFT}\{x(n) \circledast y(n)\} = X_k Y_K$$

where

$$X_k = \text{DFT}\{x(n)\}$$
$$Y_k = \text{DFT}\{y(n)\}, \qquad n = 0, , 1, \ldots, N-1$$

and

$$\text{DFT}\{x(n)y(n)\} = \frac{1}{N}\sum_{k=0}^{N-1} X_{n-k}Y_k$$

### 1.2.4 Summary

The $Z$-transform is widely used in sampled data systems for representation and analysis. Based on a familiarity with the Laplace transform it is convenient to introduce the $Z$-transform as the transformation $z = e^{sT}$. This serves as a familiar base for exploring its properties; however, this transform could equally well be studied in its own right and related to the Laplace transform as an interesting afterthought. It is convenient to approach it as we did, based on an assumed familiarity with the $S$-plane. Figure 1.11 summarizes the overall relationships between the time domain and frequency domain representation of both continuous-time and discrete-time signals.

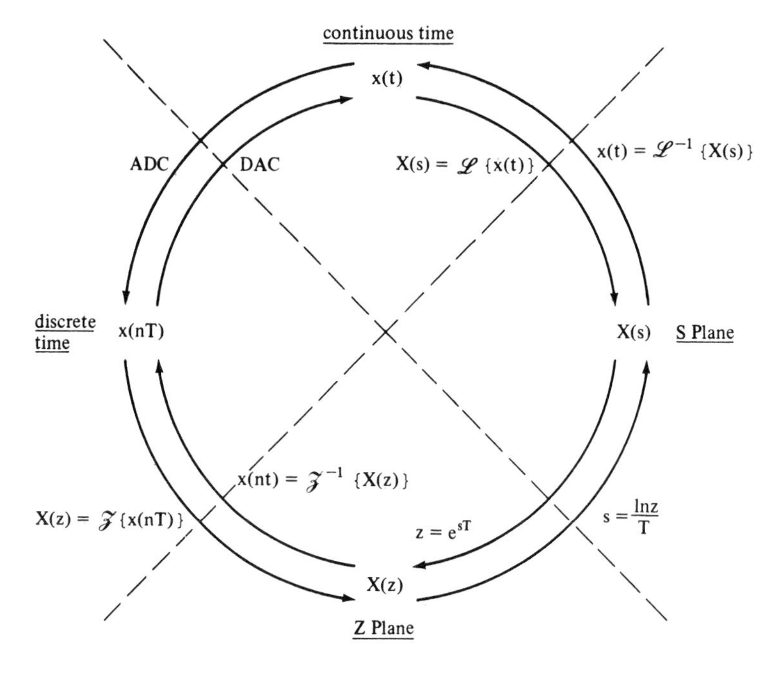


**Figure 1.11:** Regions, relations, and transforms.

In the $Z$-plane the unit circle is the equivalent of the $j\omega$ axis of the $S$ plane and therefore the Fourier transform has a discrete counterpart in the $Z$-plane as would be expected. The DFT forms an important aspect of the spectral analysis of signals.

For signal processing, the major concern is with sequences of sample values representing a signal. The $Z$-transform of such a sequence is easily represented and derived. $z^{-1}$ is a delay (as expected from the attributes of $e^{sT}$ in the $S$ plane). This property will be widely used in describing digital filtering in later chapters.

At this point we encounter a familiar problem if we begin to consider real signal environments; i.e., random noise or signals with random parameters. These concepts are studied in analog processing, and the equivalent concepts must be integrated into our digital processing arsenal.

## 1.3 A STATISTICAL SIGNAL MODEL

### 1.3.0 Introduction

In this section we review the concepts of statistical modeling and analysis of discrete-time signals. Signals for which no predictable time dependent function is known are common in all fields of communications and control. A knowledge of the statistical character of the signal gained through observation (or assumption) is, however, sufficient for many processing applications. (It is usual for noise or any other corrupting influence on a transmitted signal to be modeled statistically.)

There are three major aspects of this model to be considered here. First, it is necessary to establish the concepts of a random variable and a random or stochastic process; second, to use first and second order statistics to compute the frequency spectrum, and finally, to relate the physical samples of such signals, obtained by an ADC, to the statistical model.

In Section 1.3.1, the concepts of a random variable are introduced. The key issues here are the first and second order statistics which are used to compute moments of a random variable. In addition the concept of statistical independence and statistical correlation are described.

In Section 1.3.2 the concept of a stochastic process is introduced. The stochastic process forms the basic mathematical model for nondeterministic or random signals. Such signals are characterized by both first and second order statistics.

In Section 1.3.3, the crucial issue of replacing statistical averages with time averages is introduced. This method of computing statistical parameters is the only way available in practice; therefore it is an urgent matter to determine conditions under which such a computation is valid.

In Section 1.3.4 the computation of power spectral densities from autocorrelation sequences is discussed. This is an important problem faced in many areas of signal processing.

### 1.3.1 Random Variables

The most basic concept of probability theory is that of a probability space, denoted $S$, which contains elements denoted $e$. These elements can be considered as elementary experimental outcomes or "samples" and $S$ defines the space of all possible samples. Subsets of the elements of $S$ are called events, denoted $E$. The probability of a particular event $E$ will be denoted $P\{E\}$ and by convention is normalized to a number between 0 and 1.

#### Real Random Variables

A real random variable $\mathbf{x}$ is defined as a function of the elements of a probability space such that for any real number $x$ the inequality

$$\mathbf{x}(e) \leqslant x \tag{1.25}$$

defines a set of elements (i.e., an event), denoted $\{\mathbf{x} \leqslant x\}$ whose probability of occurrence is defined. We shall use boldface to represent random variables and standard type to represent specific sample values or realizations of the random variable.

#### Complex Random Variables

A "complex random variable" is defined as the assignment of a complex number $\mathbf{z}(e)$ to the elements of a probability space, i.e.,

$$\mathbf{z}(e) = \mathbf{x}(e) + j\mathbf{y}(e) \tag{1.26}$$

where the functions $\mathbf{x}$ and $\mathbf{y}$ are real random variables.

#### Distribution Functions

The probability of the event $\{\mathbf{x} \leqslant x\}$ defines a function of the real number $x$ with respect to the random variable $\mathbf{x}$ known as the "probability distribution function" of the random variable, i.e.,

$$F_{\mathbf{x}}(x) \triangleq P\{\mathbf{x} \leqslant x\}, \quad -\infty < x < \infty \tag{1.27}$$

The distribution function of a random variable has the following properties:

1. $$\lim_{x \to -\infty} F_{\mathbf{x}}(x) = 0 \tag{1.28}$$

2. $$\lim_{x \to +\infty} F_{\mathbf{x}}(x) = 1 \tag{1.29}$$

3. $$F_{\mathbf{x}}(x_1) \leqslant F_{\mathbf{x}}(x_2), \quad x_1 < x_2 \tag{1.30}$$

4. $$P\{x_1 < \mathbf{x} \leqslant x_2\} = F_{\mathbf{x}}(x_2) - F_{\mathbf{x}}(x_1) \tag{1.31}$$

### Continuous Random Variables and Density Functions

A "continuous random variable" is defined as one whose distribution function is a continuous function of $x$. In this case $F_{\mathbf{x}}(x)$ is differentiable at all but a finite number of points and the derivative of $F_{\mathbf{x}}(x)$ is called the "probability density function," denoted

$$f_{\mathbf{x}}(x) \triangleq \frac{d}{dx} F_{\mathbf{x}}(x) \tag{1.32}$$

The density function has the following properties:

1. $$f_{\mathbf{x}}(x) \geqslant 0 \tag{1.33}$$

2. $$\int_{-\infty}^{\infty} f_{\mathbf{x}}(x)\,dx = F_{\mathbf{x}}(+\infty) - F_{\mathbf{x}}(-\infty) = 1 \tag{1.34}$$

3. $$F_{\mathbf{x}}(x_o) = \int_{-\infty}^{x_o} f_{\mathbf{x}}(x)\,dx \tag{1.35}$$

4. $$P\{x_1 < \mathbf{x} \leqslant x_2\} = \int_{x_1}^{x_2} f_{\mathbf{x}}(x)\,dx \tag{1.36}$$

### Discrete Random Variables and Mass Functions

A discrete random variable is defined as one that can assume only a countable number of values. As such, its distribution function is of a staircase form as shown in Figure 1.12. The discontinuities appear at the allowable realization values of the random variable and the magnitude of the discontinuity represents the probability of the random variable taking on that value. Thus

$$F_{\mathbf{x}}(x) = \sum_i P\{\mathbf{x} = x_i\}, \quad x_i \leqslant x \tag{1.37}$$

The probability $P\{\mathbf{x} = x\}$ is often called the probability mass function of a discrete random variable, denoted

$$\begin{aligned} m_{\mathbf{x}}(x) &\triangleq \lim_{\epsilon \to 0}\left[F_{\mathbf{x}}(x) - F_{\mathbf{x}}(x - \epsilon)\right] \\ &= F_{\mathbf{x}}(x) - F_{\mathbf{x}}(x^-) \\ &= P\{\mathbf{x} = x\} \end{aligned} \tag{1.38}$$

Thus

$$F_{\mathbf{x}}(x) = \sum_{x \leqslant x'} m_{\mathbf{x}}(x') \tag{1.39}$$

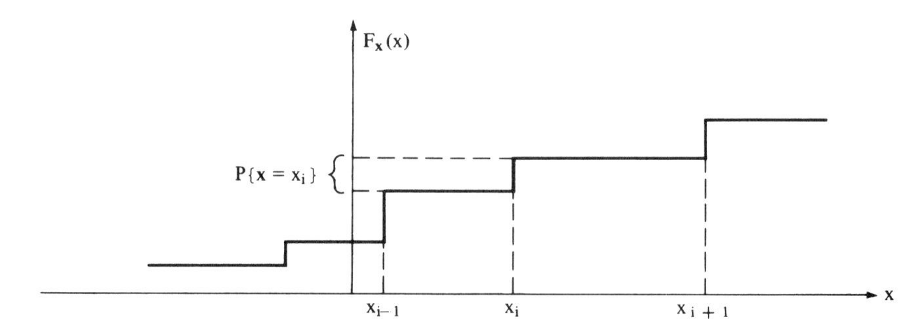


**Figure 1.12:** Probability Distribution function of a discrete random variable.

### Moments

A random variable is completely characterized by its distribution function or equivalently by its density function. However, it is often useful to describe the statistics of a random variable in terms of its moments $m_n$ defined as

$$m_n = E\{\mathbf{x}^n\} \triangleq \int_{-\infty}^{\infty} x^n f_{\mathbf{x}}(x)\,dx \tag{1.40}$$

where the subscript $n$ is called the order of the moment. The notation $E$ is usually called the "expectation operator." An important property of the expectation operator is the fact that the expectation of a sum is equal to the sum of the individual expectations.

If the random variable is discrete, equation (1.40) may be written as the summation

$$m_n = E\{\mathbf{x}^n\} = \sum_i x_i^n P\{\mathbf{x} = x_i\} \tag{1.41}$$

Clearly, for $n = 0$ we have

$$m_o = 1$$

For $n = 1$, we have

$$m_1 = E\{\mathbf{x}\} \triangleq \int_{-\infty}^{\infty} x f_{\mathbf{x}}(x)\,dx = \eta_{\mathbf{x}} \tag{1.42}$$

which is known as the "expectation," or the "mean," or the "average," or the "first-moment" of the random variable **x**. The "second-moment" or "mean square" value is

$$E\{\mathbf{x}^2\} \triangleq \int_{-\infty}^{\infty} x^2 f_{\mathbf{x}}(x)\,dx \tag{1.43}$$

The "$n^{th}$ central moment" of $\mathbf{x}$ is defined as

$$\mu_n = E\{(\mathbf{x} - \eta_{\mathbf{x}})^n\} = \int_{-\infty}^{\infty} (\mathbf{x} - \eta_{\mathbf{x}})^n f_{\mathbf{x}}(x)\,dx \tag{1.44}$$

The "second central moment" is called the "variance" of $\mathbf{x}$, i.e.,

$$\begin{aligned} var\{\mathbf{x}\} &= E\{(\mathbf{x} - \eta_{\mathbf{x}})^2\} = E\{\mathbf{x}^2\} - 2\eta_{\mathbf{x}} E\{\mathbf{x}\} + \eta_{\mathbf{x}}^2 \\ &= E\{\mathbf{x}^2\} - \eta_{\mathbf{x}}^2 \end{aligned} \tag{1.45}$$

and is a measure of the dispersion of the random variable.

The variance is often denoted $\sigma^2$ and its positive square root is known as the "standard deviation."

$$\sigma_{\mathbf{x}} \triangleq + \sqrt{var\{\mathbf{x}\}} \tag{1.46}$$

### Jointly and Marginally Distributed Random Variables

When two or more random variables are defined on the same probability space they are said to be "jointly distributed" and their characteristics can be described by their "joint distribution or density functions" and their "joint moments." If we are interested in only some of the random variables defined on a space we are concerned with marginal distribution and density functions.

The "joint probability distribution" function of a set of random variables $\mathbf{x}_1, \mathbf{x}_2, \ldots \mathbf{x}_n$ defined on the same probability space is defined as the probability of the event $\mathbf{x}_1 < x_1, \mathbf{x}_2 < x_2, \ldots, \mathbf{x}_n < x_n$. i.e.,

$$F_{\mathbf{x}_1, \mathbf{x}_2, \ldots, \mathbf{x}_n}(x_1, x_2, \ldots, x_n) = P\{\mathbf{x}_1 \leqslant x_1, \mathbf{x}_2 \leqslant x_2, \ldots, \mathbf{x}_n \leqslant x_n\} \tag{1.47}$$

The "joint density function" is defined as

$$f_{\mathbf{x}_1, \ldots, \mathbf{x}_n}(x_1, \ldots, x_n) = \frac{\partial^n F_{\mathbf{x}_1, \ldots, \mathbf{x}_n}(x_1, \ldots, x_n)}{\partial x_1 \partial x_2 \ldots \partial x_n} \tag{1.48}$$

and for discrete random variables

$$m_{\mathbf{x}_1, \ldots, \mathbf{x}_n}(x_1, \ldots, x_n) = P\{\mathbf{x}_1 = x_1, \ldots, \mathbf{x}_n = x_n\} \tag{1.49}$$

If we are interested only in the set of random variables $\mathbf{x}_1, \ldots, \mathbf{x}_m$, $m < n$, out of the total set of jointly distributed random variables $\mathbf{x}_1, \ldots, \mathbf{x}_n$, then their "marginal distribution" is

$$F_{\mathbf{x}_1, \ldots, \mathbf{x}_m}(x_1, \ldots, x_m) = F_{\mathbf{x}_1, \ldots, \mathbf{x}_n}(x_1, \ldots, x_m, \infty, \ldots, \infty) \tag{1.50}$$

Their "marginal density" function is

$$f_{\mathbf{x}_1, \ldots, \mathbf{x}_m}(x_1, \ldots, x_m) = \int_{-\infty}^{\infty} \ldots \int_{-\infty}^{\infty} f_{\mathbf{x}_1, \ldots, \mathbf{x}_n}(x_1, \ldots, x_m)\,dx_{m+1} \ldots dx_n \tag{1.51}$$

**Independence**

The jointly distributed random variables $\mathbf{x}_1, \mathbf{x}_2, \ldots, \mathbf{x}_n$ are said to be independent (mutually independent) if

$$f_{\mathbf{x}_1,\ldots,\mathbf{x}_n}(x_1,\ldots,x_n) = f_{\mathbf{x}_1}(x_1)f_{\mathbf{x}_2}(x_2)\cdots f_{\mathbf{x}_n}(x_n) \tag{1.52}$$

**Joint Moments**

The "joint moments" $m_{kr}$ of two jointly distributed random variables $\mathbf{x}$ and $\mathbf{y}$ are defined as

$$m_{kr} = E\{\mathbf{x}^k\mathbf{y}^r\} = \int_{\infty}^{\infty}\int_{\infty}^{\infty} x^k y^r f_{\mathbf{xy}}(x,y)\,dxdy \tag{1.53}$$

for continuous random variables and as

$$m_{kr} = E\{\mathbf{x}^k\mathbf{y}^r\} = \sum_i\sum_i x_i^k y_j^r P\{\mathbf{x} = x_i, \mathbf{y} = y_j\} \tag{1.54}$$

for discrete random variables.

The sum of the subscripts $k + r = n$ is the order of the moment (i.e., $m_{11}$, $m_{20}$, $m_{02}$ are all second order joint moments).

The first order joint moments $m_{10}$ and $m_{01}$ are the mean values of $\mathbf{x}$ and $\mathbf{y}$.

$$m_{10} = E\{\mathbf{x}\} = \eta_{\mathbf{x}} \tag{1.55}$$

$$m_{01} = E\{\mathbf{y}\} = \eta_{\mathbf{y}} \tag{1.56}$$

The second order moments $m_{20}$ and $m_{02}$ are the mean square values of $\mathbf{x}$ and $\mathbf{y}$, i.e.,

$$m_{20} = E\{\mathbf{x}^2\} \tag{1.57}$$

$$m_{02} = E\{\mathbf{y}^2\} \tag{1.58}$$

and $m_{11}$ is often denoted $R_{\mathbf{xy}}$, i.e.,

$$m_{11} = R_{\mathbf{xy}} = E\{\mathbf{xy}\} \tag{1.59}$$

If $R_{\mathbf{xy}} = 0$, $\mathbf{x}$ and $\mathbf{y}$ are said to be orthogonal.

The "joint central moments" $\mu_{kr}$ are defined as

$$\mu_{kr} \triangleq E\{(\mathbf{x} - \eta_{\mathbf{x}})^k(\mathbf{y} - \eta_{\mathbf{y}})^r\} = \int_{-\infty}^{\infty}\int_{-\infty}^{\infty} (x - \eta_{\mathbf{y}})^r f_{\mathbf{x},\mathbf{y}}(x,y)\,dxdy \tag{1.60}$$

for continuous random variables and as

$$\begin{aligned}\mu_{kr} &= E\{(\mathbf{x} - \eta_{\mathbf{x}})^k(\mathbf{y} - \eta_{\mathbf{y}})^r\} \\ &= \sum_i\sum_j (x_i - \eta_{\mathbf{x}})^k(y_{yj} - \eta_{\mathbf{y}})^r P\{\mathbf{x} = x_i, \mathbf{y} = y\}\end{aligned} \tag{1.61}$$

The second order central moments $\mu_{20}$ and $\mu_{02}$ are the variances of **x** and **y**, i.e,

$$\mu_{20} = \sigma_{\mathbf{x}}^2 \tag{1.62}$$

$$\mu_{02} = \sigma_{\mathbf{y}}^2 \tag{1.63}$$

The second order central moment $\mu_{11}$ is called the covariance of **x** and **y**

$$\begin{aligned}\mu_{11} &= E\{(\mathbf{x} - \eta_{\mathbf{x}})(\mathbf{y} - \eta_{\mathbf{y}})\} \\ &= E\{\mathbf{xy}\} - \eta_{\mathbf{x}}\eta_{\mathbf{y}}\end{aligned} \tag{1.64}$$

Note that if **x** and **y** are both zero mean random variables, then we have $\mu_{11} = R_{\mathbf{xy}}$.

### Correlation Coefficient

The "correlation coefficient" of **x** and **y** is defined as

$$r \triangleq \frac{\mu_{11}}{\sigma_{\mathbf{x}}\sigma_{\mathbf{y}}} = \frac{\mu_{11}}{(\mu_{20}\mu_{02})^{1/2}} \tag{1.65}$$

and $r \leqslant 1$.

If

$$E\{\mathbf{xy}\} = E\{\mathbf{x}\}E\{\mathbf{y}\} \tag{1.66}$$

then $\mu_{11} = 0$, $r = 0$ and **x** and **y** are said to be "uncorrelated."

If **x** and **y** are independent, then **x** and **y** are also uncorrelated. The converse is not generally true.

### Conditional Distributions, Densities and Expectations

For two jointly distributed random variables **x** and **y** the "conditional probability distribution function" of **x** given that $\mathbf{y} = y$ is defined as

$$F_{\mathbf{x}|\mathbf{y}}(x|y) \underset{\lim \Delta y \to 0}{=} \frac{P\{\mathbf{x} \leqslant x, y \leqslant \mathbf{y} \leqslant y + \Delta y\}}{P\{y \leqslant \mathbf{y} \leqslant y + \Delta y\}} \tag{1.67}$$

The "conditional density function" $f_{\mathbf{x}|\mathbf{y}}(x,y)$ of **x** given that $\mathbf{y} = y$ is defined as

$$f_{\mathbf{x}|\mathbf{y}}(x|y) = \frac{f_{\mathbf{x},\mathbf{y}}(x,y)}{f_{\mathbf{y}}(y)} \tag{1.68}$$

Since the definitions apply equally to either **x** or **y** equation (1.68) may be written as

$$f_{\mathbf{x}|\mathbf{y}}(x|y) = \frac{f_{\mathbf{y}|\mathbf{x}}(y|x)f_{\mathbf{x}}(x)}{f_{\mathbf{y}}(y)} \tag{1.69}$$

This equation is known as Baye's theorem. We shall see in Chapter 3 that it forms the foundation of detection and estimation theory.

The "conditional expectation" of a random variable **x** given **y** is defined as

$$E\{\mathbf{x}|\mathbf{y}\} \triangleq \int_{-\infty}^{\infty} x f_{\mathbf{x}|\mathbf{y}}(x|y)\,dx \tag{1.70}$$

The following properties apply to conditional densities and expectations.

1.
$$f_{\mathbf{x}|\mathbf{y}}(x|y) \geqslant 0 \tag{1.71}$$

2.
$$\int_{-\infty}^{\infty} f_{\mathbf{x}|\mathbf{y}}(x|y)\,dx = 1 \tag{1.72}$$

3. If **x** and **y** are dependent, then
$$f_{\mathbf{x}|\mathbf{y}}(x|y) = f_{\mathbf{x}}(x) \text{ and } E\{\mathbf{x}|\mathbf{y}\} = E\{\mathbf{x}\} \tag{1.73}$$

4.
$$f_{\mathbf{x}}(x) = E\{f_{\mathbf{x}|\mathbf{y}}(x|y)\} = \int_{-\infty}^{\infty} f_{\mathbf{x}|\mathbf{y}}(x|y) f_{\mathbf{y}}(y)\,dy \tag{1.74}$$

### 1.3.2 Stochastic Processes

In the previous section we reviewed the concepts of random variables. If the assignment of a number $x$ to an element of the probability space $S$ is time dependent, we create a two-dimensional process

$$\mathbf{x}(t,e) = x \tag{1.75}$$

known as a random or stochastic process. We shall use the terms random and stochastic interchangeably.

For any fixed time $t$ the function $\mathbf{x}(t,e)$ is simply a random variable which may be either continuous or discrete. This random variable may be considered as the value of a nondeterministic signal at any time $t$. The time dependence of $\mathbf{x}(t,e)$ may also be defined as continuous or discrete. In the case of a discrete-time dependence the process may be considered as a time-indexed set of random variables. Thus the concept of a "stochastic process" may be employed to model nondeterministic signals of either analog or discrete-time form. The time dependence of the process, continuous or discrete, corresponds to the continuous or discrete-time nature of the signal, and the type of random variable corresponds to continuous or quantized signal values.

A nondeterministic digital signal is conveniently modeled as a discrete type and discrete time-dependent stochastic process, or equivalently as a time-indexed set of discrete type random variables.

### First and Second Order Distributions

Because a stochastic process is a time dependent random variable, its probability distribution function is also time dependent, i.e.,

$$F(x,t) = P\{\mathbf{x}(t) \leqslant x\} \tag{1.76}$$

Equation (1.76) is known as the first-order distribution of the stochastic process $\mathbf{x}(t)$. Note we have dropped the explicit inclusion of $e$ in the expression of the stochastic process, and we have also omitted the subscript from the distribution function. The argument of the distribution (or density) function will identify the random variable or process to which the function applies.

At two distinct times, $t_1$ and $t_2$, we have two random variables, $\mathbf{x}(t_1)$ and $\mathbf{x}(t_2)$, that are characterized by their joint probability distribution function,

$$F(x_1,x_2;t_1,t_2) = P\{\mathbf{x}(t_1) \leqslant x_1, \mathbf{x}(t_2) \leqslant x_2\} \tag{1.77}$$

This distribution is in general dependent on the two times $t_1$ and $t_2$ and is known as the "second order distribution" of the process.

### First and Second Order Density Functions

The first and second order density functions are similarly time dependent.

The "first order density function" of $\mathbf{x}(t)$ is defined as

$$f(x,t) \triangleq \frac{\partial F(x,t)}{\partial x} \tag{1.78}$$

and the "second order density function" is given by

$$f(x_1,x_2;t_1,t_2) \triangleq \frac{\partial^2 F(x_1,x_2;t_1,t_2)}{\partial x_1 \partial x_2} \tag{1.79}$$

Higher order distributions and densities are similarly defined. The real stochastic process $\mathbf{x}(t)$ is "statistically determined" if its distribution functions (or equivalently its density functions) are known for every order $n$.

### First Order Statistics

The "mean value" of a stochastic process is defined as

$$\eta_{\mathbf{x}}(t) = E\{\mathbf{x}(t)\} = \int_{-\infty}^{\infty} x\, f(x,t)\, dx \tag{1.80}$$

or

$$\eta_{\mathbf{x}}(t) = \sum_i x_i P\{\mathbf{x}(t) = x_i\} \tag{1.81}$$

for the case of continuous or discrete random variables respectively.

The variance is also a function of time, i.e.,

$$\sigma_{\mathbf{x}}^2(t) = E\{(\mathbf{x}(t) - \eta_{\mathbf{x}}(t))\} \tag{1.82}$$

The mean and variance are known as first order statistics of the process.

### Second Order Statistics

The "autocorrelation" of a process $\mathbf{x}(t)$ denoted $R(t_1,t_2)$ is the joint moment of the random variables $\mathbf{x}(t_1)$ and $\mathbf{x}(t_2)$, i.e.,

$$R_{\mathbf{xx}}(t_1,t_2) \triangleq E\{\mathbf{x}(t_1)\mathbf{x}(t_2)\} \tag{1.83}$$

The autocorrelation of a stochastic process is a measure of the dependence between values of the process at different times, i.e., it describes the time variation of the signal modeled by the process.

The "autocovariance" of $\mathbf{x}(t)$ denoted $C_{\mathbf{xx}}(t_1,t_2)$ is the joint central moment of $\mathbf{x}(t_1)$ and $\mathbf{x}(t_2)$, i.e.,

$$C_{\mathbf{xx}}(t_1,t_2), = E\{(\mathbf{x}(t_1) - \eta_{\mathbf{x}}(t_1))(\mathbf{x}(t_2) - \eta_{\mathbf{x}}(t_2))\} \tag{1.84}$$

The autocorrelation and autocovariance functions are second order statistics of the process and are two-dimensional functions of the times $t_1$ and $t_2$.

From equations (1.83) and (1.84) we have

$$C_{\mathbf{xx}}(t_1,t_2) = R_{\mathbf{xx}}(t_1,t_2) - \eta_{\mathbf{x}}(t_1)\eta_{\mathbf{x}}(t_2) \tag{1.85}$$

### Complex Processes

A complex stochastic process is defined as a time dependent complex random variable

$$\mathbf{z}(t) = \mathbf{x}(t) + j\mathbf{y}(t) \tag{1.86}$$

where $\mathbf{x}(t)$ and $\mathbf{y}(t)$ are real stochastic processes.

For a complex process the auto correlation is defined as

$$R_{\mathbf{xx}}(t_1,t_2) \triangleq E\{\mathbf{x}(t_1)\mathbf{x}^*(t_2)\} \tag{1.87}$$

where $\mathbf{x}^*(t_2)$ represents the complex conjugate of $\mathbf{x}(t_2)$. Similarly the auto covariance of a complex process is given as

$$C_{\mathbf{xx}}(t_1,t_2) = E\{(\mathbf{x}(t_1) - \eta_{\mathbf{x}}(t_1))(\mathbf{x}^*(t_2) - \eta_{\mathbf{x}}^*(t_2))\} \tag{1.88}$$

The cross-correlation of two complex processes is defined as

$$R_{\mathbf{xy}}(t_1,t_2) = E\{\mathbf{x}(t_1)\mathbf{y}^*(t_2)\} \tag{1.89}$$

The cross-covariance of two complex processes is defined as

$$C_{\mathbf{xy}}(t_1,t_2) = E\{(\mathbf{x}(t_1) - \eta_{\mathbf{x}}(t_1))(\mathbf{y}^*(t_2) - \eta_{\mathbf{y}}^*(t_2))\}$$
$$= R_{\mathbf{xy}}(t_1,t_2) - \eta_{\mathbf{x}}(t_1)\eta_{\mathbf{y}}^*(t_2) \tag{1.90}$$

**Random Vectors**

For modeling a discrete-time signal, a stochastic process can be viewed as a time-indexed set of random variables defined on the same probability space. Thus, the probabilistic or statistical nature of the signal can be described in terms of the joint distributions and joint moments of this set of random variables.

When dealing with such a set of jointly distributed random variables it is convenient to introduce the concept of a random vector,

$$\underline{\mathbf{x}} = \begin{bmatrix} \mathbf{x}_1 \\ \mathbf{x}_2 \\ \vdots \\ \mathbf{x}_n \end{bmatrix} \tag{1.91}$$

and its transpose conjugate,

$$\underline{\mathbf{x}}^t = \begin{bmatrix} \mathbf{x}_1^*, \mathbf{x}_2^*, \ldots, \mathbf{x}_n^* \end{bmatrix} \tag{1.92}$$

The elements of the random vector $\underline{\mathbf{x}}$ are the various random variables defined on the same probability space.

The expectation vector is defined as

$$E\{\underline{\mathbf{x}}\} = \begin{bmatrix} E\{\mathbf{x}_1\} \\ E\{\mathbf{x}_2\} \\ \vdots \\ E\{\mathbf{x}_n\} \end{bmatrix} \tag{1.93}$$

The autocovariance matrix of an $n$-dimensional random vector is the $n \times n$ random matrix

$$C_{\mathbf{xx}} = \begin{bmatrix} cov\{\mathbf{x}_i\mathbf{x}_j\} \end{bmatrix} = E\{\underline{\mathbf{x}} - \underline{\eta}_{\mathbf{x}})(\underline{\mathbf{x}} - \underline{\eta}_{\mathbf{x}})^t\} \tag{1.94}$$
$$= \begin{bmatrix} var\{\mathbf{x}_1\} & cov\{\mathbf{x}_1\mathbf{x}_2\} & \ldots & cov\{\mathbf{x}_1\mathbf{x}_n\} \\ cov\{\mathbf{x}_2\mathbf{x}_1\} & var\{\mathbf{x}_2\} & \ldots & cov\{\mathbf{x}_2\mathbf{x}_n\} \\ \vdots & & & \\ cov\{\mathbf{x}_n\mathbf{x}_1\} & cov\{\mathbf{x}_n\mathbf{x}_2\} & \ldots & var\{\mathbf{x}_n\} \end{bmatrix}$$

The autocorrelation matrix of an $n$-dimensional random vector $\underline{\mathbf{x}}$ is defined as the $n \times n$ random matrix

$$R_{\mathbf{xx}} = \left[R_{\mathbf{x}_i\mathbf{x}_j}\right] = \left[E\{\underline{\mathbf{x}}\underline{\mathbf{x}}^t\}\right] \tag{1.95}$$

$$\begin{bmatrix} E\{|\mathbf{x}_1|^2\} & E\{\mathbf{x}_1\mathbf{x}_2^*\} & \dots & E\{\mathbf{x}_1\mathbf{x}_n^*\} \\ E\{\mathbf{x}_2\mathbf{x}_1^*\} & E\{|\mathbf{x}_2|^2\} & & \\ \vdots & & & \\ E\{\mathbf{x}_n\mathbf{x}_1^*\} & E\{\mathbf{x}_n\mathbf{x}_2^*\} & & E\{|\mathbf{x}_n|^2\} \end{bmatrix}$$

Clearly both the covariance and correlation matrices are symmetric for real processes.

If we are concerned with two stochastic processes $\mathbf{x}(n)$ and $\mathbf{y}(n)$ we define the cross-correlation matrix as

$$R_{\mathbf{xy}} = E\{\underline{\mathbf{x}}(n)\underline{\mathbf{y}}^t(n)\} \tag{1.96}$$

where $\underline{\mathbf{x}}$ and $\underline{\mathbf{y}}$ are random vectors representing the stochastic process as a time-indexed set of random variables.

Similarly the cross-covariance matrix is given as

$$C_{\mathbf{xy}} = E\left\{(\underline{\mathbf{x}} - \underline{\eta}_{\mathbf{x}})(\underline{\mathbf{y}} - \underline{\eta}_{\mathbf{y}})^t\right\} \tag{1.97}$$

**Gaussian, Markov and White Noise Processes**

Of particular importance in the modeling of nondeterministic signals are the concepts of Gaussian (or Normal) and Markov stochastic processes and the concept of white noise.

A "Gaussian process" is a stochastic process that is statistically determined by a normal distribution. A random variable is said to be normally distributed if its density function is of the form

$$f_{\mathbf{x}}(x) = \frac{1}{\sqrt{2\pi\sigma_{\mathbf{x}}^2}} e^{-\frac{1}{2}\left(\frac{x - \eta_{\mathbf{x}}}{\sigma_{\mathbf{x}}}\right)^2} \tag{1.98}$$

where $\eta_{\mathbf{x}}$ (mean) and $\sigma_{\mathbf{x}}$ (variance) are constants. These two parameters completely characterize a Gaussian density. Thus, a Gaussian process can be statistically determined from its mean and autocorrelation or covariance functions which can be determined from the first and second order densities of the process.

A process is said to be "first order Markov" if the statistics of the process at any future time $t > \tau$, where $\tau$ represents a given time (e.g., the present), are unaffected by the past history of the process for all $t < \tau$. This property can be stated in terms of density functions as

$$f(x(t_n)|x(t_{n-1}), \dots, x(t_1)) = f(x(t_n)|x(t_{n-1})) \tag{1.99}$$

Thus the statistical determination of a Markov process can be expressed by specifying $f(x(\tau))$ and $f(x(t)|x(\tau))$ for $t > \tau$.

The conditional densities $f(x(t)|x(\tau))$ are known as the "transitional probability densities" of the process. These transitional densities satisfy the Chapman-Kolmogoroff (*C-K*) equation

$$f(x(t_n)|x(t_s)) = \int_{-\infty}^{\infty} f(x(t_n)|x(t_r))f(x(t_r)|x(t_s))\,dx(t_r) \qquad (1.100)$$

where $n > r > s$. In terms of discrete random variables the *C-K* equation can be written as

$$P_{ij}(n,s) = \sum_k P_{ik}(n,r)P_{kj}(r,s)\,, \quad n > r > s \qquad (1.101)$$

where the notation

$$P_{ij}(n,s) = P\{\mathbf{x}(t_n) = x_i|\mathbf{x}(t_s) = x_j\}\,, \quad n > r > s$$

is called the conditional transition probability. Thus, the *C-K* equation tells us that the conditional transition probability is equal to the sum of all possible transition paths from state $j$ at time $s$ to state $i$ at time $n$ and is thus independent of any particular path. The significance of this property of Markov processes will be explored further in Chapter 3 with respect to recursive estimation.

Since a discrete time stochastic process can be viewed as a time-indexed set of random variables, it is convenient to consider this set as a sequence. If the process is Markov then this sequence of random variables is called a Markov sequence. A white random sequence or "white noise" is a Markov sequence displaying the property

$$f(x(t_n)|x(t_m)) = f(x(t_n))\,, \quad n > m \qquad (1.102)$$

This implies that all the random variables of the sequence are mutually independent and therefore uncorrelated. If the individual random variables are also Gaussian then the sequence is called a "Gaussian white noise" sequence.

### 1.3.3 Time Averages, Stationarity and Ergodicity

Thus far we have been primarily concerned with ensemble averages of stochastic processes, that is, the statistical averages of the random variable represented by the process at any specific point in time. In a practical sense we usually have only a single sample sequence representing a discrete-time signal. Each sample value represents only a single realization of each time indexed random variable of the stochastic process model (i.e., a specific measurement at a specific period of time). In this section we show how the connection is made between the ensemble properties of the stochastic process model and the data available from a sequence of samples obtained over a period of time.

**Time Averages**

We have seen that a stochastic process is a two-dimensional function of the elements of a probability space and of time. We have introduced some concepts of the ensemble averages of a process. Clearly, we can form time averages of a process as well. In the case of a continuous-time stochastic process the concept of time averages is developed through the vehicle of stochastic integrals. We do not pursue this concept, but refer the reader to the references for a discussion on stochastic integrals (8). In the case of a discrete-time signal modeled as a discrete-time random process we form the time average of the infinite sample sequence $x(n)$ as

$$<x(n)> = \lim_{N\to\infty} \frac{1}{2N+1} \sum_{n=-N}^{N} x(n) \tag{1.103}$$

Similarly we can form other time averages analogous to the previously developed ensemble averages, i.e.,

1. Mean square time average

$$<x^2(n)> = \lim_{N\to\infty} \frac{1}{2N+1} \sum_{n=-N}^{N} x^2(n) \tag{1.104}$$

2. Time variance

$$<(x(n) - <x(n))>)^2> \tag{1.105}$$

$$= \lim_{N\to\infty} \frac{1}{2N+1} \sum_{n=-N}^{N} (x(n) - <x(n)>)^2$$

3. Time autocorrelation sequence

$$R_x(m) = <x(n)x(n+m)> \tag{1.106}$$

$$= \lim_{N\to\infty} \frac{1}{2N+1} \sum_{n=-N}^{N} x(n)x*(n+m)$$

4. Time autocovariance sequence

$$C_x(m) = <(x(n) - <x(n)>)(x^*(n+m) - <x(n)>^*)> \tag{1.107}$$

which by substituting the appropriate sum is equivalent to:

$$C_x(m) = R_x(m) - |<x(m)>|^2 \tag{1.108}$$

These averages are functions of an infinite set of random variables and as such are random variables themselves. We shall now show that under certain conditions these time averages can be used in place of ensemble averages.

### Stationary Processes

The limits indicated in equations (1.103) to (1.108) can be shown to exist if the sequence $\mathbf{x}(n)$ is stationary and has a finite mean (8).

A process is said to be "stationary in the strict sense" if all its distribution (or density) functions are independent of the position of the time origin. If only those distributions of order $k$ or less are independent of a shift in the time origin, then the process is said to be "stationary of order $k$".

A stationary process has a time independent first order density and therefore its mean is a constant, i.e.,

$$\eta_x(t) = \eta_x = \text{a constant} \tag{1.109}$$

The second order density of a stationary process depends only on the time difference $t_1 - t_2$, and its autocorrelation and autocovariance functions are also dependent only on the time difference, i.e.,

$$R_{\mathbf{xx}}(t_2,t_1) = R_{\mathbf{xx}}(\tau)\,;\; t_2 - t_1 = \tau \tag{1.110}$$

and

$$C_{\mathbf{xx}}(t_2,t_1) = C_{\mathbf{xx}}(\tau) = R_{\mathbf{xx}}(\tau) - |\eta_x|^2 \tag{1.111}$$

Furthermore, since the choice of time origin position is arbitrary we have

$$R_{\mathbf{xx}}(\tau) = R^*_{\mathbf{xx}}(-\tau) \tag{1.112}$$

It can also be shown that

$$|R_{\mathbf{xx}}(\tau)| \leqslant R(0) \tag{1.113}$$

A somewhat weaker degree of stationarity known as "wide sense stationarity" is defined if equations (1.109) and (1.110) are valid. Clearly, a process that is stationary of order 2 or greater is wide sense stationary; the converse, however, is not generally true.

Two processes are said to be "jointly stationary" in the wide sense if equations (1.109) and 1.110 hold for each process and their cross-correlation also depends only on the time difference $\tau$, i.e.,

$$E\{\mathbf{x}(t + \tau)y(t)\} = R_{\mathbf{xy}}(\tau) \tag{1.114}$$

Thus, for stationary processes the mean is a constant and the correlation and covariance functions are one-dimensional functions of the time difference $\tau$, or for discrete-time processes, the difference in time indices $n_1 - n_2 = m$.

### Ergodic Processes

The "Ergodic hypothesis" assumes that the time averages of equations 1.103 to 1.108 are equal to the ensemble averages of a stationary process. The definition of an ergodic process is often stated as:

> The process $\mathbf{x}(t)$ is ergodic if time averages equal ensemble expected values.

This is often very difficult to prove in practice.

The determination of the specific criteria for ergodicity is based on the concept of taking time averages over a finite period [-T,T] and seeing under what conditions the variance of the resulting stochastic integral approaches zero as T approaches infinity. For a more detailed discussion of ergodicity the reader is referred to Papoulis (8).

**Estimates of Time Averages**

The infinite limits expressed in equations (1.103) to (1.108) cannot in general be computed practically. It is common practice to estimate such time averages from a finite sequence of sample values and utilize these estimates as expected values under an ergodic hypothesis, i.e.,

$$E\{\mathbf{x}(n)\} = <x(n)> \simeq \frac{1}{2N+1} \sum_{n=-N}^{N} x(n) \tag{1.115}$$

and

$$R_{\mathbf{xx}}(m) = <x(n)x(n+m)> \simeq \frac{1}{2N+1} \sum_{n=-N}^{N} x(n)x*(n+m) \tag{1.116}$$

We shall have more to say about such estimations in Chapter 3.

### 1.3.4 Power Spectral Density

We have seen that the autocorrelation of a stationary stochastic process is a function only of the time difference $\tau$, i.e.,

$$E\{\mathbf{x}(t+\tau)\mathbf{x}^*(t)\} = R(\tau) \tag{1.117}$$

If $\tau = 0$ we have

$$R(0) = E\{|\mathbf{x}(t)|^2\} \geqslant 0 \tag{1.118}$$

Thus $R(0)$ represents the "mean square or average power" of the process.

The Fourier transform of $R(\tau)$, denoted $S(\omega)$, is known as the "power density spectrum" or "spectral density" of the process, i.e.,

$$S(\omega) = \int_{-\infty}^{\infty} R(\tau)e^{-j\omega\tau}d\tau \tag{1.119}$$

and

$$R(\tau) = \frac{1}{2\pi}\int_{-\infty}^{\infty} S(\omega)e^{j\omega\tau}d\omega \tag{1.120}$$

Clearly, for $\tau = 0$

$$R(0) = \frac{1}{2\pi}\int_{-\infty}^{\infty} S(\omega)d\omega = E\{|\mathbf{x}(t)|^2\} \geqslant 0 \tag{1.121}$$

and we see that the total area under $S(\omega)$ is just $2\pi$ times the average power.

Furthermore, it can be shown that $S(\omega) \geqslant 0$ for all $\omega$ [8]. It can also be shown that

$$\Delta = \frac{1}{\pi} \int_{\omega-}^{\omega+} S(\omega)\, d\omega \tag{1.122}$$

represents the average power in the incremental frequency band $\omega-$ to $\omega+$.

If $\mathbf{x}(t)$ is a real process then $R(\tau)$ is real and even and thus

$$S(-\omega) = S(\omega) \tag{1.123}$$

The cross spectral density of two stationary processes $\mathbf{x}(t)$ and $\mathbf{y}(t)$ is defined as

$$S_{\mathbf{xy}}(\omega) = \int_{-\infty}^{\infty} R_{\mathbf{xy}}(\tau) e^{-j\omega\tau} d\tau = S_{\mathbf{xy}}^{*}(\omega) \tag{1.124}$$

assuming both $\mathbf{x}(t)$ and $\mathbf{y}(t)$ are wide sense stationary. If $\mathbf{x}(t)$ and $\mathbf{y}(t)$ are orthogonal, then

$$R_{\mathbf{xy}}(\tau) = 0 \text{ and } S_{\mathbf{xy}}(\omega) = 0 \tag{1.125}$$

### 1.3.5 Summary

We have presented the concept of a stochastic process as a model for non-deterministic signals and introduced some statistical quantities for characterizing these signals. We have shown that under certain constraints (i.e., stationarity and ergodicity) the time averages of a single sample sequence can be used in place of ensemble averages to describe a random signal. In practice, time averages formed from a finite length sample sequence are utilized as estimates of the mean, autocorrelation and autocovariance sequences of a random signal. With these estimates, a variety of processing algorithms can be formulated: for example, the spectral filtering of a known signal with additive random noise so as to minimize the noise energy. We will approach the formulation of such problems in later sections.

The power spectral density $S(\omega)$ of a real random signal is a real, even and non-negative function obtained by taking the Fourier transform of the autocorrelation function.

We conclude this section with a summary of important properties of the correlation and covariance functions for two stationary discrete-time signals $x(n)$ and $y(n)$, which may, in general, be real or complex.

$$R_{\mathbf{xx}}(m) = E\{\mathbf{x}(n)\mathbf{x}^{*}(n+m)\} \tag{1.126}$$

$$\begin{aligned} C_{\mathbf{xx}}(m) &= E\{(\mathbf{x}(n) - \eta_{\mathbf{x}})(\mathbf{x}(n+m) - \eta_{\mathbf{x}}^{*})\} \\ &= R_{\mathbf{xx}}(m) - |\eta_{\mathbf{x}}|^2 \end{aligned} \tag{1.127}$$

$$\begin{aligned} C_{\mathbf{xy}}(m) &= E\{(\mathbf{x}(n) - \eta_{\mathbf{x}})(\mathbf{y}^{*}(n+m) - \eta_{\mathbf{y}}^{*})\} \\ &= R_{\mathbf{xy}}(m) - \eta_{\mathbf{x}}\eta_{\mathbf{y}}^{*} \end{aligned} \tag{1.129}$$

$$R_{xx}(0) = E\{|\mathbf{x}(n)|^2\} = \text{mean square value} \tag{1.130}$$

$$C_{xx}(0) = \sigma_x^2 = \text{variance} \tag{1.131}$$

$$R_{xx}(m) = R_{xx}^*(-m) \tag{1.132}$$

$$C_{xx}(m) = C_{xx}^*(-m) \tag{1.133}$$

$$R_{xy}(m) = R_{xy}^*(-m) \tag{1.134}$$

$$C_{xy}(m) = C_{xy}^*(-m) \tag{1.135}$$

If $x(n)$ and $y(n)$ are real

$$|R_{xx}(m)| \leqslant R_{xx}(0) \tag{1.136}$$

$$|C_{xx}(m)| \leqslant C_{xx}(0) \tag{1.137}$$

## 1.4 CHAPTER SUMMARY

In this chapter we have taken the first step toward a review of signal processing theory from a purely digital perspective by establishing the basic concepts of digital signals. This was done in three distinct sections.

First we established our definition of a digital signal as an ordered sequence of finite precision (quantized) numbers and considered the issues of transforming analog signals to digital signals and vice versa. Two important concepts of analog to digital conversion were discussed.

1. The Sampling Theorem requires that we sample a band-limited analog signal at a rate of at least twice that of the highest frequency to avoid aliasing.
2. Sample quantization introduces irreversible errors into the digital signal representation.

Several common digital number representations were reviewed and the basic concepts of digital to analog conversion and signal generation were presented.

Second, because a digital signal is a discrete-time signal, the remainder of the chapter was devoted to discrete-time signal analysis. In Section 1.2 we introduced the *Z*-transform as the discrete-time counterpart to the familiar Laplace and Fourier transforms. The discrete Fourier transform was shown to provide a unique frequency domain representation of periodic and finite duration digital signals that is equivalent to evaluating the *Z*-transform at uniformly spaced intervals around the unit circle of the *Z*-plane. We also saw that both the *Z*-transform and the discrete Fourier transform exhibit a number of properties, somewhat similar to those of the continuous time Laplace and Fourier transforms, that are very useful in signal modeling and analysis.

Finally, in Section 1.3 we reviewed the basic concepts of random variables and stochastic processes which lead to a statistical model of a nondeterministic digital signal as a discrete-time discrete-valued stochastic process which can be thought of as a time-indexed set of random variables. The concepts of stationarity and ergodicity were introduced which, when assumed for a random signal, allow the use of time averages in place of ensemble averages. It is through these assumptions that a measurable sequence of sample values (i.e., a digital signal) can be manipulated and interpreted in a statistical sense. The first and second order moments of the statistical signal model (i.e., mean, variance, autocorrelation and autocovariance) are of particular importance in signal analysis. We saw that the Fourier transform of the autocorrelation function defines the power spectral density of the signal.

The prime purpose for reviewing the mathematical concepts of digital signal modeling and analysis in this chapter has been to establish a familiarity with the terminology and basic properties of these models in a discrete-time sense, upon which the discussions on digital signal processing theory and techniques of the following chapters can be based. Having established these basic concepts we now proceed to Chapter 2 where we begin our discussion of the processing of these signals.

## 1.5 EXERCISES

### Basic Concepts

**1.** Construct a reference table of converter characteristics as follows:

Column 1: resolution in bits (4, 6, 8, 10, 12, 14, 16)

Column 2: states ($2^n$)

Column 3: Binary weight ($2^{-n}$)

Column 4: Quantization size for 10 volts full scale (VFS)

Column 5: S/N Ratio (dB)

Column 6: Dynamic Range (dB)

Column 7: Maximum output for 10VFS

**2.** Consider sampling a signal by modulating it with a uniform pulse train having pulse width $P$. Suppose the signal is band-limited at 20 KHz and that it is sampled at the Nyquist rate.

(a) Compute a relationship between width of $P$ and the maximum value of $n$ so that the coefficients $c_n$ are within 5% of those for an ideal delta function.

(b) Suppose the clock period $T$ of the sample interval is uncertain by an amount $\pm\tau$. Assume a uniform probability within the region. Compute on the effect of this uncertainty on the maximum frequency components of the original band-limited signal that can be recovered by ideal low pass filtering with zero aliasing error.

**3.** The sampling theorem stipulates conditions for sampling and recovering a signal. Suppose instead of a low pass filter at the output that a bandpass filter is proposed. Discuss

(a) Is this feasible?

(b) If it is, propose a mathematical description of the filter.

(c) If it is not, why not?

**4.** Assume the following signals are sampled by an ideal sampler.

$$x(t) = te^{-at}$$

$$x(t) = e^{-at}\sin\omega t$$

(a) Determine the Laplace transform $X(s)$ of the sample sequence.

(b) Derive the $Z$-transform of the sample sequence.

(c) Can the Laplace transform be converted to a $Z$-transform? Demonstrate.

**5.** Consider a 16 bit A/D converter with input range $\pm 10$ VFS.

(a) Compute the maximum $S/N$ ratio.

(b) Recompute the $S/N$ ratio for each 2 volt increment up to full scale.

## Transform Analysis

**6.** Write a HOL program to evaluate the DFT (equation (1.24)). Structure the program so that the number of points and their value can be input.

(a) Estimate (or measure) the run time for 8, 16, 32, 64,... points.

(b) If possible, assemble and examine the complexity of the data references.

(c) Assuming you do not yet know about the FFT, can you see from your program any mechanisms for reducing the execution time?

**7.** Suppose a 16 bit computer is to be used to process data from a 12 bit A/D converter. The required computation is defined as:

$$z(n) = \sum_{k=0}^{N-1} a_k x(n-k)$$

This computation must be completed on the last $N$ samples before the next sample time.

(a) Either

(i) Write the required code in assembler, or

(ii) Write the required code in some HOL.

(b) Estimate the real-time capability of the system (i.e., what is the limit on $N$ as a function of $T$).

(c) Discuss the considerations for the word length imposed by the machine as the computations progress.

**8.** Suppose the spectrum of a time-limited signal is to be sampled in order to reconstruct $s(t)$. If $s(t) = 0$ for $|t| > T$.

(a) How many samples of $S(w)$ are required to preserve all information about $s(t)$?

(b) From the information given, can the frequency spacing of the samples be derived? If not, what further information is needed?

**9.** For example 4 in Section 1.2.1, derive the original signal on the assumption it was band-limited.

**10.** Consider two arrays $x(i)$ and $y(i)$, with $i = 0$ to 99. Write a HOL program to compute

$$z(n) = \sum_{k=0}^{99} x(k)y(n-k)$$

Examine this program for the following features:

(a) The complexity of data addressing.

(b) The ratio of code devoted to manipulating data as compared to operating on it.

(c) Estimate the total data storage required.

(d) How many multiplications are required?

(e) Estimate the minimum run time if possible.

**11.** Suppose the numbers in question 7 are complex. Assume the real and imaginary parts are stored in successive locations in memory.

(a) Write the code for a subroutine to multiply complex numbers.

(b) Rewrite the program for question 7 and compare the features required in this case to real arrays.

**12.** Suppose the numbers in question 10 were obtained from a 16 bit A/D converter.

(a) How many bits contain significant data?

(b) For the multiplications in question 10, how many bits of $z(n)$ are significant?

(c) Suggest a mechanism for discarding bits and defend it; discuss the other alternatives.

**13.** Prove the DFT properties proposed at the end of Section 1.2.3.

**14.** Compute the DFT of the following sequences:

$$s(n) = \delta(n) \quad n \leqslant N$$

$$s(n) = \delta(n - n_0) \quad n \leqslant N$$

In the latter case assume $n_0$ is less than $N$. Note in both cases that the sequences are of finite length.

CHAPTER 2

# Linear Systems and Digital Filters

## 2.0 INTRODUCTION

In Chapter 1 we considered the acquisition, representation, modeling, and analysis of digital signals. We now turn our attention to the processing of these signals. In the simplest terms, this involves the manipulation of sequences of digital numbers. Our overall objective is to examine these manipulations in a manner that exposes the fundamental computational requirements so that we can understand, and eventually design, signal processors.

The exact nature of a specific processing requirement is highly dependent on the particular application and on how the information content of the signal is modeled. We shall consider this application dependence in more detail in Chapter 5. In the present chapter we are more concerned with establishing the basic characteristics of digital signal processing, which to a large extent are based on the concepts of digital filtering, which are in turn based on the fundamentals of linear systems theory. To do this we shall first review the properties of discrete-time linear systems and then introduce the basic concepts of digital filters.

Two general approaches to the representation and analysis of linear systems exist. The first is in terms of their frequency domain characteristics arrived at through transform anaylsis. We shall examine these in Section 2.1 for the one-dimensional case and then extend these results to two-dimensions. The second approach emphasizes time domain analysis. Section 2.2 examines this approach. Perhaps the most significant impact of VLSI components is to bring implementation techniques based on time domain analysis within the reach of an increasing range of applications. This approach will become increasingly impoıtant, perhaps completely replacing the frequency domain approach in many applications in the near future.

A large portion of digital signal processing theory is concerned with digital filtering. In this chapter we introduce digital filters as an important subclass of discrete-time linear time-invariant systems that obey a general linear difference equation. The approach taken here is to present digital filters from the perspective of four basic areas:

1. the mathematical and graphical representation of digital filters,
2. the specification of digital filter performance,
3. the design of digital filters,
4. the implementation issue.

Each of these areas can be approached from either the time domain or the frequency domain.

In Section 2.3 we review the representations of digital filters in both the time domain and frequency domain and show how these representations lead to various realization structures. Two common methods of subdividing digital filters result. The first breaks digital filters into two major categories: infinite impulse response (IIR) filters, and finite impulse response (FIR) filters. The second method is to classify digital filters as either recursive or non-recursive. While both IIR and FIR filters can be implemented (or at least approximated) by either recursive or non-recursive techniques, it is common to associate IIR filters with recursive realizations while FIR filters are generally associated with non-recursive realizations. The reasons for this will become apparent from our examination of filter representations.

In Section 2.4 we consider some design techniques and implementation issues for digital filters whose performance specifications are given in terms of frequency domain characteristics. The general concepts of time domain filter specification and design are considered in Chapter 3 within the context of detection and estimation theory.

## 2.1 LINEAR SYSTEMS

### 2.1.0 Introduction

Before we begin our discussion of digital filters, we shall review some basic concepts of discrete-time linear systems. The reader is assumed to have some familiarity with linear transformations and linear systems theory. Thus our coverage serves mostly to establish the notation and convention for the discussion of digital filters in the latter sections of the chapter.

In this section we restrict our discussion to a review of the properties and methods of representation of discrete-time linear time-invariant (LTI) systems. We review the basic concepts of impulse response and transfer functions and the response to random signals for the one-dimensional case, and then discuss the extension of these concepts to two dimensions.

The basic concepts and resulting properties of linear digital processing are similiar to those associated with analog processing. These properties are enumerated in the following section.

### 2.1.1 Basic Concepts

**Linearity**

A discrete-time system can be described as a transformation of an input sequence $x(n)$ into an output sequence $y(n)$, represented mathematically as

$$y(n) = T[x(n)] \tag{2.1}$$

or graphically as shown in Figure 2.1.

A system is said to be linear if and only if the transformation $T[.]$ is such that

$$T[ax_1(n) + bx_2(n)] = ay_1(n) + by_2(n) \tag{2.2}$$

where

$$y_1(n) = T[x_1(n)]$$

and

$$y_2(n) = T[x_2(n)]$$

**Time Invariance**

A system is said to be time-invariant (or shift-invariant) if

$$y(n-k) = T[x(n-k)] \tag{2.3}$$

where

$$y(n) = T[x(n)]$$

**Unit Sample Response**

If an arbitrary input sequence

$$x(n) = \sum_{k=-\infty}^{\infty} x(k)\delta(n-k) \tag{2.4}$$

is applied to a linear time-invariant (LTI) system the output $y(n)$ can be expressed as

$$\begin{aligned} y(n) &= T\left[\sum_{k=-\infty}^{\infty} x(k)\delta(n-k)\right] \\ &= \sum_{k=-\infty}^{\infty} x(k)\,T\left[\delta(n-k)\right] \end{aligned} \tag{2.5}$$

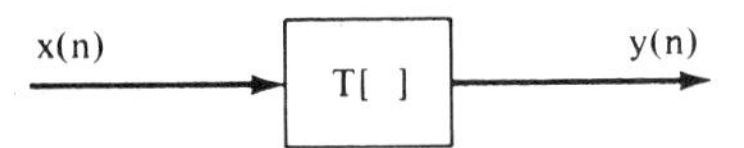


**Figure 2.1:** A linear transformation.

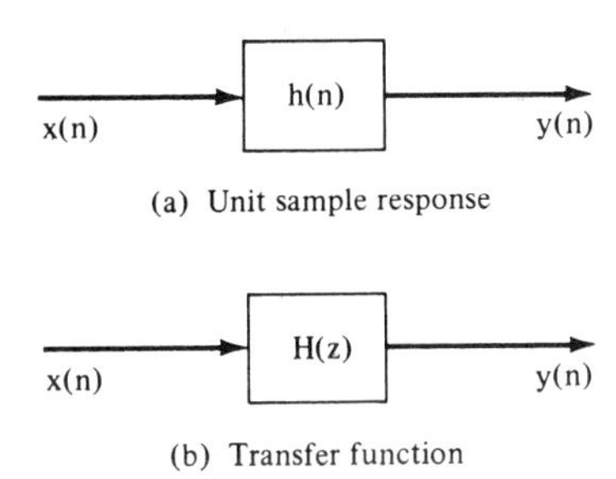


**Figure 2.2:** Graphical representation of a linear system.

Since the system is time-invariant the sequence $T[\delta(n-k)]$ is simply a shifted version of $T[\delta(n)]$. If $T[\delta(n)]$ is denoted $h(n)$, then we obtain

$$y(n) = \sum_{k=-\infty}^{\infty} x(k)\, h(n-k) = x(n) * h(n) \tag{2.6}$$

which we recognize as the convolution sum.

The sequence $h(n)$ is the response of the system to a unit sample sequence and is therefore known as the "unit sample response" of the system. Thus we have the important result:

> The output of a linear time-invariant system is the convolution of the input sequence with the system unit sample response.

The system is completely characterized by its unit sample response and is often represented graphically as in Figure 2.2(a).

**Properties of Linear Systems**

There are several properties of the convolution operation that are useful in the analysis of linear time-invariant systems. Convolution is:

1. Commutative:

$$y(n) = x(n) * h(n) = h(n) * x(n)$$

2. Associative:

$$x(n) * [h_1(n) * h_2(n)] = [x(n) * h_1(n)] * h_2(n)$$

3. Distributive:

$$x(n) * [h_1(n) + h_2(n)] = [x(n) * h_1(n)] + [x(n) * h_2(n)]$$

4. Shift-invariant:
   If $y(n) = x(n) * h(n)$
   then $y(n-k) = x(n-k) * h(n) = x(n) * h(n-k)$

Figure 2.3 illustrates these properties in block diagram form.

**Stability and Causality**

Two additional concepts of linear systems are important:

1. Stability: A system is said to be stable if and only if every bounded input sequence results in a bounded output sequence. This implies that the unit sample response of the system must be either a finite sequence or a convergent infinite sequence. Hence, a necessary and sufficient condition for a system to be stable is

$$\sum_{n=-\infty}^{\infty} |h(n)| < \infty$$

2. Causality: A system is said to be causal if the output for any $n = n_o$ depends only on inputs for $n \leqslant n_o$. A linear shift-invariant system is causal if and only if the unit sample response is a causal sequence, that is, $h(n) = 0$ for $n < 0$.

We note that the requirement for an LTI system to be both causal and stable implies that the $Z$-transform of the unit sample response has poles only on the inside of the unit circle of the $Z$-plane.

The foregoing properties can be viewed in their correspondence to familiar time domain concepts. Unless stated explicitly otherwise signals and systems are assumed to have the properties described here. The next requirement is to relate the frequency domain with the same set of concepts.

### 2.1.2 Frequency Domain Representation

Because LTI systems are completely characterized by their unit sample response they can be analyzed with $Z$-transforms. The first concept is the "transform function" of a system and then the "frequency response."

**Transfer Function**

Consider the LTI system with unit sample response $h(n)$. The response of the system to an arbitrary input sequence $x(n)$ is (from equation (2.6)):

$$y(n) = \sum_{k=-\infty}^{\infty} x(k)h(n-k) \tag{2.7}$$

If we take the $Z$-transform of both sides of this equation we get

$$Y(z) = \mathcal{Z}\{x(n)*h(n)\} = X(z)H(z) \tag{2.8}$$

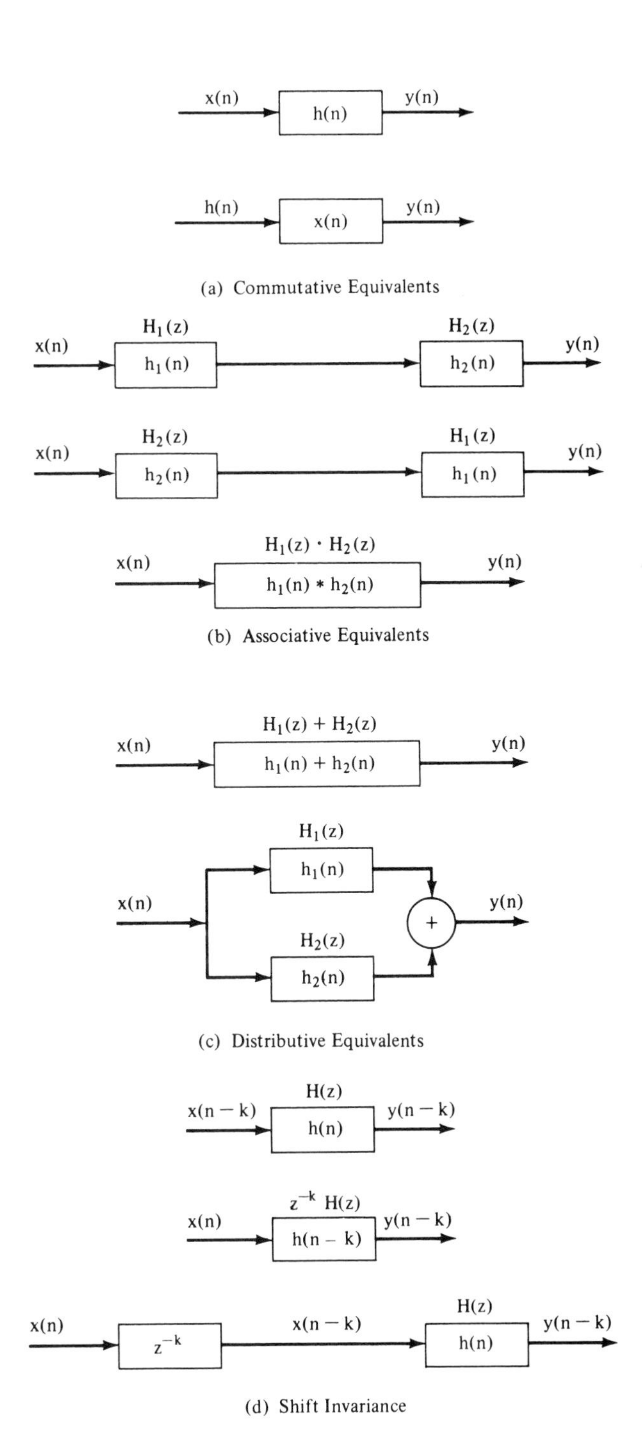


**Figure 2.3:** Equivalent systems.

From equation (2.8) it is clear that the $Z$-transform of the unit sample response of the system is equal to the ratio of the $Z$-transforms of the output sequence and the input sequence, i.e.,

$$H(z) = \frac{Y(z)}{X(z)} \tag{2.9}$$

$H(z)$ is known as the "system transfer function." The output of an LTI system can be calculated by taking the inverse $Z$-transform of the product of the transfer function and the $Z$-transform of the input sequence, i.e.,

$$y(n) = \mathcal{Z}^{-1}\{H(z)X(z)\} \tag{2.10}$$

Since an LTI system is completely characterized by its unit sample response, $h(n)$, it is also completely characterized by its transfer function $H(z)$. Thus, a linear system is often represented by its transfer function, as shown in Figure 2.2(b).

**Frequency Response**

When $H(z)$ is evaluated as $z = e^{j\omega}$ (i.e., the Fourier transform of $h(n)$), it is known as the frequency response of the system. To see why, consider a sinusoidal input $x(n) = e^{j\omega n}$.

$$\begin{aligned} y(n) &= h(n)*x(n) \\ &= \sum_{k=-\infty}^{\infty} h(k)e^{j\omega(n-k)} \\ &= e^{j\omega n}\sum_{k=-\infty}^{\infty} h(k)e^{-j\omega k} \\ &= e^{j\omega n}H(e^{j\omega}) \end{aligned} \tag{2.11}$$

Therefore $y(n) = H(e^{j\omega})x(n)$ when $x(n)$ is a pure sinusoid of the form $x(n) = e^{j\omega n}$.

Thus $H(e^{j\omega})$ completely characterizes the response of an LTI system to a pure sinusoid. The magnitude of $H(e^{j\omega})$ gives the amplitude response as a function of angular frequency $\omega$, and the angle gives the phase response as a function of $\omega$. Similar concepts are available to analyze the response to random signals, which we explore in the next section.

In specifying the frequency response of a discrete-time LTI system, several units of frequency are commonly used. Recall from Section 1.2 that $H(e^{j\omega})$ will be periodic with period $2\pi/T$, where $T$ is the sample period. Thus, the frequency response of a system is often normalized to a range of 0 to $2\pi$ giving a frequency representation in radians per sample. If the frequency response is expressed in Hertz (Hz), then the range represented by once around the unit circle is $0 < f \leqslant f_s$ where $f_s = 1/T$ is the sample frequency.

### 2.1.3 Response to Random Signals

We have seen that the output of a linear time-invariant system is given by the convolution of the input sequence with the system unit sample response, i.e.,

$$y(n) = \sum_{k=-\infty}^{\infty} x(k)h(n-k) = \sum_{k=-\infty}^{\infty} h(k)x(n-k) \tag{2.12}$$

We now wish to consider the effect of passing a random signal through a linear system. In particular we shall consider the case of a stationary random signal characterized by its mean and autocorrelation. If we pass this signal through an LTI system we wish to know the mean and autocorrelation of the output.

For a stable LTI system with a stationary random input signal we can write

$$\begin{aligned} E\{y(n)\} &= E\{\sum_{k=-\infty}^{\infty} h(k)x(n-k)\} \\ &= \sum_{k=-\infty}^{\infty} h(k)E\{x(n-k)\} \\ &= \eta_x H(e^{j0}) \qquad (2.13) \\ &= \text{a constant} \end{aligned}$$

Thus the mean of the output sequence of an LTI system whose input is a stationary random signal is a constant, equal to the product of the mean of the input sequence and the dc response of the system.

To determine the autocorrelation of the output sequence $y(n)$ we shall first determine the cross-correlation of the input and output sequences, i.e.,

$$\begin{aligned} R_{xy}(n,n+m) &= E\{x(n)y(n+m)\} \\ &= E\{x(n)\sum_{k=-\infty}^{\infty} h(k)x(n+m-k)\} \\ &= \sum_{k=-\infty}^{\infty} h(k)E\{x(n)x(n+m-k)\} \qquad (2.14) \\ &= \sum_{k=-\infty}^{\infty} h(k)R_{xx}(m-k) \end{aligned}$$

Noticing that this last expression is just the convolution of the autocorrelation of the stationary input with the system unit sample response, we conclude that the cross-correlation of the input with the output is also dependent only on the time difference $m$., i.e.,

$$R_{xy}(n,n+m) = R_{xy}(m) = R_{xx}(m)*h(m) \tag{2.15}$$

We now write the autocorrelation of the output as

$$\begin{aligned} R_{yy}(n+m,n) &= E\{y(n+m)\sum_{k=-\infty}^{\infty} h(k)x(n-k)\} \\ &= \sum_{k=-\infty}^{\infty} h(k)E\{y(n+m)x(n-k)\} \\ &= \sum_{k=-\infty}^{\infty} h(k)R_{xy}(m+k) \end{aligned} \tag{2.16}$$

Again noting that the last equation is independent of $n$, we conclude that the autocorrelation of the output is dependent only on the time difference $m$, i.e.,

$$R_{yy}(m) = R_{xx}(m)*h(m)*h(-m) \tag{2.17}$$

Hence we have the important result:

> If the input to an LTI system is a stationary random sequence then the output is also a stationary sequence.

Further, we note that the sequence

$$h(n)*h(-n) = \sum_{k=-\infty}^{\infty} h(k)h(n+k) \tag{2.18}$$

is simply the autocorrelation sequence of $h(n)$, and we have a second important observation:

> The autocorrelation of the output of an LTI system whose input is a stationary random sequence is the convolution of the autocorrelation of the input with the autocorrelation of the system unit sample response.

From equation (2.17) we conclude that

$$S_{yy}(\omega) = S_{xx}(\omega)|H(e^{j\omega})|^2 \tag{2.19}$$

It follows from this equation that the spectral density of any process, real or complex, is non-negative, i.e.,

$$S(\omega) \geqslant 0 \tag{2.20}$$

and thus, for a real signal, the spectral density is real, even and positive.

From equation (2.15) we also note that the cross-spectral density of the input and output is

$$S_{xy}(\omega) = S_{xx}(\omega)H(e^{j\omega}) \tag{2.21}$$

This relationship is useful in the case of a white noise input sequence $w(n)$ characterized by a constant spectral density

$$S_{ww}(\omega) = \sigma^2 \tag{2.22}$$

Thus the cross-power spectrum of input and output is proportional to the frequency response of the system and may be used to estimate the frequency response of the system, i.e.,

$$S_{yw}(\omega) = \sigma^2 H(e^{j\omega}) \tag{2.23}$$

The significant result here is that the spectral characteristics of a random signal after linear processing is predictable based on the transfer funcion. This property is identical to that for analog systems.

In some applications the incoming information can be considered as two signals. This dimensionality can be modeled conveniently in terms of our previous concepts. A two-dimensional case will be considered next.

### 2.1.4 Two-Dimensional Signals and Systems

Thus far we have considered only one-dimensional signals and systems. There are, however, several signal processing applications (e.g., sonar, geophysics and image processing) that are concerned with multi-dimensional problems. Two-dimensional signals are fairly common, e.g., the range and bearing of an aircraft. We now present an extension of some of our previous concepts to the two-dimensional case. In the following, we first define a two-dimensional sequence and show that for a certain separable class of sequences our concepts of $Z$-transform and of linear systems can be extended to accommodate both a description and an analysis.

#### Two-Dimensional Sequences

A two-dimensional signal is a function of two variables such as the coordinates on a plane, or time and position. When sampled, the result is a two-dimensional sequence that is a function of two integer indices. Figure 2.4 is a graphical representation of a two-dimensional sequence.

The unit sample, unit step and exponential sequences are defined for the two-dimensional case as follows:

1. Unit Sample Sequence

$$\delta(m,n) = \begin{cases} 1, m = n = 0 \\ 0 \text{ otherwise} \end{cases} \tag{2.24}$$

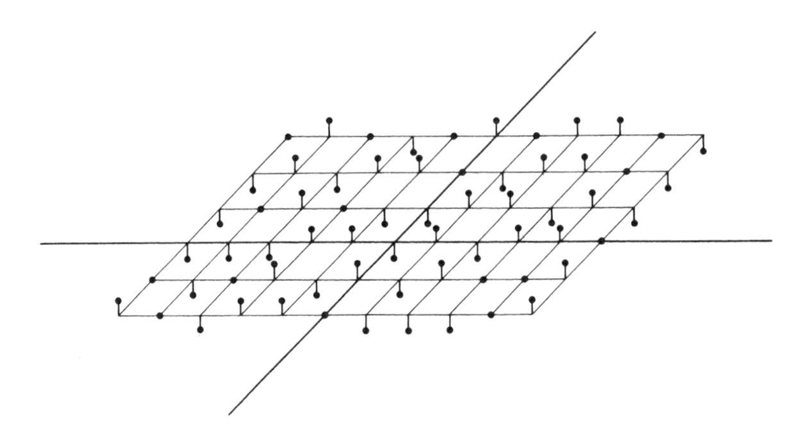

**Figure 2.4:** A two-dimensional sequence.

2. Unit Step Sequence

$$u(m,n) = \begin{cases} 1, m,n \geqslant 0 \\ 0 \text{ otherwise} \end{cases} \tag{2.25}$$

3. Exponential Sequence

$$x(m,n) = a^m b^n \tag{2.26}$$

**Separable Sequences**

A two-dimensional sequence is said to be 'separable' if it can be expressed as the product of one-dimensional sequences, i.e.,

$$x(m,n) = x_1(m)x_2(n) \tag{2.27}$$

The elementary sequences defined above are all separable.

Similar to the one-dimensional case an arbitrary two-dimensional sequence can be expressed as a summation of weighted, shifted unit samples, i.e.,

$$x(m,n) = \sum_{k=-\infty}^{\infty} \sum_{i=-\infty}^{\infty} x_{ki}\delta(m-k, n-i) \tag{2.28}$$

**Two-Dimensional** $Z$**-Transform**

The two-dimensional $Z$-transform of a sequence, $x(m,n)$, is defined as

$$X(z_1,z_2) = \sum_{m=-\infty}^{\infty} \sum_{n=-\infty}^{\infty} x(m,n) z_1^{-m} z_2^{-n} \tag{2.29}$$

and the inverse relation as

$$x(m,n) = \frac{1}{(2\pi j)} \int_{c_1}\int_{c_2} X(z_1,z_2) z_1^{m-1} z_2^{n-1} dz_1 dz_2 \tag{2.30}$$

where $C_1$ and $C_2$ are suitable closed curves in the $z_1$ and $z_2$ planes.

**Two-Dimensional Convolution**

As in the one-dimensional case, the product of the $Z$-transforms of two sequences, $x(m,n)$ and $y(m,n)$, is equal to the $Z$-transform of the two-dimensional convolution of these sequences, defined as

$$x(m,n) * y(m,n) = \sum_{k=-\infty}^{\infty} \sum_{i=-\infty}^{\infty} x(k,i) y(m-k, n-i) \tag{2.31}$$

and

$$\mathcal{Z}\{x(m,n) * y(m,n)\} = X(z_1,z_2) Y(z_1,z_2) \tag{2.32}$$

**Two-Dimensional DFT**

The Fourier transform of a two-dimensional sequence is, as in one dimension, a special case of the $Z$-transform, i.e.,

$$\mathcal{Z}\{x(m,n)\} = X(z_1,z_2)\Big|_{z_1=e^{j\omega_1}, z_2=e^{j\omega_2}} \tag{2.33}$$

and the inverse relation is given as

$$x(m,n) = \frac{1}{4\pi^2} \int_{-\pi}^{+\pi} \int_{-\pi}^{+\pi} X(e^{j\omega_1}, e^{j\omega_2}) e^{j\omega_1 m} e^{j\omega_2 n} d\omega_1 d\omega_2 \tag{2.34}$$

The two dimensional DFT of a finite duration two dimensional sequence is defined as

$$\begin{aligned} DFT\{x(m,n)\} = X_{k_1,k_2} &= X(z_1,z_2)\Big|_{\substack{z_1=e^{j(2\pi/M)k_1} \\ z_2=e^{j(2\pi/N)k_2}}} \\ &= \sum_{m=j}^{M-1} \sum_{n=0}^{N-1} x(m,n) e^{-j(2\pi/M)k_1 m} e^{-j(2\pi/N)k_2 n} \end{aligned} \tag{2.35}$$

and

$$x(m,n) = IDFT\{X_{k_1,k_2}\} = \frac{1}{MN} \sum_{m=0}^{M-1} \sum_{n=0}^{N-1} X_{k_1,k_2} e^{j(2\pi/M)k_1 m} e^{j(2\pi/N)k_2 n} \tag{2.36}$$

**Two-Dimensional Linear Systems**

A two-dimensional LTI system is characterized by its response to a unit sample input, as in the one-dimensional case. Its output is given by the two-dimensional convolution of the unit sample response and the input sequence, i.e.,

$$y(m,n) = \sum_{k=-\infty}^{\infty} \sum_{i=-\infty}^{\infty} x(k,i) h(m-k, n-i) \tag{2.37}$$

where

$$h(m,n) = T[\delta(m,n)] \tag{2.38}$$

The transfer function of the system is

$$H(z_1,z_2) = \mathfrak{Z}\{h(m,n)\}$$

and the frequency response is

$$H(e^{j\omega_1},e^{j\omega_2}) = \sum_{k=-\infty}^{\infty}\sum_{i=-\infty}^{\infty} h(k,i)e^{j\omega_1 k}e^{j\omega_2 i} \tag{2.39}$$

If $h(m,n)$ is separable, then

$$H(e^{j\omega_1},e^{j\omega_2}) = H_1(e^{j\omega_1})H_2(e^{j\omega_2}) \tag{2.40}$$

and the inverse relation is given as

$$x(m,n) = \frac{1}{4\pi^2}\int_{-\pi}^{+\pi}\int_{-\pi}^{+\pi} X(e^{j\omega_1},e^{j\omega_2})e^{j\omega_1 m}e^{j\omega_2 n}d\omega_1 d\omega_2 \tag{2.41}$$

The concepts of two-dimensional signals and systems can be extended to multiple dimensions. Such an extension is beyond our scope. For a complete discussion of two-dimensional signal processing the reader is referred initially to (10), and to further reading suggested in the bibliography.

### 2.1.5 Summary

In this section we have introduced the elementary concepts of linear systems, primarily from a frequency domain point of view. The description and analysis rests on the same conceptual framework as continuous-time systems. Indeed, corresponding concepts exist in both. For a stationary random input sequence, the output sequence is also a stationary random sequence whose mean is the product of the mean of the input sequence and the dc response of the system. The autocorrelation of the output is obtained by convolving the autocorrelation of input with the autocorrelation of the system unit sample response. For a white noise input the output signal spectrum is that of the filter. This result forms a mechanism for generating signals with a specified spectrum and proves useful in adaptive processing (Section 4.4) and in other applications.

A discrete-time LTI system is completely characterized by its unit sample response sequence or by its transfer function. In this respect also, a correspondence exists with similar concepts in continuous time systems. The correlation function and the spectral density function form the prime application of the statistical concepts developed in Chapter 1.

For a reasonable class of two-dimensional signals, the transform techniques can be easily extended for both representation and analysis.

In the next section we shall concentrate on extending the concepts of time domain representations.

## 2.2 TIME DOMAIN REPRESENTATIONS OF DISCRETE-TIME LINEAR SYSTEMS

### 2.2.0 Introduction

Discrete-time linear systems can be represented and analyzed in the time domain as well as in the frequency domain. In this section we shall examine some elementary concepts of time-series analysis models and state-space representations for discrete-time linear systems. These models are perhaps more important than frequency domain models because they deal directly with the samples obtained for processing from an ADC, which are intrinsically in this domain.

The basic concepts of time series analysis and the general autoregressive moving average model are introduced in Section 2.2.1. This model will be extended in Section 2.2.2 and shown to be equivalent to the most general digital filter.

The concept of the state of a system leads to a state-space representation and analysis of linear systems in terms of a pair of first order vector-matrix equations. We shall see that this approach provides a convenient method of representing multiple-input multiple-output systems. The concepts of state-space representations will be used again in Chapter 3 when we discuss the Kalman filter.

In this section we shall change notation slightly to comply with the conventions of this field. The notation $x(n)$ will be used for representing the state variables of a system rather than the input sequence, which will be represented as $v(n)$.

### 2.2.1 Basic Concepts of Time Series Analysis

In this section, fundamental concepts are developed which yield an "autoregressive moving average" model of the input-output relation of linear processing. This model yields a difference equation which defines the most general digital filter. From this equation we will later develop digital filter realizations. As we proceed, note that we do not go into the $Z$-domain. Rather the concepts presented here represent the basics of time domain linear processing.

We begin by introducing a general moving average process model in which the output samples of a linear processing operation are derived from a linear weighted sum of the present and past input samples. In terms of actually computing such an output process, we may consider that our system has memory for storing input samples which are appropriately combined to produce successive output samples.

Next, we introduce a general autoregressive model in which the output samples are derived from only the current input sample and a linear weighted sum of past output samples. This model corresponds to systems that have memory for storing only output samples which are weighted and summed with the current input to create the next output sample.

Finally, we shall examine the equivalence of representing a process as either a moving average or as an autoregressive process. This leads to a general autoregressive moving average model representing a system with storage for past values of both input and output samples which are combined to create the next output sample.

**Moving Average Process Model**

Consider passing an input sequence $v(n)$ through an LTI system to produce an output sequence $y(n)$. If we denote the linear operator that transforms $v(n)$ to $y(n)$ as $L$ then we can write

$$y(n) = L[v(n)] \tag{2.42}$$

For an LTI system, the steady-state output sequence $y(n)$ can be represented by a weighted linear combination of the input samples, i.e.,

$$y(n) = \alpha_o v(n) + \alpha_1 v(n-1) + \alpha_2 v(n-2) + \ldots \tag{2.43}$$

where we assume the initial conditions $v(n) = o$ for $n < o$.

We can define a new operator $B$, called the "backward operator," such that $B[x(n)] = x(n-1)$ and rewrite equation (2.43) as:

$$\begin{aligned} y(n) &= \alpha_o v(n) + \alpha_1 B[v(n)] + \alpha_2 B^2[v(n)] + \ldots \\ &= L(B)[v(n)] \end{aligned} \tag{2.44}$$

where

$$L(B) = \alpha_o + \alpha_1 B + \alpha_2 B^2 + \ldots$$

The backward operator $B$ can be identified with the unit delay element having transfer function $z^{-1}$, and the operator $L(B)$ represents the transfer function of an LTI system.

In general the coefficient sequence $\alpha_1$ may be either finite or infinite. In the case of an infinite sequence, the sequence must be convergent if the system is to be stable.

If the coefficient sequence $\alpha_i$ is a convergent infinite geometric series of the general form

$$\alpha_n = r(1-r)^n \qquad 0 < r < 1 \tag{2.45}$$

then

$$y(n) = r\sum_{i=0}^{\infty}(1 - r)^i v(n - i) \tag{2.46}$$

This model is known as an "exponential or geometric smoothing process."

In the case of a finite sequence of coefficients of length $N + 1$, the output $y(n)$ is given by

$$y(n) = \alpha_o v(n) + \alpha_1 v(n - 1) + \ldots \alpha_N v(n - N) \tag{2.47}$$

This is known as a "moving average model of order $N + 1$" for the process $y(n)$, and $y(n)$ is referred to as a "moving average process," (i.e., it is computed as a weighted sum of the last $N + 1$ input samples).

### Autoregressive Process Model

Consider passing an input signal sequence $v'(n)$ through an LTI system to produce the output sequence

$$y'(n) = v'(n) + b_1 y'(n - 1) + b_2 y'(n - 2) + \ldots + b_N y'(n - N) \tag{2.48}$$

Equation (2.48) represents a general autoregressive model of the output process $y'(n)$ given the input process $v'(n)$.

We can rearrange equation (2.48) such that the input process $v'(n)$ is described as a moving average process in terms of the output process $y'(n)$, i.e.,

$$v'(n) = \phi(B)[y'(n)] \tag{2.49}$$

where

$$\phi(B) = 1 + \beta_1 B + \beta_2 B^2 + \ldots + \beta_N B^N \tag{2.50}$$

and $\beta_i = -b_i$.

We note that an autoregressive process as defined by equation (2.48) is the inverse of a moving average process as defined by equation (2.49). No loss of generality occurs in claiming equation (2.49) to be a general moving average process such as introduced in equation (2.44) since the assumption that $\beta_0 = 1$ implies only a constant scale factor for the process $v'(n)$.

As can be seen from equation (2.50), the coefficients of the autoregressive model for the process $v'(n)$, given the process $y'(n)$, are simply the negative values of the coefficients of moving average model for the process $y'(n)$, given the process $v'(n)$. This inverse relationship between the autoregressive and moving average models is illustrated in Figure 2.5(b).

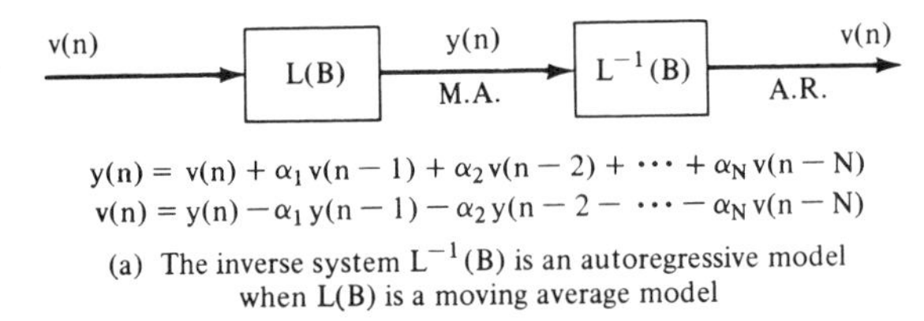


$$y(n) = v(n) + \alpha_1 v(n-1) + \alpha_2 v(n-2) + \cdots + \alpha_N v(n-N)$$
$$v(n) = y(n) - \alpha_1 y(n-1) - \alpha_2 y(n-2 - \cdots - \alpha_N v(n-N)$$

(a) The inverse system $L^{-1}(B)$ is an autoregressive model when L(B) is a moving average model

y(n) → $\phi$ (B) → v(n) A.R. → $\phi^{-1}$(B) → y(n) M.A.

$$v(n) = y(n) + \alpha_1 v(n-1) + \alpha_2 v(n-2) + \cdots + \alpha_N v(n-N)$$
$$y(n) = v(n) - \alpha_1 v(n-1) - \alpha_2 v(n-2) - \cdots - \alpha_N v(n-N)$$

(b) The inverse system $\phi^{-1}(B)$ is a moving average model when $\phi(B)$ is an autoregressive model

**Figure 2.5:** Autoregressive and moving average models.

### Equivalent Autoregressive and Moving Average Models

The inverse relationship between the autoregressive and moving average models also leads to an equivalence relationship. Since we can express $v'(n)$ as

$$v'(n) = \phi(B)[y'(n)] \tag{2.51}$$

we can also write

$$y'(n) = \phi^{-1}(B)[v'(n)] \tag{2.52}$$

This equation implies that the autoregressive process $y'(n)$, as defined in equation (2.48), can be represented by an equivalent moving average model $\phi^{-1}(B)$. We shall now outline an intuitive procedure for determining an equivalent moving average model for the autoregressive process $y'(n)$.

Noting the recursive nature of equation (2.48), we see that each output sample $y'(n)$ is defined in terms of the present input sample $v'(n)$ and past outputs. By working backward, we can express each past output in terms of a past input and the $N$ previous outputs. For example, $y'(n-1)$ can be expressed as

$$y'(n-1) = v'(n-1) + b_1 y'(n-2) + b_2 y'(n-3) + \cdots + b_N y'(n-N-1) \tag{2.53}$$

In the same fashion $y'(n-2)$ is given as

$$y'(n-2) = v'(n-2) + b_1 y'(n-3) + b_2 y'(n-4) + \cdots + b_N y'(n-N-2) \tag{2.54}$$

and similarly for $y'(n-3)$, $y'(n-4)$, and so on. Thus, each delayed output term (i.e., $y'(n-\eta)$) can be represented in terms of a similarly delayed input (i.e., $v'(n-\eta)$) plus a combination of past outputs with greater delays (each of which can in turn be represented in terms of greater delayed inputs and outputs). By continuing this process we eventually arrive at an infinite moving average model for $y'(n)$ of the form

$$\begin{aligned} y'(n) &= v'(n) + \tau_1 v'(n-1) + \tau_2 v'(n-2) + \cdots \\ &= T(B)[v'(n)] \end{aligned} \tag{2.55}$$

where $T(B)$ will actually be of some finite order $k$ assuming $v'(n)$ and $y'(n)$ are causal sequences (i.e., $v'(n), y'(n) = 0, n < 0$).

Comparing equations (2.52) and (2.55) we see that

$$T(B) = \phi^{-1}(B) \tag{2.56a}$$

or

$$\phi(B) = T^{-1}(B) \tag{2.56b}$$

**Autoregressive Moving Average Model**

The moving average and autoregressive models can be combined to form a general autoregressive moving average (ARMA) model. Let the moving average process $y(n)$ of equation (2.44) be the input to a system whose output is the autoregressive process $y'(n)$ as defined in equation (2.52). This system is shown in Figure 2.5(c) and can be represented as

$$\begin{aligned} y'(n) &= T(B)[y(n)] \\ &= \phi^{-1}(B)[L(B)[v(n)]] \end{aligned} \tag{2.57}$$

or

$$\phi(B)[y'(n)] = L(B)[v(n)] \tag{2.58}$$

corresponding to the general linear difference equation

$$\sum_{k=0}^{N-1} \beta_k y'(n-k) = \sum_{k=0}^{N-1} \alpha_k v(n-k) \tag{2.59}$$

We shall consider linear systems that can be represented by this general difference equation in some detail in this chapter, for it defines the most general digital filter.

### 2.2.2 Basic Concepts of State-Space Representations

The concept of the state of a system can be generalized as follows:

> The state of a system can be represented by a set of parameters that summarize the past behavior of the system such that only the knowledge of

the present state and the inputs to the system are necessary to determine future system behavior.

Thus, a knowledge of past behavior is unnecessary and the manner in which the system arrived at its present state has no effect on the future behavior of the system. From our discussion on statistical signal modeling we see that a state-determined system may also be described as one whose output is a first order Markov sequence.

Linear systems that can be represented by the general Nth order difference equation (2.59), or by a system of such equations, can be conveniently described by a pair of first-order linear vector-matrix equations: one describing the behavior of an $n$-dimensional state-vector and another relating the output of the system to the present state and the input. These equations are known as the "state equation" and the "output equation" respectively, or collectively as the "state-space representation" of the system.

Equation (2.59) does not provide a unique description of the input-output relation of the system without the specification of a set of initial conditions. These initial conditions can be considered as specifying the initial state of the system. Given this specification and a knowledge of the inputs, the future outputs of the system can be uniquely determined.

### State Equations

A general description of a state-determined, time-varying system is usually given in terms of a pair of equations, called the "state equations" having the following form

$$x(t) = f\{x(t_0); v(t_0,t)\}, \text{ the next state} \tag{2.60}$$

$$y(t_0,t) = g\{x(t_0); v(t_0,t)\}, \text{ the output} \tag{2.61}$$

Here $x(t)$ is a vector function of time whose elements are the set of parameters that determine the state of the system at time $t$; while $y(t_0,t)$ is a vector function of time whose elements represent the outputs of the system over the interval $t_0$ to $t$. Both $f$ and $g$ are single-valued functions of the initial state of the system $x(t_0)$, and the input, $v(t_0,t)$, over the interval $t_0$ to $t$. Equation (2.60) is called the state equation of the system and equation (2.61) is called the output equation. The following example illustrates the description of a simple single-input single-output system by a pair of state equations.

***Example:***

Consider a discrete-time system with input $v(n)$ and output $y(n)$ related by the difference equation

$$y(n) + c_1 y(n-1) + c_2 y(n-2) = dv(n) \tag{2.62}$$

This second order difference equation can be expressed as a first order matrix equation. If initial conditions specifying $y(n_o - 1)$, $y(n_o - 2)$ are given and $v(n)$ is known for $n > n_o$ then the output $y(n)$ can be determined recursively from equation (2.62). Therefore if the parameters $y(n-1)$ and $y(n-2)$ are chosen as representing the state of the system, and the state vector $\underline{x}(n)$ of the two-dimensional system is defined as

$$\underline{x}(n) = \begin{bmatrix} x_1(n) \\ x_2(n) \end{bmatrix} = \begin{bmatrix} y(n-2) \\ y(n-1) \end{bmatrix} \tag{2.63}$$

Thus by examining successive indexes

$$x_1(n+1) = y(n-1) = x_2(n)$$

and by appropriate substitutions

$$x_2(n+1) = y(n) = dv(n) - c_1 x_2(n) - c_2 x_1(n) \tag{2.64}$$

In matrix form these equations become

$$\begin{bmatrix} x_1(n+1) \\ x_2(n+1) \end{bmatrix} = \begin{bmatrix} 0 & 1 \\ -c_2 & -c_1 \end{bmatrix} \begin{bmatrix} x_1(n) \\ x_2(n) \end{bmatrix} + \begin{bmatrix} 0 \\ d \end{bmatrix} v(n) \tag{2.65}$$

or

$$\underline{x}(n+1) = A\underline{x}(n) + B\underline{v}(n)$$

and

$$y(n) = [-c_2, -c_1] \begin{bmatrix} x_1(n) \\ x_2(n) \end{bmatrix} + \begin{bmatrix} 0 \\ d \end{bmatrix} v(n) \tag{2.66}$$

or

$$\underline{y}(n) = C\underline{x}(n) + D\underline{v}(n)$$

Thus we have described the system in terms of a first order matrix state equation (2.65) and an output equation (2.66), rather than the second order difference equation (2.62). It is intuitively obvious at this point that the dimensions of the state equations are determined by the order of the original difference equation.

For a general multiple input-output linear system with $p$ inputs $v_1(n)$, . . . , $v_p(n)$ and $q$ outputs $y_1(n)$, . . ., $y_q(n)$ represented by the vectors

$$\underline{v}(n) = \begin{bmatrix} v_1(n) \\ v_2(n) \\ \vdots \\ v_p(n) \end{bmatrix} \quad \text{and} \quad \underline{y}(n) = \begin{bmatrix} y_1(n) \\ y_2(n) \\ \vdots \\ y_q(n) \end{bmatrix}$$

we have the general state space representation

$$\underline{x}(n+1) = A(n)\underline{x}(n) + B(n)\underline{v}(n) \tag{2.67}$$

$$\underline{y}(n) = C(n)\underline{x}(n) + D(n)\underline{v}(n) \tag{2.68}$$

where

$\underline{x}(n)$ is an N—dimensional column state vector

$\underline{v}(n)$ is a p—dimensional column input vector

$\underline{y}(n)$ is a q—dimensional column output vector

$A(n)$ is an NxN nonsingular 'feedback' matrix

$B(n)$ is an Nxp input matrix

$C(n)$ is a qxN output matrix

$D(n)$ is a qxp feed forward matrix

The general block diagram of the system is shown in Figure 2.6.

**State-Transition Matrix**

Consider the system as shown in Figure 2.6. The state vector $x(n)$ can be expressed in terms of a transient and steady-state response (similar to any linear system). The state-transition matrix for a sequence of state changes is defined as the matrix

$$\Phi(n_2,n_1) = \begin{cases} A(n_2-1)A(n_2-2)\ldots.A(n_1), & n_2 > n_1 \\ I & , n_2 = n_1 \\ A^{-1}(n_1)\ldots A^{-1}(n_2-1) & , n_2 < n_1 \end{cases} \tag{2.69}$$

which relates the states of the system under zero input conditions ($\underline{v}(n) = 0$) at any two times $n_2$, $n_1$. The state vector at $n_2$ is related to the state vector at $n_1$ as

$$\underline{x}(n_2) = \Phi(n_2,n_1)\underline{x}(n_1) \tag{2.70}$$

From equations (2.67) and (2.69) we obtain the general relation for the state of the system at an arbitrary time $n$ starting at time $n_o$ as

$$\underline{x}(n) = \Phi(n,n_o)\underline{x}(n_o) + \sum_{k=n_o}^{n-1} \Phi(n,k+1)B(k)\underline{v}(k) \tag{2.71}$$

The second term shows the state by state changes over the interval where $\underline{x}(n_o)$, $B(n_o)$ and $\underline{v}(n_o)$ represent the initial conditions.

In the previous discussion the matrices $A(n)$, $B(n)$, $C(n)$ and $D(n)$ were assumed to be functions of the time-index $n$, corresponding to a time-varying system. If the system is time-invariant these matrices are constant and can be written simply as $A$, $B$, $C$ and $D$. Thus the state equations for LTI systems become

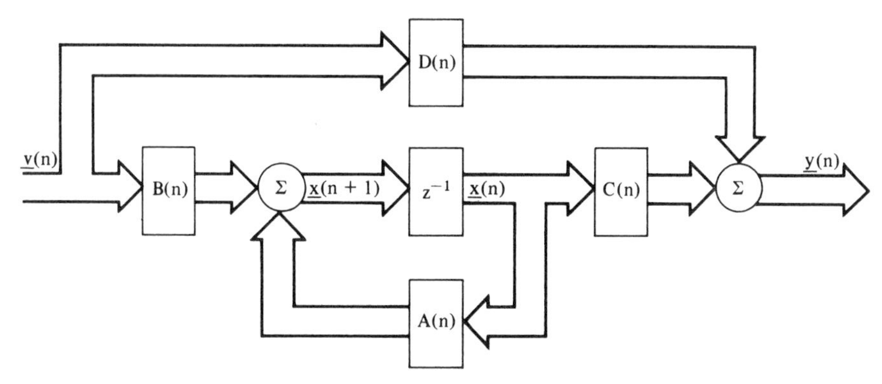


**Figure 2.6:** Block diagram of state space representation of a linear system.

$$\underline{x}(n+1) = A\underline{x}(n) + B\underline{v}(n) \tag{2.72}$$

$$\underline{y}(n) = C\underline{x}(n) + D\underline{v}(n) \tag{2.73}$$

and the state transition matrix depends only on the difference $n_2 - n_1$, i.e.,

$$\Phi(n_2,n_1) = \phi(k) = A^k, \qquad k = n_2 - n_1 \tag{2.74}$$

For LTI systems

$$\phi(k_1 + k_2) = \phi(k_1)\phi(k_2) \tag{2.75}$$

and

$$\phi(k) = \phi^{-1}(-k) \tag{2.76}$$

The general solution of the state equation (2.71) becomes

$$\underline{x}(n) = \phi(n - n_o)\underline{x}(n_o) + \sum_{k=n_o}^{N-1} \phi(n - k - 1)B\underline{v}(k), n > n_o \tag{2.77}$$

The state-space representation of LTI systems can easily be analyzed using familiar $Z$-transform techniques. It can be shown [16] that the state transition matrix is:

$$\phi(n) = A^n = \mathcal{Z}_+^{-1}\{(zI - A)^{-1}z\}, n > 0 \tag{2.78}$$

where $I$ is the unit matrix and the one-sided $Z$-transform of a matrix is a matrix whose elements are the one-sided $Z$-transforms of the corresponding elements of the original matrix, i.e.,

$$\Phi(z) = \mathcal{Z}_+\{A^n\} = (zI - A)^{-1}z \tag{2.79}$$

The transfer function matrix $\underline{H}(z)$ of the overall system, can be derived as:

$$\underline{H}(z) = C(zI - A)^{-1}B + D \tag{2.80}$$

and therefore

$$\underline{Y}(z) = (C(zI - A)^{-1}B + D)\underline{V}(z) \tag{2.81}$$

where $\underline{Y}(z)$ and $\underline{V}(z)$ are column vectors whose elements are the transforms of the outputs and inputs of the system. This representation is now in familiar format despite the appearances of vectors as terms and coefficients.

### 2.2.3 Summary

In this section we have reviewed the basic concepts of time domain representations of linear systems. The moving average and autoregressive process models were shown to be related in that one is the inverse of the other. When combined they form a general autoregressive moving average model that is equivalent to a general Nth order linear difference equation. Systems that can be represented by such equations can also be represented in state space by a pair of first order linear vector-matrix equations. The use of such a representation is convenient for handling multiple-input multiple-output systems. The computations implicit in equation (2.81) are extensive. The powerful generality of this representation has not often been exploited because of the computational problems imposed by the matrix arithmetic. We shall find the concepts of both time series analysis and state-space representations useful in our discussions of linear estimation in Chapter 3.

## 2.3 DIGITAL FILTERS

### 2.3.0 Introduction

An important class of LTI systems satisfies the general Nth order constant coefficient linear difference equation

$$\sum_{k=0}^{M} b_k y(n-k) = \sum_{k=0}^{N} a_k x(n-k) \tag{2.82}$$

A system satisfying this equation is not necessarily a causal system. In addition, the input-output relation of the system is not uniquely specified by equation (2.82). As in the case of an ordinary differential equation, the specification of initial conditions is necessary for a unique solution. However, if we assume a causal system with initial rest conditions we obtain an explicit representation of the system input-output relation, i.e.,

$$y(n) = \sum_{k=0}^{N} \alpha_k x(n-k) - \sum_{k=1}^{M} \beta_k y(n-k) \tag{2.83}$$

where

$$\alpha_k = a_k/b_o \text{ and } \beta_k = b_k/b_o$$

LTI systems that satisfy this equation are commonly known as digital filters. The constants $\alpha_k$ and $\beta_k$ are called the coefficients of the filter.

If the coefficients $\beta_k$ are not all zero, then the unit sample response of the filter is of infinite duration due to the recursive nature of the defining equation. Filters of this type are generally known as infinite impulse response (IIR) filters. If the coefficients $\beta_k$ are all identically zero, then equation (2.83) reduces to a finite convolution sum, i.e.,

$$y(n) = \sum_{k=0}^{N} \alpha_k x(n-k) \tag{2.84}$$

In this case the coefficients $\alpha_k$ represent the values of the unit sample response. Since these coefficients represent a finite sequence, filters of this type are known as finite impulse response (FIR) filters. FIR filters are also commonly referred to as transversal filters or tapped delay line filters (when the implementation is through the use of a tapped delay line) and the coefficients $\alpha_k$ are referred to as tap gains.

Since digital filters are a subclass of linear systems, they can be analyzed and represented in either the time domain or the frequency domain. In each case a number of important concepts and equivalent realization structures are exposed.

The actual implementation of a digital filtering operation can also be carried out in either the time domain or the frequency domain. The choice of the domain in which to implement a filter, and the particular realization structure chosen will determine the actual computational requirements of the filtering operation. This in turn impacts the required architectural features of the digital computing system utilized to implement the filter, whether it be a dedicated hardware structure or a programmable processor. These issues will be examined in Section 2.4. In the present section we concentrate on the mathematical and graphical representations of digital filters from both a frequency domain and a time domain perspective.

### 2.3.1 Frequency Domain Representations

In Section 2.1.2 we saw that a linear system can be represented in the frequency domain by its transfer function $H(z)$. The transfer function of the general digital filter represented by equation (2.83) is easily obtained by taking the $Z$-transform of both sides of the equation, i.e.,

$$Y(z) = \sum_{k=0}^{N} \alpha_k z^{-k} X(z) - \sum_{k=1}^{M} \beta_k z^{-k} Y(z) \tag{2.85}$$

Rearranging this equation we have

$$H(z) = \frac{Y(z)}{X(z)} = \frac{\sum_{k=0}^{N} \alpha_k z^{-k}}{1 + \sum_{k=1}^{M} \beta_k z^{-k}} \tag{2.86}$$

This equation represents the most general transfer function of a digital filter. The roots of the numerator polynomial represent the zeros of the filter and the roots of the denominator polynomial are the poles. For the case of an FIR filter where $\beta = 0, k \geqslant 1$, the filter has only zeros (i.e., the poles of an FIR filter are restricted to occur only at $z = 0$ or at $z = \infty$). For the case of a purely recursive filter where $\alpha_k = 0, k \geqslant 1$, the filter has only poles.

Since equation (2.86) is a ratio of polynomials in $z^{-1}$ it can generally be factored to a ratio of products of the form

$$H(z) = G\frac{\prod_{k=0}^{N}(1 - c_k z^{-1})}{\prod_{k=0}^{M}(1 - d_k z^{-1})} \tag{2.87}$$

where $G$ represents a constant gain factor and the coefficients $c_k$ and $d_k$ correspond to the zeros and poles respectively.

Equation (2.87) can be written as a product of lower order transfer functions, i.e.,

$$H(z) = GH_1(z)H_2(z).H_3(z) \cdots \tag{2.88}$$

This is known as a "cascade realization" as illustrated in Figure 2.7. In this manner digital filters of an arbitrary order can be broken down into a cascaded series of first and second order filter sections of the following form:

**First Order Cascade Section:**

$$H_{f_c}(z) = \frac{1 + \alpha_1 z^{-1}}{1 + \beta_1 z^{-1}} \tag{2.89}$$

**Second Order Cascade Section:**

$$H_{S_c}(z) = \frac{1 + \alpha_1 z^{-1} + \alpha_2 z^{-2}}{1 + \beta_1 z^{-1} + \beta_2 z^{-2}} \tag{2.90}$$

Equation (2.87) may also be expanded in partial fractions to a sum of lower order transfer functions giving a parallel realization, as illustrated in Figure 2.8, i.e.,

$$H(z) = H_1(z) + H_2(z) + \cdots \tag{2.91}$$

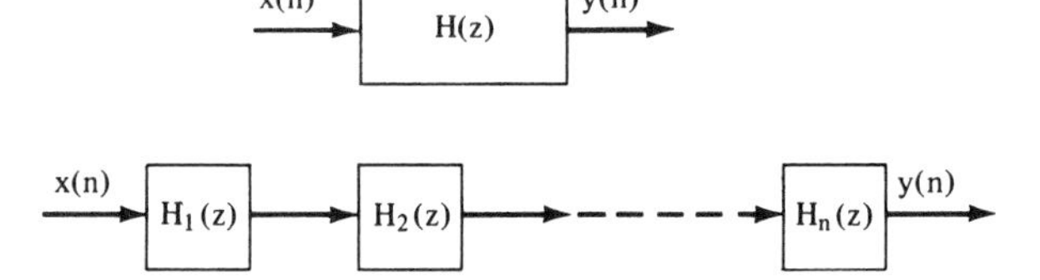


**Figure 2.7:** Cascade realization of $H(z)$.

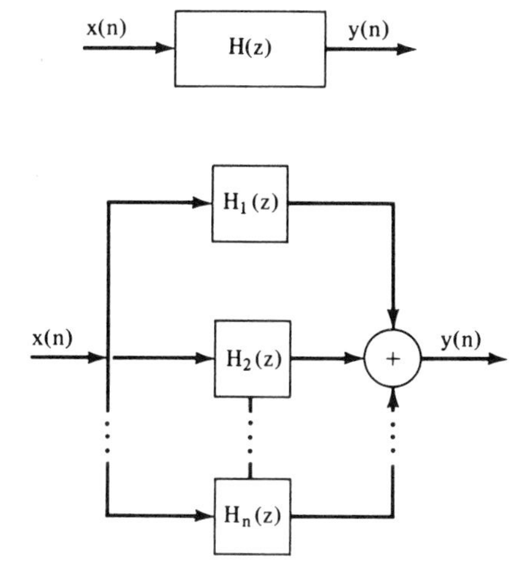


**Figure 2.8:** Parallel form realization.

Again a filter of arbitrary order can be created by combining first and second order sections of the following form:

**First Order Parallel Section:**

$$H_{f_p}(z) = \frac{\alpha_o}{1 + \beta z^{-1}} \tag{2.92}$$

**Second Order Parallel Section:**

$$H_{S_p}(z) = \frac{\alpha_o + \alpha_1 z^{-1}}{1 + \beta_1 z^{-1} + \beta_2 z^{-2}} \tag{2.93}$$

We note that each of these subfilter sections can be achieved from a common second order section of the form

$$H(z) = \frac{\alpha_o + \alpha_1 z^{-1} + \alpha_2 z^{-2}}{1 + \beta_1 z^{-1} + \beta_2 z^{-2}} \tag{2.94}$$

by assigning zero to the appropriate coefficients. Hence, this general second order section transfer function serves as a basic building block for virtually all digital filters.

Care must be taken when combining multiple first and second-order sections that the required transfer function is not altered by pole-zero cancelling. Further, the order of combination of the various sub-filter sections can have severe effects on the dynamic range and noise error properties of the final filter.

The general second order section transfer function of equation (2.94) corresponds to the second order difference equation:

$$y(n) = \alpha_0 + \alpha_1 x(n-1) + \alpha_2 x(n-2) - \beta_1 y(n-1) - \beta_2 y(n-2) \quad (2.95)$$

Hence, equation (2.95) defines a basic computational requirement for the implementation of an arbitrary digital filter. The implications of this requirement will be examined further in the next section. In that section we shall carry on to examine other representations of digital filters.

### 2.3.2 Digital Filter Network Representation

Equations (2.83) and (2.86) provide equivalent descriptions of the most general digital filter: in the time domain as a difference equation, and in the frequency domain as a transfer function. From the difference equation representation (2.83) a direct realization of the filter can be implemented either in dedicated hardware or in software. Figure 2.9 is a block diagram of such a structure, sometimes referred to as a direct form 1 realization. Such block diagrams are commonly referred to as "digital filter networks" or "flow graphs." The blocks labeled $\alpha_k$ and $\beta_k$ represent multiplications and these coefficients are taken directly from the difference equation (2.83), or a transfer function in the form of equation (2.86). The blocks labeled $z^{-1}$ represent unit delay elements of one sample period. While this direct realization is easily obtained, it is not particularly efficient due to the large number of delay elements, or equivalently, the number of memory elements and store/fetch operations required. By noting that the transfer function of the filter, as represented by equation (2.86), can be broken up as cascade of an FIR and an IIR section, i.e.,

$$H(z) = H_1(z)H_2(z) = \sum_{k=0}^{N} \alpha_k z^{-k} \cdot \frac{1}{1 + \sum_{k=1}^{M} \beta_k z^{-k}} \quad (2.96)$$

we see that Figure 2.9 can be transformed to the equivalent structures illustrated in Figures 2.10 and 2.11. As can be seen in Figure 2.10 the transfer function $H_1(z)$ represents a purely non-recursive (or FIR) structure representing the zeros of $H(z)$. $H_2(z)$ is a purely recursive structure representing the poles of $H(z)$. Since the sequence of cascading these structures is not important they can be rearranged as in Figure 2.10 and combined as in Figure 2.11 to reduce the number of delay elements needed. Figure 2.11 represents a Direct Form II realization of the filter defined by equation (2.83) or (2.86).

While filters of arbitrary order can be implemented in either direct form I or II realizations, it is more common to realize higher order filters as a cascade and/or parallel combination of first and second-order filter sections realized in direct form.

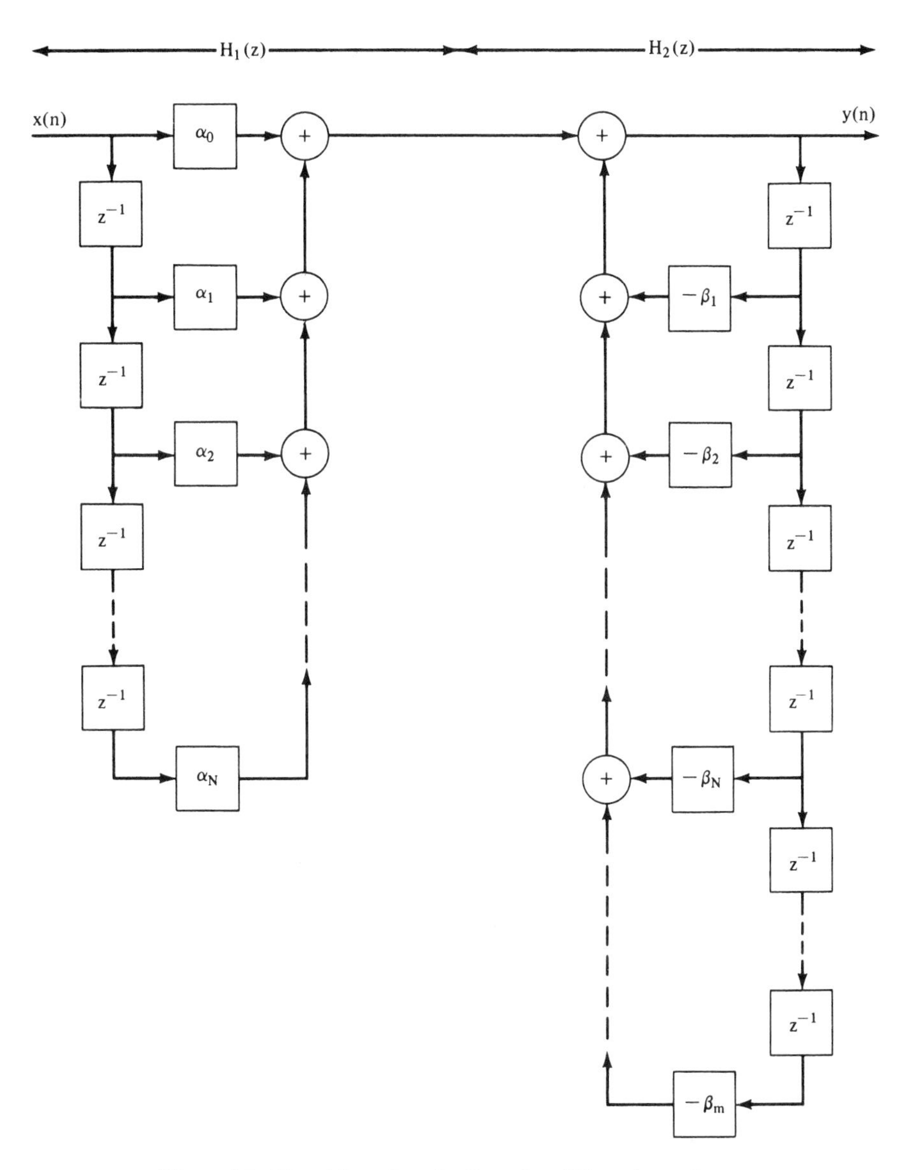


**Figure 2.9:** Direct form I realization of an Mth order system.

The basic direct form II second order filter section, or "biquad filter" as it is often called, is shown in Figure 2.12. This filter section represents the realization of equation (2.95). The coefficient $\alpha_0$ represents an interstage gain factor which is often omitted from the standard second-order section.

As noted at the end of Section 2.2.1 the general autoregressive moving average (ARMA) model corresponds to the general Nth order linear difference equation (2.83). Hence, the ARMA model corresponds to the most

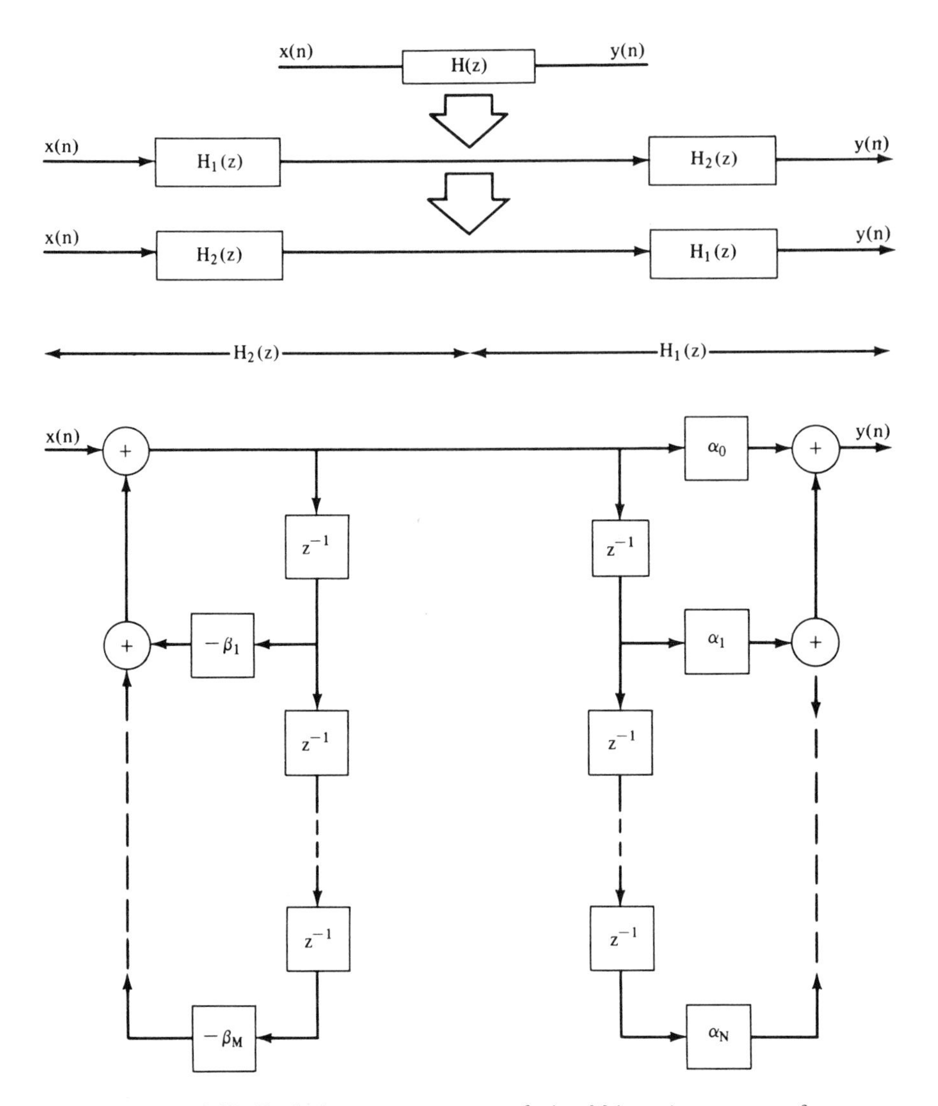


**Figure 2.10:** Equivalent arrangements of the Mth order system of Figure 2.9.

general digital filter structure of Figure 2.9. The moving average process corresponds exactly to an all-zero non-recursive FIR filter structure represented by $H_1(z)$ in Figures 2.9 and 2.10 while the autoregressive process corresponds exactly to an all-pole purely recursive IIR filter structure represented by $H_2(z)$ in Figures 2.9 and 2.10.

It is worth noting here that some variation exists in the literature regarding the use of the terms recursive and non-recursive for describing classes of digital filters. It is often suggested that the terms refer to the realization of

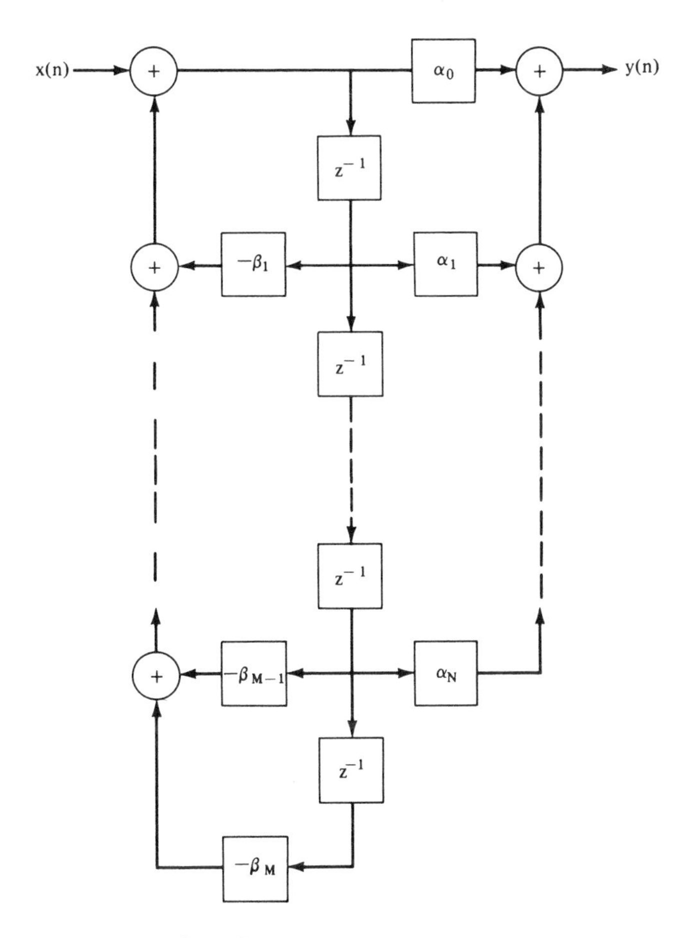


**Figure 2.11:** Direct form II realization of an Mth order system of Figures 2.9 and 2.10.

the filter and not to whether the filter is FIR or IIR. It is, however, common for IIR filters to be realized recursively and for FIR filters to be realized non-recursively. In this chapter we have used the terms recursive and non-recursive to refer to the mathematical representation of the filter. A difference equation such as equation (2.83) is a recursive equation and thus a recursive filter is implied. Equation (2.84) represents a convolution sum which is a non-recursive equation for an FIR filter. If the impulse response is extended to infinity, the convolution sum becomes an IIR filter equation (although non-realizable) but is still a non-recursive description of that filter. The concept of recursive and non-recursive implementations of IIR and FIR filters is discussed further in Section 2.4 where we discuss design and implementation issues.

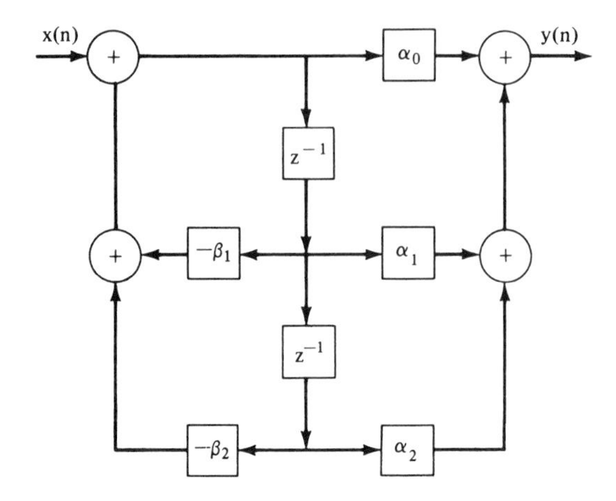


**Figure 2.12:** Second order section (biquad filter).

### 2.3.3 State-Space Representations

In Section 2.3.2 we introduced some realization structures for digital filters. In this section we shall re-examine realization structures from a state-space point of view.

Consider again the general Nth order difference equation for the most general digital filter, i.e.,

$$y(n) = \sum_{k=0}^{N} a_k x(n-k) - \sum_{k=1}^{N} b_k y(n-k) \tag{2.97}$$

where we have chosen $N = M$ for convenience sake.

The general Direct Form II realization of equation (2.97) is illustrated in Figure 2.13 for the case of $M = N$. If we choose as state variables the outputs of the delay elements then from the block diagram we can write

$$\begin{aligned}
x_1(n+1) &= x_2(n)\\
x_2(n+1) &= x_3(n)\\
&\vdots\\
x_{N-1}(n+1) &= x_N(n)\\
x_N(n+1) &= -b_N x_1(n) - b_{N-1}x_2(n) - \ldots\ldots - b_1 x_N(n) + v(n)
\end{aligned} \tag{2.98}$$

and

$$\begin{aligned}
y(n) &= a_N x_1(n) + a_{N-1}x_2(n) + \ldots. + a_1 x_N(n) + a_0 x_N(n+1)\\
&= (a_N - a_0 b_N)x_1(n) + (a_{N-1} - a_0 b_{N-1})x_2(n) + \ldots\\
&+ (a_1 - a_0 b_1)x_N(n) + a_0 v(n)
\end{aligned} \tag{2.99}$$

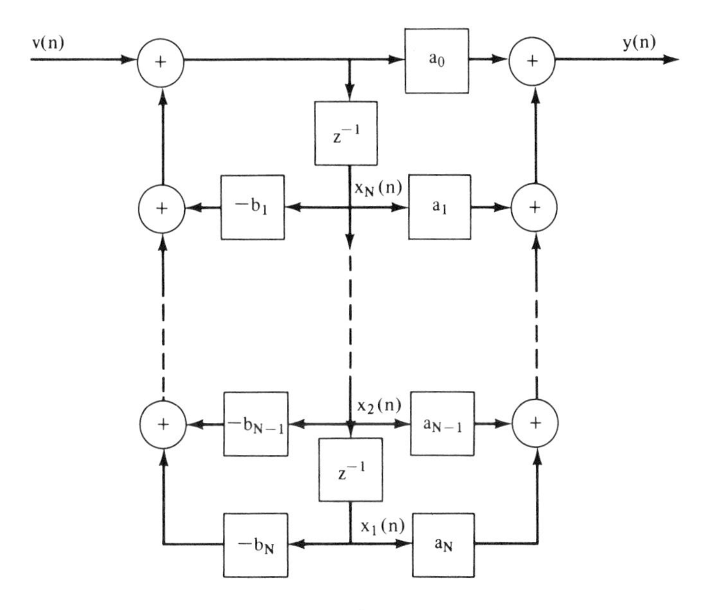


**Figure 2.13:** Direct form II realization.

From equations (2.98) and (2.99) we can write the matrix state equation:

$$\begin{bmatrix} x_1(n+1) \\ x_2(n+1) \\ \vdots \\ \\ x_{N-1}(n+1) \\ x_N(n+1) \end{bmatrix} = \begin{bmatrix} 0 & 1 & 0 & \dots & 0 \\ 0 & 0 & 1 & \dots & 0 \\ \vdots & \vdots & & & \vdots \\ & & & 1 & 0 \\ 0 & 0 & & 0 & 1 \\ -b_N & -b_{N-1} & \dots & -b_2 & -b_1 \end{bmatrix} \begin{bmatrix} x_1(n) \\ x_2(n) \\ \vdots \\ \\ \\ x_N(n) \end{bmatrix} + \begin{bmatrix} 0 \\ 0 \\ \vdots \\ \\ 0 \\ 1 \end{bmatrix} \underline{v}(n) \quad (2.100)$$

and the output equation

$$y(n) = [(a_N - a_o b_N), (a_{N-1} - a_o b_{N-1}), \dots, (a_1 - a_o b_1)] \begin{bmatrix} x_1(n) \\ x_2(n) \\ \vdots \\ x_N(n) \end{bmatrix} + a_o v(n) \quad (2.101)$$

Equations (2.100) and (2.101) provide a state-space representation of the filter.

Another realization structure for equation (2.97), commonly known as a standard form realization, is shown in Figure 2.14. The coefficients $\alpha_k$ and

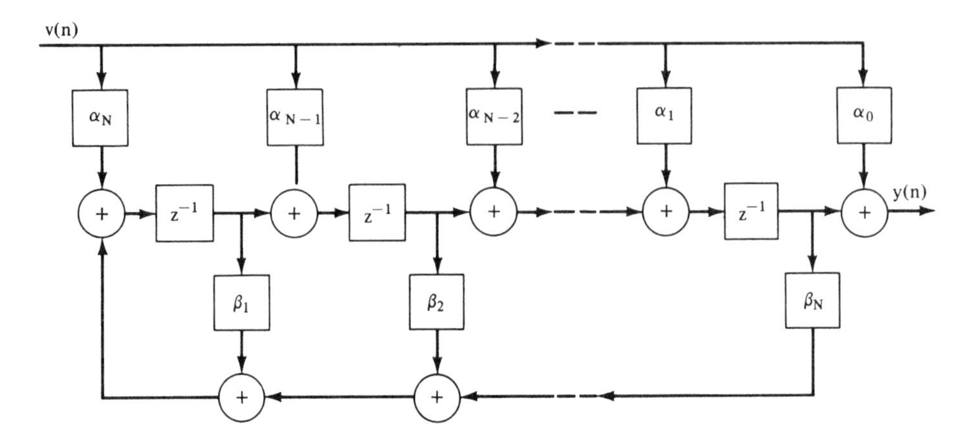


**Figure 2.14:** Standard form realization.

$\beta_k$ of Figure 2.14 are related to the coefficients $a_k$ and $b_k$ of equation (2.97) as follows:

$$\beta_k = b_k \qquad k = 1, \ldots, N \tag{2.102}$$

and

$$\begin{aligned}
\alpha_0 &= a_0 \\
\alpha_1 &= a_1 - b_1\alpha_o \\
\alpha_2 &= a_2 - b_2\alpha_o - b_1\alpha_1 \\
&\vdots \\
\alpha_N &= a_N - b_N\alpha_o - b_{N-1}\alpha_1 - \ldots - b_1\alpha_{N-1}
\end{aligned}$$

or equivalently

$$\begin{bmatrix} a_c \\ a_1 \\ \vdots \\ a_{N-1} \\ a_N \end{bmatrix} = \begin{bmatrix} 1 & 0 & 0 \ldots & & & & 0 \\ b_1 & 1 & 0 \ldots & & & & 0 \\ \vdots & & & & & & \vdots \\ b_{N-1} & b_{N-2} & & 0 & 1 & 0 & \\ b_N & b_{N-1} & b_{N-2} & \ldots & b_2 & b_1 & 1 \end{bmatrix} \begin{bmatrix} \alpha_0 \\ \alpha_1 \\ \\ \\ \alpha_N \end{bmatrix} \tag{2.103}$$

The state equation and output equation for the standard form realizations have the form:

$$\begin{bmatrix} x_1(n+1) \\ x_2(n+1) \\ \vdots \\ \\ x_N(n+1) \end{bmatrix} = \begin{bmatrix} 0 & 1 & 0 \ldots & & 0 & x_1(n) \\ 0 & 0 & 1 \ldots & & 0 & x_2(n) \\ \vdots & & & & \vdots & \vdots \\ 0 & & & 0 & 1 & \\ -b_N & -b_{N-1} & \ldots & -b_2 & -b_1 & x_N(n) \end{bmatrix} + \begin{bmatrix} \alpha_1 \\ \alpha_2 \\ \vdots \\ \\ \alpha_N \end{bmatrix} \underline{v}(n) \tag{2.104}$$

$$y(n) = [1,0,0,\ldots,0]\begin{bmatrix} x_1(n) \\ x_2(n) \\ \vdots \\ x_N(n) \end{bmatrix} + \alpha_o \underline{v}(n) \tag{2.105}$$

$$= x_1(n) + \alpha_o \underline{v}(n)$$

State-space representations can also be derived for both parallel and cascade realizations. Of particular interest is the "normal form" representation corresponding to an all pole parallel realization. Provided that the filter has only $N$ simple poles occurring at locations $a_1, a_2, \ldots, a_n$, then a partial fraction expansion of the transfer function results in the $A$ matrix being diagonal, i.e.,

$$\begin{bmatrix} x_1(n+1) \\ \vdots \\ x_N(n+1) \end{bmatrix} = \begin{bmatrix} a_1 & 0 & 0 & \cdots & 0 \\ 0 & a_2 & 0 & \cdots & 0 \\ \vdots & & \vdots & \cdots & \vdots \\ & & & & 0 \\ 0 & 0 & \cdots & 0 & a_N \end{bmatrix} \begin{bmatrix} x_1(n) \\ \vdots \\ x_N(n) \end{bmatrix} + \begin{bmatrix} 1 \\ 1 \\ \vdots \\ 1 \end{bmatrix} v(n) \tag{2.106}$$

and

$$y(n) = [d_1, d_2, \ldots, d_N]\begin{bmatrix} x_1(n) \\ \vdots \\ x_N(n) \end{bmatrix} + a_o v(n) \tag{2.107}$$

where the constants $d_i$ are the numerator coefficients of the partial fraction expansion of the transfer function and $a_o$ is a constant feed forward gain factor for the input.

We note finally that an alternate network known as the "transpose configuration" can be achieved for all the preceding networks by simply reversing the directions of all signal flow arrows and interchanging all branch and summation points. These transpose networks will have the same transfer function as the original but may differ with respect to finite word length error effects.

An even larger number of forms can be developed for realizing digital filters. The important point to note is that the general matrix formulation of digital filters involves the recursive calculation of the next state of the system based on the present state and the present input. The present output is calculated from the present state and the present input. This technique is of central importance in the realization of recursive estimation techniques, such as the Kalman filter to be discussed in Chapter 3.

### 2.3.4 Summary

In this section we have examined the basic concepts of the mathematical and graphical representation and analysis of digital filters. Digital filters were introduced as an important subclass of LTI systems. As such, they can be represented and analyzed in either the time domain or the frequency domain. The most general digital filter can be represented as the Nth order linear difference equation (2.82) or equivalently by the general transfer function equation (2.86).

By examining the general transfer function of equation (2.86) we saw that filters of an arbitrary order can be equivalently represented as cascaded and/or parallel combinations of first and second order sections.

Based on the general difference equation (2.83) for the most general digital filter two important subclasses were identified: infinite impulse response (IIR) and finite impulse response (FIR). IIR filters are characterized by an infinite unit sample response and are generally represented recursively in the time domain. In the frequency domain IIR filters are distinguished as those with poles in their transfer function. FIR filters have finite unit sample response sequences and are generally represented in the time domain by a discrete convolution operation. An FIR filter is characterized in the frequency domain by a transfer function with only zeros (i.e., poles can exist only at zero or infinity).

In the next section we begin the discussion of specification, design and implementation issues of digital filtering.

## 2.4 DIGITAL FILTER DESIGN TECHNIQUES AND IMPLEMENTATION ISSUES

### 2.4.0 Introduction

We have seen that digital filters can be represented and analyzed in either the time domain or the frequency domain. It is possible to approach the specification, design and implementation of digital filters from either of these domains as well.

Thus far we have considered only the general Nth order linear difference equation for specifying the most general digital filter. In practice the design and implementation of a digital filter must be based not on a specific difference equation but on a filter performance specification given in terms of its frequency response, or perhaps, in terms of some predetermined or required characteristics of its time domain output sequence for a given input. In this section we restrict our discussion to filters specified in terms of their frequency domain characteristics. In Chapter 3 we discuss the topics of detection and estimation theory, and show how these concepts lead to the specification and design of a broader class of digital filters from a time domain approach.

The design and implementation of a digital filter primarily involves the determination of a realization structure and the identification of the associated filter coefficient values. The manner in which filter performance is specified will in general affect the design techniques and implementation approach chosen.

The synthesis of frequency domain specified filter response characteristics is constrained by the intrinsic limitations of the Fourier transform on the arbitrary specification of gain and of phase shift. Thus, at the extremes, either could be specified if the other could be ignored. Much of the challenge faced by early designers revolved around finding amplitude functions which yielded acceptable phase characteristics within a cost constraint of using as few components as possible.

In this section we shall consider a number of techniques commonly used for designing digital filters which are based on a required frequency response filter performance specification. Such filters correspond directly to familiar analog filtering operations for spectral shaping. Indeed, we shall see that a common method of digital filter design is based on the transformation of a suitable analog filter that meets the performance requirements, into a digital realization.

The discussion of filter design techniques is brief intentionally. Its purpose is to indicate that systematic approaches to filter design problems do exist. Numerous volumes deal exclusively with this topic, and a guide to further reading is available in the bibliography.

Of more immediate interest to us in this volume is the identification of the actual computational requirements of digital filters and the implications for hardware architectures. A number of implementation issues arise in connection with the realization of digital filters. Indeed, one motive for studying realization structures is to identify those structures that provide better immunity to the effects of truncation, rounding and finite word lengths. The chosen implementation approach will also impact the required hardware characteristics for dedicated hardware realizations. If the filter is to be implemented on a programmable processor there are implications for the instruction set and hardware architecture for the efficient performance of filter operations. These considerations are examined in the latter part of this section.

### 2.4.1 Digital Filter Specification

The specification of performance requirements for a specific digital filtering operation represents the first step in the move from theory to implementation. A wide variety of filters are commonly specified in terms of their required frequency domain characteristics. Figure 2.15 illustrates several parameters that form part of such a specification.

Filters specified by their frequency domain characteristics are generally spectral shaping filters (i.e., low pass, high pass, band pass). The essential parameters for specifying such a filter are:

1. Pass-band bandwidth: The range of frequencies that are passed unattenuated (or with some specified gain) by the filter (commonly measured between the -3 dB attenuation points).
2. Stop-band(s): The range(s) of signal frequencies that are significantly attenuated by the filter.
3. Minimum stop-band attenuation: The ratio of minimum stop-band to mean pass-band attenuation (usually specified in dB).
4. Transition-band: The frequency range in which the filter response changes from being within the pass band to the stop band.
5. Pass-band ripple: The variation of the filter response characteristic about a designated characteristic. By specifying a maximum allowable ripple (e.g., about the desired pass-band characteristic) the degree of accuracy with which the desired response is approximated may be specified within the band.

### 2.4.2 FIR vs IIR Filter

There are several advantages and disadvantages associated with either IIR or FIR filter implementations. The choice of one approach over the other will usually be highly dependent on the specific requirements of the application. Some of the general factors that will influence such a choice are listed below.

**IIR Advantages:**

1. In general, IIR filters can be implemented recursively with a small number of coefficients to realize a sharp frequency cutoff characteristics. This leads to reduced hardware requirements in dedicated realizations, particularly where amplitude response is the major concern and phase response is of lesser importance.
2. Several systematic design techniques exist for mapping well known analog fitler designs to an IIR digital filter realization. This allows the digital filter designer to take advantage of well established closed form analog designs.

**IIR Disadvantages**

1. IIR filters implemented recursively can be unstable, while finite word length error effects can lead to such problems as "limit cycles" (these error effects are discussed further at the end of this section).

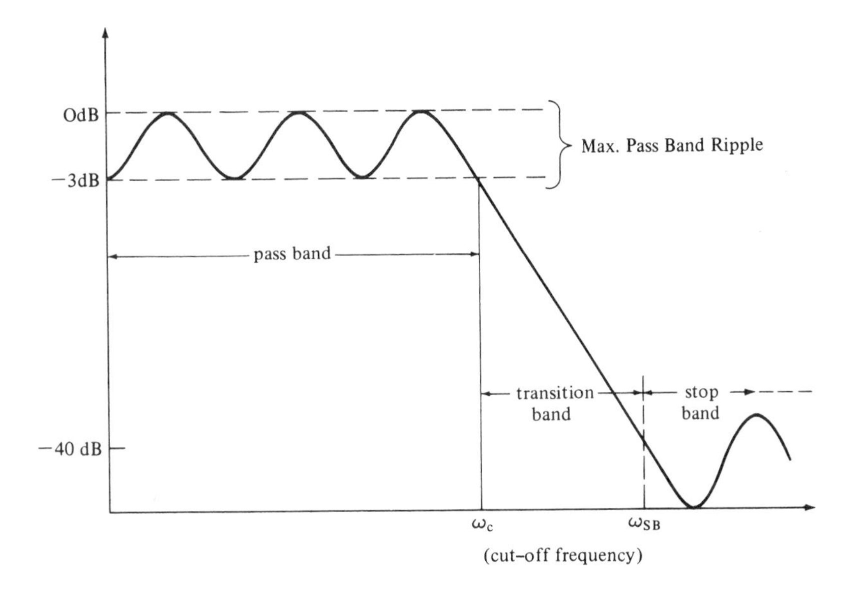


**Figure 2.15:** Filter domain filter characteristics.

2. Closed form design approaches for IIR filters tend to be limited to low pass, band pass and high pass filters.
3. IIR filters exhibit nonlinear phase characteristics which may make them a poor choice for some applications (e.g., speech synthesis).

**FIR Advantages**

1. FIR filters can be implemented with exactly linear phase.
2. FIR filters implemented by direct convolution are always stable.
3. The error effects due to round-off or truncation can be easily minimized in nonrecursive FIR implementations.
4. FIR filters can be more easily altered in a dynamic fashion to implement adaptive filtering functions.
5. Existing approximation techniques for FIR filter design can be applied easily to a wider range of filtering problems than is the case with closed form solutions for IIR design.

**FIR Disadvantages**

1. FIR filters tend to require a very long unit sample response (a large number of coefficients) to realize a sharp frequency cutoff characteristic. This implies an increased computational requirement to implement highly frequency selective filters as nonrecursive structures.
2. Closed form design formulas for FIR filters do not exist. Thus, FIR filter design is always an approximation problem.

3. Linear phase FIR filters do not always exhibit delay characteristics that are an integral number of sample periods, which may be necessary for some applications.

### 2.4.3 IIR Filter Design Techniques

The most common methods for designing IIR spectral shaping filters are based on the concept of first designing or finding an analog filter that meets the specified requirements and then transforming that filter to a digital realization. Before reviewing these methods we review the basic characteristics of the common analog filters from which IIR filters are often derived.

One of the following analog filter types often forms the starting point of an IIR design:

1. Butterworth Filter: Butterworth filters are specified by the magnitude squared relationship

$$|H(j\omega)|^2 = \frac{1}{1 + (\omega/\omega_C)^{2n}} \tag{2.108}$$

The typical response of these filters as a function of the order $n$ is shown in Figure 2.16(a). The pole locations in the $S$ plane are shown in Figure 2.16(b). For an *nth* order filter the poles are equally spaced around a circle of radius $\omega_c$ in the $S$ plane. These filters have a monotonically decreasing amplitude function.

2. Chebyshev Filters: Chebyshev filters are specified by the magnitude squared relationship

$$|H(j\omega)^2 = \frac{1}{1 + \epsilon^2 C_N^2(\omega/\omega_c)} \tag{2.109}$$

where $C_N(\omega)$ is a Chebyshev polynomial of order $N$ having the property of equal ripple over a particular range of $\omega$. The parameter $\epsilon$ allows the specification of a magnitude function with equal ripple in the pass band and monotonic decay in the stop band. The ripple amplitude is given as

$$\delta = 1 - \frac{1}{\sqrt{1 + \epsilon^2}} \tag{2.110}$$

Figure 2.17 shows the magnitude squared response and the $S$-plane pole locations for a Chebyshev filter. The poles lie on the ellipse determined from the parameters $\epsilon$, $N$, and $\omega_c$.

3. Elliptic Filters: Elliptic filters are specified by the magnitude squared relationship

$$|H(j\omega)|^2 = \frac{1}{1 + \epsilon^2 C_N^2(\omega)} \tag{2.111}$$

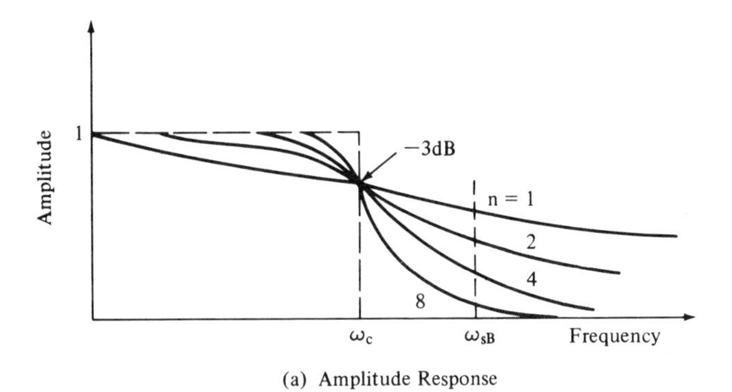


(a) Amplitude Response

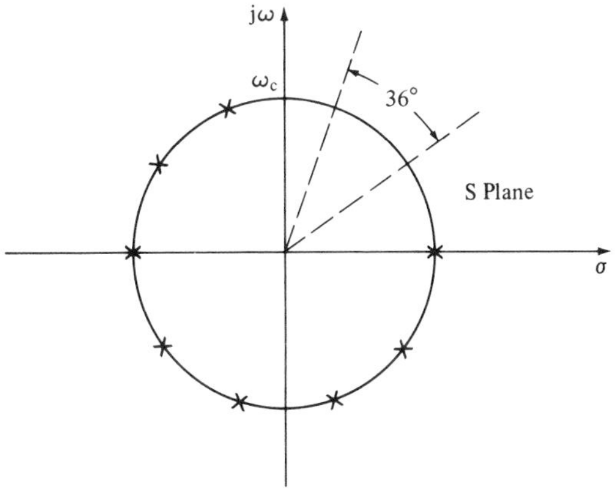

(b) Poles of a Fifth-Order Butterworth Filter

**Figure 2.16:** Butterworth filter example.

where $C_N(\omega)$ is a rational Chebyshev function involving elliptical functions. These filters have the property of equal ripple in both the pass band and the stop band. A magnitude squared response is illustrated in Figure 2.18.

Several methods of obtaining the digital filter coefficients from a given analog transfer function have been utilized. Some of the more common methods are:

1. Impulse Invariance Method: This technique is based on deriving a digital filter with a unit sample response that is equivalent to a sampled version of the impulse response of an analog filter that meets the performance specification, i.e.,

$$h(nT) = h(t)\,|_{t\,=\,nT} \tag{2.112}$$

Thus if

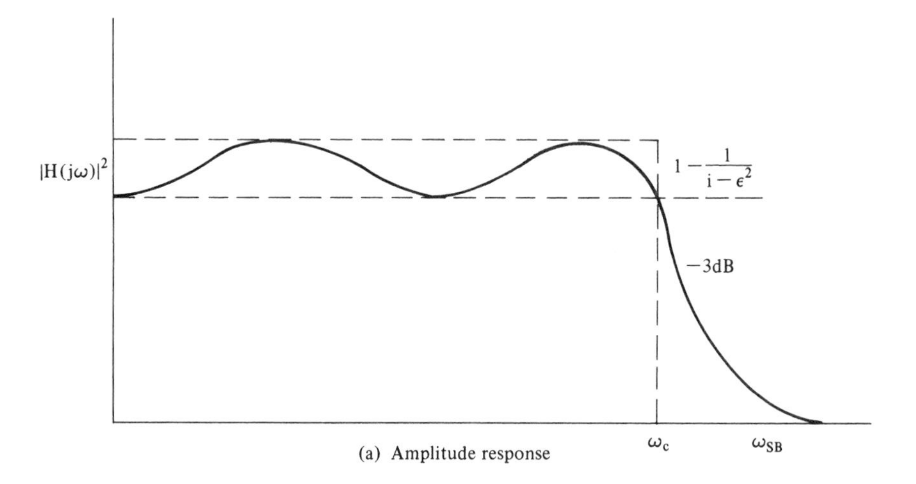


(a) Amplitude response

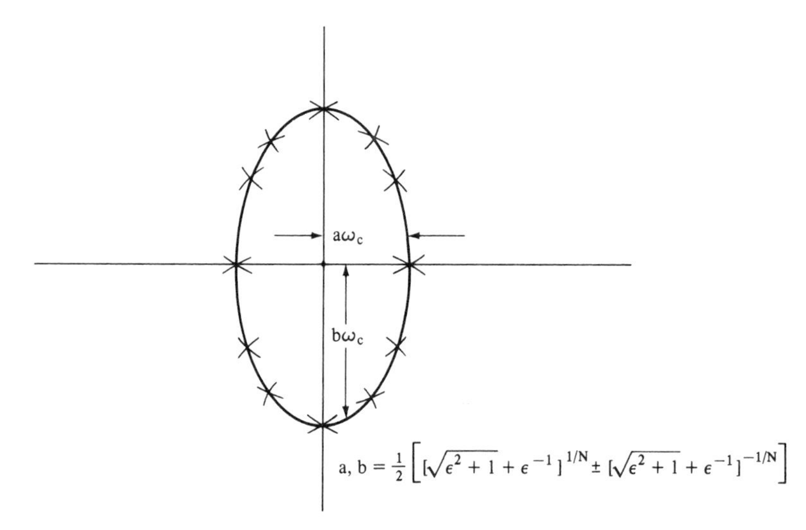


(b) Pole location in the S plane

**Figure 2.17:** Chebyshev filter.

$$H(s) = \sum_{i=1}^{M} Ai/(s + \alpha_i) \tag{2.113}$$

we have

$$H(z) = \sum_{i=1}^{M} Ai/(1 - e^{-\alpha_i T} z^{-1}) \tag{2.114}$$

2. Bilinear $Z$-Transform Method: The bilinear $Z$-transform technique is based on the the algebraic transformation

$$s = \sigma + j\omega = \frac{z-1}{z+1} = \frac{e^{\sigma T}e^{j\omega T} - 1}{e^{\sigma T}e^{j\omega T} + 1} \tag{2.115}$$

which maps the entire left half $S$-plane inside and the entire right half $S$-plane outside the unit circle in the $Z$-plane.

This mapping produces a frequency axis warping described by the relation

$$\omega_A = \tan \omega_D T/2 \tag{2.116}$$

where $\omega_A$ is the continuous time frequency variable along the $j\omega$ axis of the $S$-plane and $W_D$ is the digital frequency variable around the unit circle of the $Z$-plane. This frequency warping must be taken into account when the digital transfer function is derived as:

$$H(z) = H(s)\Big|_{s = \frac{Z-1}{Z+1}} \tag{2.117}$$

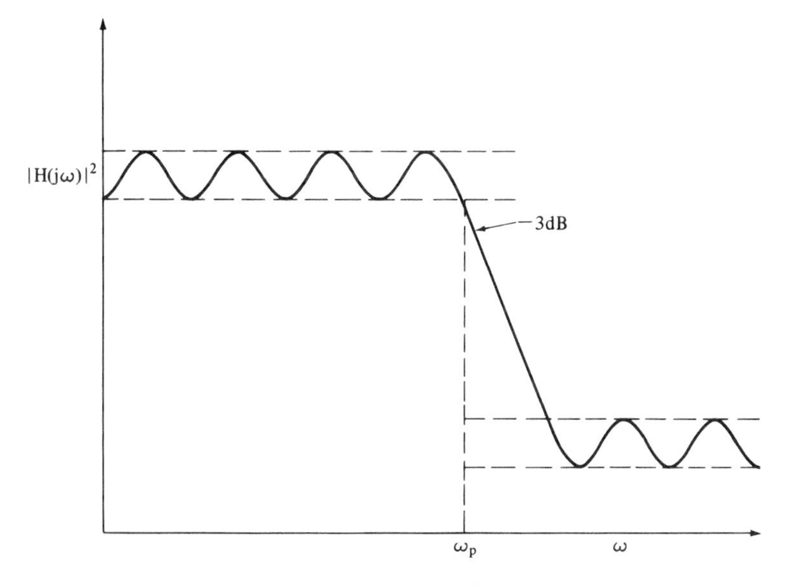


**Figure 2.18:** Elliptic filter.

3. Matched $Z$-Transform: The matched $Z$-transform technique is based on the concept of mapping the poles and zeros of a continuous time transfer function to the discrete time transfer function. That is, for $H(s)$ of the form

$$H(s) = \frac{1}{\prod_{i=1}^{M} (s + s_i)} \tag{2.118}$$

then

$$H(z) = \frac{1}{\prod_{t=1}^{M} (1 - e^{-S_t T} z^{-1})} \tag{2.119}$$

This results in the poles of $H(z)$ being identical to those of the impulse invariant method, while the zero locations will be different.

***Example*** *(Butterworth Filter; Bilinear Transform):*

*Filter Specifications*

Sample rate: 10 KHz;
Low Pass: 0 to 1000 Hz
Transition band: 1000 to 2000 Hz
Stop band: -10dB (starting at 2000 Hz)
Filter must be monotonic in pass and stop band

*Design*

- A Butterworth filter is indicated by the monotonicity requirement
- Use the bilinear transformation

$$W_c T = 2\pi \times 1000 x 10^{-4} = .2\pi$$

$$W_{sB} T = 2\pi \times 2000 \times 10^{-4} = .4\pi$$

from which

$$W_{A1} = \tan .2\pi/2 = 0.325$$

$$W_{A2} = \tan .4\pi/2 = 0.726$$

- The filter requires a cut-off (-3dB point) at $W_{A1} = 0.325$. To find the order $n$ set

$$1 + (W_{A2}/W_{A1})^{2n} = 10$$

which yields

$$n = 1.74 \text{ (use } n = 2 \text{ as the filter order)}$$

- The transfer function is

$$H(s) = \frac{s_1 s_2}{(s + s_2)(s + s_1)} = \frac{2X(0.23)^2}{(s + 0.23)^2 + (0.23)^2}$$

$$= \frac{0.1058}{s^2 + 0.46s + 0.1058}$$

- Substitute the bilinear transform and obtain (after some simplification)

$$H(z) = \frac{0.067569(z^2 - 2z + 1)}{z^2 - 1.14216z - 0.412441}$$

It is not always necessary to begin IIR filter design from an analog transfer function. In particular, the magnitude squared technique, or the frequency sampling technique can be applied without explicit reference to an analog filter. In addition a number of optimization methods for designing IIR filters exist [10]. Further discussion of the details of IIR filter design techniques, the basic results of which are the specification of filter coefficient values, would not contribute significantly to the main thrust of our presentation, namely, the identification of the major computational requirements of signal processing. Thus we shall not pursue these techniques here. The reader who is interested in a study of filter design techniques is referred to the wealth of available literature on the subject. Suitable references for an initial investigation are contained in the bibliography.

### 2.4.4 FIR Filter Design

One of the major features of FIR (non-recursive) filters is the ability to realize an exactly linear phase response (since the output is not included in the processing). Two common approaches exist for the design of linear phase FIR filters:

**Windowing**

FIR filter design by windowing is based on the idea of truncating an infinite impulse response to a finite length for implementation. To avoid drastic alteration in the filter response, the original impulse response coefficients are weighted by a windowing function as part of the truncation process.

To see how this can be done, consider the case of truncating an infinite impulse response to some finite length. This is equivalent to multiplying the original infinite impulse response $h(n)$ by a window function of the form

$$w(n) = \begin{cases} 1, & 0 < n < N \\ 0, & \text{otherwise} \end{cases} \tag{2.120}$$

to give a finite unit sample response of the form

$$h'(n) = h(n)w(n) \tag{2.121}$$

The FIR filter with unit sample response $h'(n)$ has the frequency response

$$H'(e^{j\omega}) = H(e^{j\omega}) * W(e^{j\omega}) \tag{2.122}$$

where $H(e^{j\omega})$ is the frequency response of the IIR filter to be approximated and $W(e^{j\omega})$ is the Fourier transform of the window function $w(n)$. For $H'(e^{j\omega})$ to approximate $H(e^{j\omega})$ as closely as possible we wish $W(e^{j\omega})$ to approximate a delta function. This requires that $N$ in equation (2.120) approach infinity. By altering the shape of the window function $w(n)$ from the "boxcar" shape of Figure 2.19(a) (corresponding to equation 2.120) to a tapered

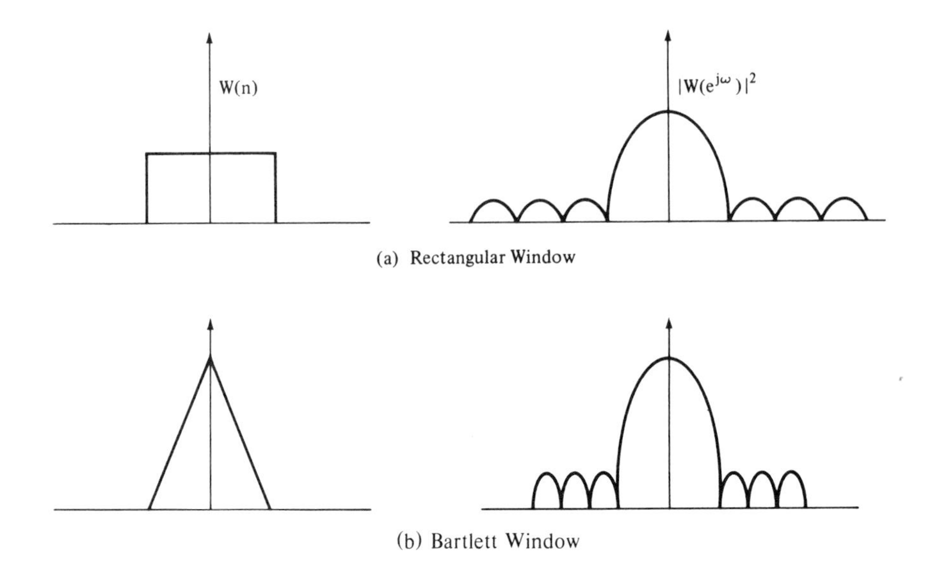


(a) Rectangular Window

(b) Bartlett Window

**Figure 2.19:** Spectral window functions.

shape, as shown in Figure 2.19(b), the shape of $W(e^{j\omega})$ can be altered to approximate more closely a sharp peak or reduce the side lobes. The fundamental tradeoff is in the length of the window. The longer the window the narrower $W(e^{j\omega})$ can be made (and hence the more closely $H'(e^{j\omega})$ will approximate $H(e^{j\omega})$). The shorter the window length the shorter the finite impulse response which minimizes computational requirements.

For the "boxcar" window a typical $(\sin x)/x$ shape for $W(e^{j\omega})$ is obtained as illustrated in Figure 2.19(a). The height of the side lobes and width of the peak can be altered by choosing window shapes that taper off smoothly to zero at the edges of the window. Several such window functions have been developed:

1. Bartlett
2. Hanning
3. Hamming
4. Blackman
5. Kaiser

The Bartlett window is simply a triangle shaped window as shown in Figure 2.19(b), while the others are more complicated smoothly tapered windows, each providing its own characteristic advantage or disadvantage with respect to main lobe width and side lobe height for a given window length.

We shall consider windowing again in Chapter 3 when we discuss spectral estimation. For a further discussion of windowing in filter design the reader is directed to the references and bibliography.

**Frequency Sampling**

FIR filter design by frequency sampling is based on the specification of the required frequency response on a point-by-point basis. Recall that the concept of "sampling in frequency" was introduced in the DFT discussion. If we specify the value of the DFT coefficients of a finite duration sequence to match a classical filter response, then the IDFT of this sequence yields a unit sample response for an FIR filter. This is the essence of FIR filter design by frequency sampling. To prevent radical deviation of the final frequency response from that desired between the specified points, the frequency sample points must often be chosen close together, resulting in a long unit sample response. In practice, frequency sampling is a useful technique for narrow-band filter design where only a few non-zero samples occur in the desired frequency response.

In Chapter 3 we shall examine a technique for designing FIR filters from a time domain approach based on a minimum mean-squared-error optimality criterion. For a complete discussion of the design techniques mentioned above, several texts are available and suitable references are indicated in the bibliography.

### 2.4.5 Implementation Issues

A wide range of issues must be faced when the mathematical structures discussed thus far are to be implemented. If performance or memory requirements are not important, then almost any computer language could be used to perform the computations. However, this is rarely the case, and stringent requirements are the usual situation faced by the processor designer. Two major issues must be initially considered. First, there is the problem of performing the required computations using a given set of data to produce results within some specified accuracy. The finite word length of any machine introduces a range of computationally important issues. Second, the architecture of the processor introduces constraints which can render the computations difficult to execute (or even program), and limit throughput rates.

Finite-word-length effects introduce a range of pathologies into any computation which in themselves form a separate area of study. It is not our intention to pursue these, except to acknowledge their existence.

In Chapter 1 (for example) it was shown that when an analog signal is sampled and quantized an irreversible error or noise component was introduced in its representation. In a similar manner, the quantization of filter coefficient values into finite length digital words can alter the performance of the filter. Additional errors may also occur due to the accumulation of truncation or round-off errors as a result of the finite precision arithmetic operations which have the effect of creating a non-linearity in the filter. This can lead to a situation known as "limit cycles" where there is an oscillation in the filter output even though the input is constant. This effect is often treated as an output "noise" effect.

The assurance of filter stability is another important consideration. A particular filter design may be such that the filter is stable if implemented with ideally accurate arithmetic. However, the effects of coefficient quantization and accumulated round-off or truncation errors may result in an unstable implementation. The statistical analysis of error effects due to finite word length is covered in detail in several other publications, a few of which are indicated in the bibliography. Of more immediate interest to us in this section is the impact of the various computational requirements of a digital filter on the architecture of signal processors.

The outstanding computational characteristic of filters is the requirement for multiplications and accumulating-type additions. A single second order digital filter requires four multiplications, four additions, plus two store operations. Given a typical 8KHz sampling rate (say for voice band data communications), these operations must be repeated every 125 microseconds. A processor for this application must have a specialized architecture, instruction set and speed to be useful in this environment.

For direct time-domain implementations the computational requirements of a general digital filter are completely summarized by the general difference equation (equation (2.83)):

$$y(n) = \sum_{k=0}^{N} \alpha_k x(n-k) - \sum_{k=1}^{M} \beta_k y(n-k) \tag{2.123}$$

As noted in Section 2.3, an arbitrary order filter can be implemented by a cascade and/or parallel combination of first and second order filter sections. The basic computational requirement is that of a standard second order section, i.e.,

$$y(n) = \sum_{k=0}^{2} \alpha_k x(n-k) - \sum_{k=1}^{2} \beta_k y(n-k) \tag{2.124}$$

or

$$y(n)/\alpha_o = \sum_{k=1}^{2} a_k x(n-k) - \sum_{k=1}^{2} b_k y(n-k) \tag{2.125}$$

where

$$a_i = \alpha_i/\alpha_o \text{ and } b_i = \beta_i/\alpha_o$$

If an FIR filter is to be implemented by direct convolution (i.e., $\beta_i = 0$ in equation (2.123)), then the required computation is

$$y(n) = \sum_{k=0}^{N} \alpha_k x(n-k) \tag{2.126}$$

where the coefficients $\alpha_k$ represent the unit sample response of the filter.

This computation represents a time-domain approach to FIR filter realization. FIR filters can also be implemented in the frequency domain. Recall that a time-domain convolution can be represented as a frequency domain multiplication, i.e.,

$$Y(z) = X(z)H(z) \tag{2.127}$$

where $H(z)$ is the $Z$-transform of the sequence of coefficients $\alpha_k$ of equation (2.126).

The computational requirements of this representation of an FIR filter can be summarized as follows:

1. Initially compute and store $H(z)$ (this need be done only once);
2. Compute $X_i(z)$ for some finite number of input samples $x_i$ (i.e., DFT operation);
3. Compute $Y_i(z) = X_i(z)H(z)$;
4. Compute the IDFT $\{Y_i(z)\} = y_i(n)$.

Several issues arise when attempting to implement this algorithm, such as determining how to segment the input sequence $x(n)$ into finite sequences $x_i(n)$ for practical computation of $X_i(z)$ and how to combine resulting outputs $y_i(n)$. The details of a technique known as "fast convolution" for performing frequency domain FIR filtering are presented in Chapter 4. Here we wish to note that this approach imposes a set of computational requirements on processor architecture different from that imposed by a direct time-domain implementation; specifically, the requirement to perform DFT and IDFT computations and cross-sequence multiplication of complex frequency-domain sequence representations.

In contrast to the frequency domain approach, consider a direct implementation of equation (2.126). This suggests that the output is computed based on the present plus the past $(N - 1)$ samples. This entire computation must be done at least before the next sample arrives. Other implications are:

- Memory must be available for storing coefficients and data; half of this memory must be RAM, while the coefficients could be in ROM.
- Coefficients and data must be supplied to a multiplier simultaneously for efficient operation.
- The results of each multiplication must be summed with previous products.
- The $N$ samples are continuously being displaced backward in time to give the $Z$ delays. A convenient mechanism is required to move the data sample to the appropriate locations in memory as the data ages.

Many other questions emerge and form the subject of later chapters and volumes. To illustrate the foregoing comments the computation of $y(n)$ can be carried out sequentially as follows:

```
At each sample interval
   begin

- For each coefficient compute as follows:
  - accumulate product from previous computation;
  - get coefficient and data, start multiplication
  - age the data, e.g., more data from previous RAM
      location to present
- When complete store final result in output data
      memory location
   end;
```

Consider as a second example the requirement to implement a general IIR filter on a programmable processor as a cascaded network of second-order sections. Computationally only complex multiplications and summations are required. In addition, the algorithm is data independent in the sense that no program modification or branching takes place because of tests on intermediate results. As a result of these features the computation can be regarded as a fixed functional unit with known response time. (This concept will be extended to other signal processing functions in Chapter 5).

The real time response imposed by the application determines the computational rate. This is proportional to both the signal bandwidth and the sampling rate. Data points are usually stored sequentially in a random access memory (RAM). A new data point overwrites the oldest one in a logical ring configuration. A pointer must be maintained for access to the record which points to the nearest sample. As computation proceeds an addressing mechanism must be in place which

1. steps through the samples;
2. tests for record limits.

In addition, if multiple channels are being processed, each must have a pointer. Addressing is therefore a significant problem if a large data flow is to be maintained.

Figure 2.20 shows two extreme possibilities of strict sequentiality and parallelism. In Figure 2.20(a) coefficients and data are assumed to be fetched serially and the AU performs a multiply and accumulates the results. In Figure 2.20(b) all coefficients and data arrive simultaneously. The AU provides full parallelism. This architecture is ideally suited for pipelining by inserting registers between each of the arithmetic operations. With earlier technologies it was necessary to pipeline the multiplication in order to maintain a uniform flow; more modern components are such that multiplication is not a bottleneck.

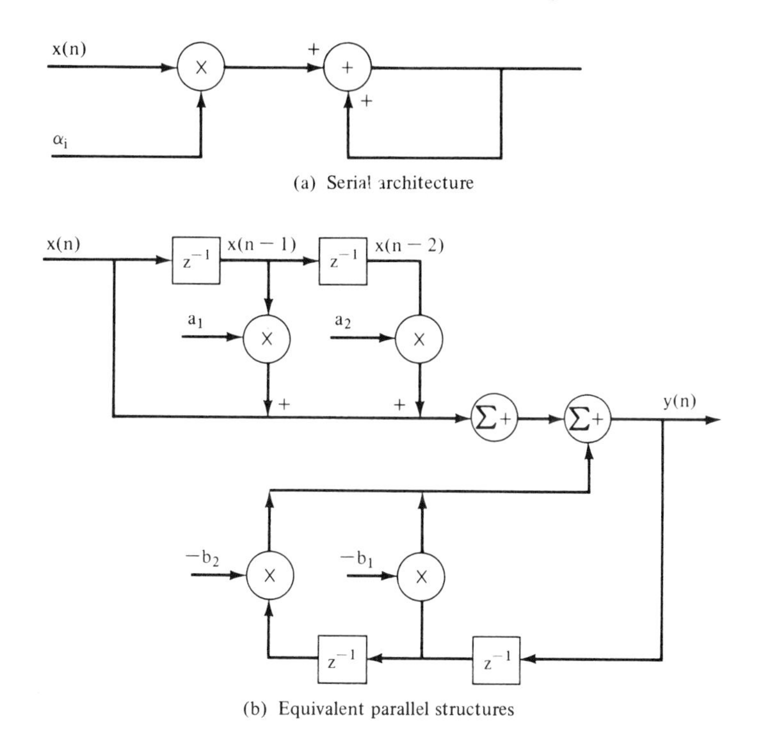


(a) Serial architecture

(b) Equivalent parallel structures

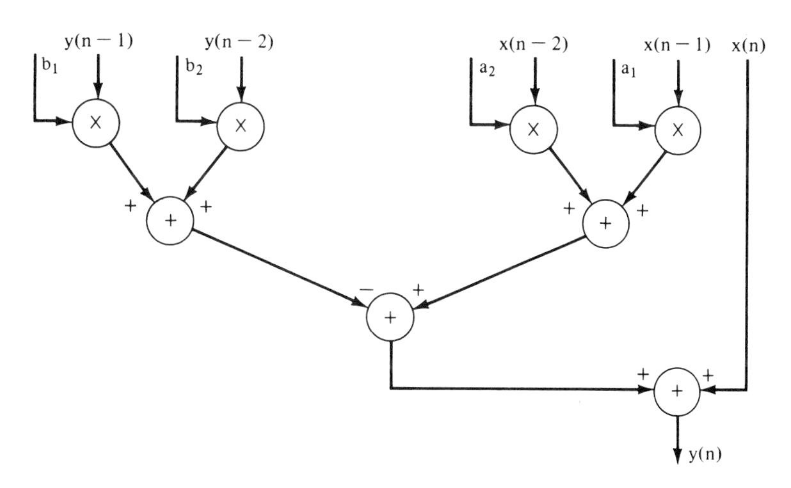


**Figure 2.20:** Recursive second order section computations.

If the filter is to be represented in a state-space manner then the state equations explicitly indicate the numerical computations in vector matrix form (equations (2.72) and (2.73)).

$$\underline{x}(n+1) = A\underline{x}(n) + B\underline{v}(n) \quad (2.128a)$$

$$\underline{y}(n) = C\underline{x}(n) + D\underline{v}(n) \quad (2.128b)$$

Matrix operations form another perplexing architectural problem. Matrix arithmetic imposes data storage problems if high speed access is required. This is particularly true if multiple processors are cooperating to produce a result. Multiplication, for example, requires fetching one array by rows and the other by columns. The results must be stored in a manner which could be dependent on the next set of computations.

Clearly, the implementations of any algorithm will be more or less efficient depending on the architecture of the hardware. If dedicated hardware is to be designed to implement a filter then a great deal of optimization can be done to obtain a specific mathematical representation that minimizes error effects, hardware complexity, and control requirements. These situations must be considered on an unstructured case by case basis. On the other hand if it is desired to implement filter operations on a programmable processor, the various possible implementation approaches and their associated computational forms place differing architectural requirements on the processor for efficient operation.

Addressing complexity and algorithm complexity are often antagonistic attributes. Consider for example the non-recursive filter

$$y(i) = \sum_{n=0}^{N-1} h_n x(i-n) \quad (2.129)$$

If the samples are real and the impulse response complex and symmetric, this can be rearranged to

$$y(i) = \sum_{n=0}^{N-1} ([x(i-n) + x(i-N+1+n)]Re(h(n))$$

$$+ [x(i-n) - x(i-N+1+n)]\mathrm{Im}(h(n))$$

$$+ [x(i-(N-1)/2)]h((N-1)/2)) \quad (2.130)$$

This changes almost half the multiplications to additions and subtractions. If multiplication takes significantly longer than addition, then this rearrangement would be attractive; note however the problem of generating the addresses for the samples. In the first case a sequential index through the record beginning at $n = 0$; in the second a very awkward requirment for computing the locating of the pair of samples is required. This is a another trade-off which must be addressed by the architect.

For multiple processor systems, both indexing and data interchange networks are required to maintain high performance. Complex indexing registers for computing addresses have already emerged as an architectural requirement in previous considerations of filters. Networks for routing data which must be shared by cooperating processors imposes a severe limitation on multiprocessor architectures.

In future chapters this initial discussion will be focused by considering in more detail the algorithms of typical applications and several real processor architectures.

## 2.5 CHAPTER SUMMARY

The primary purpose of this chapter has been the establishment of the basic terminology and concepts of digital filters. The approach taken was to discuss first the characteristics of discrete-time linear time-invariant systems and then consider digital filters as a major subclass of such systems.

In the first two sections of the chapter we dealt with linear systems representation and emphasized that the theoretical modeling of LTI systems can be approached from either the time domain or the frequency domain. From each of these approaches many useful concepts emerge. Of fundamental importance is the concept of the unit sample response of an LTI system and the fact that this sequence completely characterizes the system. For an arbitrary input sequence the output sequence is the convolution of the input sequence with the unit sample response. By applying the transform techniques of Chapter 1, we obtained the transfer function of the system as the $Z$-transform of the unit sample response and the frequency response of the system as the Fourier Transform of the unit sample response; these completely characterize the system in the frequency domain. For the case of random input sequences we showed that a stationary random input signal applied to an LTI system results in a stationary random output sequence. Section 2.1 concluded with a discussion of how these fundamental concepts can be extended to the case of two-dimensional signals and systems.

In Section 2.2 we discussed the representation and modeling of LTI systems from a time-domain approach. The basic concepts of time series analysis lead to the moving average, autoregressive, and general autoregressive moving average linear system models. The general ARMA model corresponds to a general Nth order linear difference equation. Such systems can be conveniently represented in a vector-matrix notation in a state-space representation.

Digital filters were introduced as a subclass of LTI systems. The most general digital filter is represented by a general Nth order linear difference equation and corresponds exactly to the ARMA model. The approach taken in discussing digital filters was to examine four basic areas:

1. mathematical and graphical representation
2. performance specification
3. design
4. implementation

Each of these areas can be considered from either the frequency domain or the time domain. In Section 2.3 we considered the mathematical and graphical representation of digital filters in each of these domains. We saw that digital filters are generally subdivided into categories according to either their unit sample response length (IIR or FIR) or their representation (recursive or non-recursive). IIR digital filters are generally associated with recursive representations while FIR filters are commonly associated with non-recursive representations.

In Section 2.4 we concentrated on design from a frequency domain point of view, and on implementation issues. The significant results of a filter design are the specification: first, of the desired realization structure; second, the associated coefficient values. Implementation considerations therefore tend to be independent of the particular attributes of the filter; they are more concerned with the structure of the architecture to support efficient sums of products type calculations. In Chapter 5 we will find that almost all signal processing reduces to a small set of computational forms. This becomes a powerful unifying force on processor design. Before this, however, we must complete our study of computational models in the time domain (Chapter 3) and explore the range of application techniques by which they are used (Chapter 4).

The field of filter design is well explored and is a speciality in its own right. Extensive reading is suggested in the bibliography.

## 2.6 EXERCISES

### Linear Systems

**1.** Prove that

$$\sum_{n=0}^{\infty} |\, h(nT) \,| < \infty$$

if $h(nT) = k^n$, $|k| < 1$.

**2.** Consider a filter with an impulse response

$$h(nt) = k^n |k| < 1$$

(a) Find the transfer function.

(b) What is the bandwidth (i.e., 3dB down from the response of $f = 0$) of the filter?

(c) Suppose samples from white Gaussian noise of variance $\sigma^2$ are applied to the filter. Compute the variance of the output sample sequence.

(d) Suppose the input signal is Gaussian with a correlation function

$$R(\tau) = \sigma^2 \epsilon^{-\tau/2\sigma^2}$$

What is the spectrum of the output if the sampling frequency is twice the bandwidth (as defined in (b))?

## Time Domain Representation

**3.** Compute the number of multiplications and additions as a function of $n$ for multiplication of $n \times n$ matrices. Write a matrix multiplication procedure using some assembler language.

(a) Comment on the addressing required to access data.

(b) Suggest improvements to the addressing modes of the language in order to facilitate the data flow.

## Digital Filters

In the following two exercises, a program in any computer language is required. If the reader is multilingual, the exercises should be done in each language. In the event that there is no facility in any language, use a pseudo-code of terse English prose.

**4.** Consider a biquad filter shown in Figure 2.12. Write a procedure (a subroutine) to compute $y(n)$. Assume suitable data structures for the samples and the coefficients. Assume the parameters passed to the procedure include the starting address of the input data and the address for the result. Coefficient data can be assumed private to the procedure. From your code:

(a) Estimate the overhead code used to move data, and fetch coefficients, etc.

(b) Estimate the intermediate storage required.

**5.** Using the biquad procedure developed in question 4:

(a) Write additional code to implement a direct form I realization with $M = 6$.

(b) Write code to implement a direct form II realization with $M = 6$.

(c) Comment on the complexity of each, and estimate the difference in execution time.

**6.** Consider a filter with a single (real) pole at $-k$. Assume the dc response is $D$ and the response at $t = 0$ is $G$.

(a) Derive an expression for the impulse response $h(t)$ and the transfer function $H(s)$.

(b) Using the impulse invariant transform compute the coefficients for a first-order section digital filter.

(c) Repeat (b) using the matched $Z$-transform.

(d) Repeat (b) using the bilinear transform.

**7.** Consider a filter with a pair of complex conjugate poles and a pair of complex conjugate zeros. Assume the poles are at $-A \pm -jB$ and the zeros at $-C \pm -jD$.

(a) Derive an expression for the impulse response $h(t)$ and the transfer function $H(s)$.

(b) Using the impulse invariant transform compute the coefficients for a first-order section digital filter.

(c) Repeat (b) using the matched $Z$-transform.

(d) Repeat (b) using the bilinear transform.

**8.** Suppose the output of an A/D produces uniformly distributed quantization noise with a variance as given by equation (1.8). Suppose the quantization $E_0 = 1$ millivolt. If

(i) the dynamic range of the signal is 60 dB,

(ii) the impulse response of a filter is:

$$h(nT) = k^n, \quad |k| < 1,$$

(iii) the output quantization noise should be at least 40 dB below the smallest signal.

(a) How many bits are required at the input?

(b) What is the smallest and the largest input signal (in volts) that can be processed (with the specifications)?

(c) Suppose a 16-bit microprocessor is to be used to implement the filter. Can the output be represented in one processor word?

**9.** Modify a second order section to compute a single real zero (i.e., set $B_1 = B_2 = \alpha_2 = 0$). Redraw the filter. Suppose the zero is located at $S = -\sigma$.

(a) Show $H(z) = 1 - z^{-1}e^{-\sigma T}$

(b) Let $\alpha_0 = 1$, show $\alpha_1 = e^{-\sigma T}$

(c) What is the value of $\alpha_1$ if the zero is at $s = 0$?

**10.** Consider a second order section to implement a complex conjugate zero at $-\sigma \pm j\omega$.

(a) Find the impulse response.

(b) Show that $H(s) = s^2 + \sigma^2 s + (\sigma^2 + \Omega_0^2)$.

(c) Show $H(z) = \alpha_0 + \alpha_1 z^{-1} + \alpha_2 z^{-2}$.

where

$$\alpha_0 = \text{a constant}$$

$$\alpha_1 = 2\alpha_0 \epsilon^{-\sigma T} \cos \omega_0 T$$

$$\alpha_2 = \alpha_0 \epsilon^{-2\sigma T}$$

(d) What is the maximum gain?

CHAPTER 3

# Detection and Estimation

## 3.0 INTRODUCTION

As noted in Chapter 2, the specification of digital filter performance requirements is not always in terms of a spectral shaping operation. In this chapter we shall examine the closely related areas of detection and estimation theory and show how this theory leads to the specification, design and implementation of digital filters from a time domain perspective. The discussion of estimation theory also leads us into the major topic of spectral estimation.

We begin with the basic concepts of detection theory in Section 3.1. This leads to the consideration of an important detection processing technique, the matched filter.

In Section 3.2 we present an overview of the basic concepts of estimation. For an estimate to be useful, it is necessary first that it can be computed from the observed data, and second that it have a reasonably close relationship to the actual parameter in question. Both of these aspects will be considered.

In Section 3.3 the concepts of linear minimum mean-squared-error estimation are explained. This estimation technique has been extensively used as the basis for time domain digital filter design and implementation in many application areas. The implementation of solutions to processing problems formulated in this manner is heavily dependent on matrix inversion techniques. Due to the high computational load associated with matrix inversion the use of time domain techniques has generally been limited to applications where real-time performance is not a major issue. With the increased performance (at lower cost) promised by VLSI component technology, this implementation constraint may soon be overcome by raw processing power.

Linear mean square estimation leads to a variety of optimum filters for extracting signal parameters. It is possible to extend this estimate in a recursive algorithm either to reduce the error or to track dynamic signal conditions in an adaptive manner. Indeed it is intuitively appealing to consider refining an estimate as a result of several tries. This leads to the concept of a Kalman filter as a recursive estimator.

Finally, in Section 3.4, the widespread problem of spectral estimation is introduced and discussed. The classical approach is used to introduce the subject and then both a single estimate and an autoregressive estimation procedure are discussed.

## 3.1 BASIC CONCEPTS OF DETECTION THEORY

### 3.1.0 Introduction

The underlying requirement in detection processing is to make a decision regarding the presence or absence of a particular signal in a noisy environment. Two important concepts arise from considering this requirement. First, there are two ways to make a correct decision and two ways to make an incorrect decision. However, due to the binary nature of the decision, any decision is either correct or incorrect. As we shall see, the performance requirements for a detection process are generally specified in terms of the probabilities associated with making correct decisions or committing errors. Second, the detection problem must ultimately rely on having a comparison parameter of some kind upon which to base the decision. Indeed, the processing problems involved stem from the determination of which quantities are to be compared and the actual comparison mechanisms.

In this section we shall examine the basic concepts of detection. First, the process of hypothesis testing, and the various approaches to specifying performance and determining comparison thresholds. We then show, by examining the case of detecting a known signal in noise, how this leads to a comparison processing operation known as correlation or matched filtering.

### 3.1.1 Hypothesis Testing

Hypothesis testing forms the framework for the fundamental yes-no decision required of detection processing. To illustrate the basic concepts we shall examine a simple example. Consider the case of searching for a signal $s(n)$ in a noisy environment. A sample of the received signal can be either:

$$r(n) = s(n) + w(n)$$

or

$$r(n) = 0 + w(n) \tag{3.1}$$

where $s(n)$ represents a source signal and $w(n)$ represents the additive random noise. By observing $r(n)$ we must decide if the source signal $s(n)$ is present or not. Intuitively, we will expect to make the decision based on some amplitude comparison. The processing problems are, "What do we compare?" and, "How do we make the comparison?", such that the decision process will be optimum (and implicitly, "What do we mean by optimum?").

To begin, we designate two mutually exclusive hypotheses $H_0$ and $H_1$ as follows:

$$H_0\colon r(n) = w(n)$$

$$H_1\colon r(n) = s(n) + w(n)$$

If we observe $r(n)$ and make an arbitrary decision on the presence or absence of $s(n)$ we can have four possible outcomes:

Two outcomes correspond to a correct decision.

1. Choose $H_1$ and $H_1$ is true
2. Choose $H_0$ and $H_0$ is true

Two outcomes correspond to committing an error:

3. Choose $H_1$ and $H_0$ is true
4. Choose $H_0$ and $H_1$ is true

There are four common approaches which lead to rules for choosing a hypothesis. Each depends on an increasing level of statistical knowledge about the environment.

**1. Maximum a Posteriori (MAP) Criterion**

One reasonable approach is to choose the hypothesis that is most likely to have resulted in the given observation. Consider the a posteriori (after the fact) conditional probabilities $P\{H_0|r\}$ and $P\{H_1|r\}$ representing the probabilities that hypothesis $H_0$ is true given the observation $r$ and that hypothesis $H_1$ is true given the observation $r$ respectively. Our decision rule is simply to choose $H_1$ if

$$P\{H_1|r\} > P\{H_0|r\} \tag{3.2(a)}$$

or

$$\frac{P\{H_1|r\}}{P\{H_0|r\}} > 1 \tag{3.2(b)}$$

This approach demands the minimum knowledge about the statistics of the environment.

**2. Maximum Likelihood (ML) Criterion**

The previous decision rule considers only the a posteriori probabilities for $H_0$ and $H_1$ with no consideration of the actual a priori probabilities of the source signal $s(n)$ being present or not. This could obviously bias our decision. Recalling Bayes theorem from Chapter 1 we can write a posteriori probabilities as

$$P\{H_1|r\} = \frac{P\{r|H_1\}P\{H_1\}}{P\{r\}} \tag{3.3}$$

and

$$P\{H_0|r\} = \frac{P\{r|H_0\}P\{H_0\}}{P\{r\}} = \frac{P\{r|H_0\}(1-P\{H_1\})}{P\{r\}} \tag{3.4}$$

where $P\{r|H_1\}$ represents the probability of an observation $r$ given $H_1$ is true.

Substituting in equation (3.2) we have the decision rule choose $H_1$ if

$$\frac{P\{r|H_1\}}{P\{r|H_0\}} > \frac{1 - P\{H_1\}}{P\{H_1\}} \tag{3.5}$$

otherwise choose $H_0$. The ratio

$$\frac{P\{r|H_1\}}{P\{r|H_0\}} \tag{3.6}$$

is known as the "likelihood ratio" and equation (3.5) represents a likelihood ratio test. The effect is to shift the level at which the decision changes based on the a priori probability $P\{H_1\}$. We note that the maximum likelihood and maximum a posteriori criterion are equivalent when $P\{H_1\} = P\{H_0\} = 1/2$.

Both the preceding criteria ignored the cost or penalty of being wrong. In many instances some errors are more costly than others. This will be considered next.

**3. Bayes Criterion**

The Bayes decision criterion is based on the assignment of a cost to each possible outcome of the decision process and performing a likelihood ratio test that minimizes the average cost.

Consider the joint probabilities for each possible outcome of the decision process, $P_1$, $P_2$, $P_3$ and $P_4$, i.e.,

$$\begin{aligned} P_1 &= \{\text{choose } H_1, \text{ and } H_1 \text{ is true}\} = P\{h_1,H_1\} \\ &= P\{h_1|H_1\}P\{H_1\} \end{aligned} \tag{3.7}$$

where $P\{h_1|H_1\}$ is the conditional probability of the event of choosing hypothesis $H_1$ given that $H_1$ is true, and

$$P\{h_1|H_1\} = \frac{P\{h_1,H_1\}}{P\{H_1\}} \tag{3.8}$$

Similarly

$$P_2 = P\{h_o \mid H_0\}P\{H_0\} \tag{3.9}$$

$$P_3 = P\{h_1 \mid H_0\}P\{H_0\} \tag{3.10}$$

$$P_4 = P\{h_0 \mid H_1\}P\{H_1\} \tag{3.11}$$

By assigning costs $C_1$, $C_2$, $C_3$ and $C_4$ to each joint probability $P_1$, $P_2$, $P_3$ and $P_4$ we form the "Bayes risk function."

$$\begin{aligned} R &= C_1P_1 + C_2P_2 + C_3P_3 + C_4P_4 \\ &= P\{H_1\}[C_1P\{h_1|H_1\} + C_4P\{h_0|H_1\}] \\ &\quad + P\{H_0\}[C_2P\{h_0|H_0\} + C_3P\{h_1|H_0\}] \end{aligned} \tag{3.12}$$

By noting that

$$P\{h_o|H_1\} = 1 - P\{h_1|H_1\} \tag{3.13}$$

and

$$P\{h_o|H_0\} = 1 - P\{h_o|H_1\} \tag{3.14}$$

we can write the risk function as

$$\begin{aligned} R = P\{H_1\}C_4 + P\{H_0\}C_2 - P\{H_1\}[C_4 - C_1]P\{h_1|H_1\} \\ + P\{H_0\}[C_3 - C_2]P\{h_1|H_0\} \end{aligned} \tag{3.15}$$

The first two terms of this equation are independent of the decision. Thus, to minimize the risk we must minimize the sum of the last two terms. We assume that the cost of a correct decision is less than that of an incorrect decision; therefore the quantities $(C_4 - C_1)$ and $(C_3 - C_2)$ should be positive. We wish to make our decision such that

$$P\{H_1\}[C_4 - C_1]P\{h_1|H_1\} \geqslant P\{H_0\}[C_3 - C_2]P\{h_1|H_0\} \tag{3.16}$$

or

$$\frac{P\{h_1|H_1\}}{P\{h_1|H_0\}} \geqslant \frac{P\{H_0\}[C_3 - C_2]}{P\{H_1\}[C_4 - C_1]} \tag{3.17}$$

To relate this equation to the likelihood ratio we simply note that $P\{h_1|H_1\}$ can be interpreted as representing the probability that the observation $r$ is one that the decision rule assigns to $H_1$ given that $H_1$ is true and similarly for $P\{h_1|H_0\}$. Hence, our decision rule is choose $H_1$ if

$$\frac{P\{r|H_1\}}{P\{r|H_0\}} \geqslant \frac{P\{H_0\}[C_3 - C_2]}{P\{H_1\}[C_4 - C_1]}, \tag{3.18}$$

otherwise, choose $H_0$.

The Bayes decision criterion results in a likelihood ratio test against a threshold determined by the a priori probabilities $P\{H_1\}$ and $P\{H_0\}$ and costs assigned to each possible outcome of the decision process.

**4. Neyman-Pearson Criterion**

The Neyman-Pearson criterion for optimum detection is based on specifying a fixed probability for false detection [i.e., fix $P\{h_1|H_0\}$ which is often denoted $P_f$ for probability of false alarm] and maximizing the probability of detection $P_D = P\{h_1|H_1\}$. This problem can be solved in terms of the Bayes approach.

Since $P\{h_1|H_1\} = 1 - P\{h_0|H_1\}$ and $P\{h_1|H_0\}$ is fixed, we can minimize the quantity

$$Q = P\{h_0|H_1\} + \alpha P\{h_1|H_0\} \tag{3.19}$$

which is equivalent to maximizing $P\{h_1|H_1\}$, where $\alpha$ represents an arbitrary constant. This problem can be put in a familiar format. If we assume zero cost for a correct decision, then the Bayes risk can be written as:

$$\begin{aligned} R &= P\{H_1\}C_4 - P\{H_1\}C_4P\{h_1|H_1\} + P\{H_0\}C_3P\{h_1|H_0\} \\ &= P\{H_1\}[1 - P\{h_1|H_1\}]C_4 + P\{H_0\}C_3P\{h_1|H_0\} \\ &= P\{H_1\}C_4P\{h_0|H_1\} + P\{H_0\}C_3P\{h_1|H_0\} \end{aligned} \tag{3.20}$$

Thus, if

$$P\{H_1\}C_4 = 1 \tag{3.21}$$

and

$$P\{H_0\}C_3 = \alpha \tag{3.22}$$

we have

$$R = Q$$

From the Bayes criterion we already know that minimizing $Q$ will give the decision rule: choose $H_1$ if

$$\frac{P\{r|H_1\}}{P\{r|H_0\}} \geqslant \frac{P\{H_0\}C_3}{P\{H_1\}C_4} = \alpha \tag{3.23}$$

Since $C_3$ is the cost associated with false alarms and is also used to determine $\alpha$ we have a mechanism for implementing the Neyman-Pearson criterion.

The obvious problem at this point is the identification of a mechanism for relating these probabilistic decision criteria to observable signal characteristics. A simple example helps to illustrate this. Assume a unit value signal in zero mean Gaussian noise with unit variance. If the signal is present then the observed signal will be the additive result of signal plus noise and will appear as a Gaussian distribution with unit mean. If the signal is not present the noise alone will appear as a Gaussian distribution with zero mean, i.e.,

$$P\{r|H_1\} = \frac{1}{\sqrt{2\pi}} e^{-(r-1)^2/2} \tag{3.24}$$

and

$$P\{r|H_0\} = \frac{1}{\sqrt{2\pi}} e^{-r^2/2} \tag{3.25}$$

The likelihood ratio, $\lambda(r)$, is

$$\frac{P\{r|H_1\}}{P\{r|H_0\}} = \lambda(r) = e^{-(r-1)/2} \tag{3.26}$$

and our decision rule becomes: choose $H_1$ if $e^{r-1/2} \geqslant T$ where $T$ is a threshold whose determination depends on which optimality criterion is being used.

We note that it is common to use a log likelihood ratio test derived from the fact that taking logarithms does not alter the threshold position but gives a linear comparison of the form: choose $H_1$ if

$$r \geqslant 1/2 + 1nT \tag{3.27}$$

The observation $r$ (possibly a voltage level measurement) represents a sufficient statistic upon which to base our decision.

In the next section a mechanism (the matched filter) is discussed which illustrates a processing mechanism for optimum detection of a signal in noise.

Figure 3.1 illustrates the effect that altering the threshold value has on the probability of making correct or incorrect decisions. If the observation $r$ lies below the threshold then $H_0$ is chosen and the probability of a "missed detection" is the shaded region marked $P_{\text{miss}}$. If the observation $r$ lies above the threshold then $H_1$ is chosen, and the probability of a false alarm is the shaded region marked $P_f$.

### 3.1.2 The Matched Filter

A basic model used to illustrate detection is that of a known signal in additive white noise. Given a received signal as in equation (3.1), where $s(n)$ is known and $w(n)$ is assumed to be white noise, it can be shown that the optimum detection receiver is a correlation filter or "matched filter" [17]. The matched filter forms an important mechanism in detection processing.

From our discussion in the previous section, it is intuitively obvious that the detection of a unit signal in white noise will be more reliable if we make several observations rather than a single observation. If we were to average the observations over a period of time and then make the threshold comparison we would smooth out the sample-by-sample fluctuations due to noise,

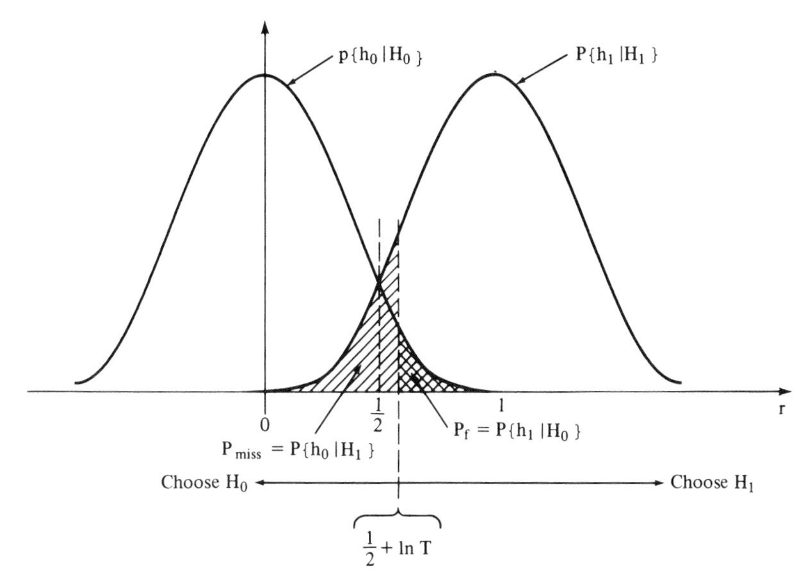


**Figure 3.1:** Effect of threshold placement on the probability of false alarm and missed detections.

and essentially have a measure of the mean value over the observation time with which to make our comparison. For the continuous time case, it can be shown that for a received signal

$$r(t) = s(t) + w(t) \tag{3.28}$$

the optimum detection receiver is a correlation receiver based on the decision rule: choose $H_1$ if

$$\int_0^T r(t)s(t)\,dt > T \tag{3.29}$$

where $T$ is the predetermined threshold value [17]. This operation can be obtained by applying $r(t)$ to a linear filter with impulse response $S(\tau - t)$. The output of this filter will be equivalent to equation (3.29) at $t = \tau$. This filter is known as a "matched filter" and its impulse response is a time reversed replica of the signal to be detected. Thus, matched filtering is equivalent to a replica correlation operation.

Consider now the digital implementation of a matched filter. The impulse response of a matched filter is the time inverse of the signal to be detected. Hence, if the received signal is either

$$r(n) = s(n) + w(n)$$

or

$$r(n) = w(n) \tag{3.30}$$

passing the received signal through a matched filter results in

$$\begin{aligned} r(n) * S(-n) &= \sum_{k=-\infty}^{\infty} s(-k)r(n-k) \\ &= \sum_{i=-\infty}^{\infty} s(i)r(n+i), \quad i=-k \\ &= \sum_{i=-\infty}^{\infty} s(i)(s(n+i)+w(n+i)) \\ &= R_{ss}(n) + R_{sw}(n) \end{aligned} \tag{3.31}$$

or

$$\begin{aligned} r(n) * s(-n) &= \sum_{i=-\infty}^{\infty} s(i)w(n+i) \\ &= R_{sw}(n) \end{aligned} \tag{3.32}$$

Thus, for detecting a known signal the output of a correlation receiver (or matched filter) represents the sufficient statistic that is compared to a threshold value. The decision rule is: at each time $\tau$ choose $H_1$ if

$$r(n) * s(-n) \geqslant T \tag{3.33}$$

This is an intuitively appealing result of our model.

### 3.1.3 Summary

In this section we have outlined the concepts of hypothesis testing as the foundation of detection theory. The most important concept to emerge is that detection processing means the computation of a sufficient statistic to be compared with a threshold value.

Solving the problem of known signal detection leads to the requirement to perform a matched filter operation which is equivalent to the cross correlation of a replica of the signal to be detected with the received signal. In the frequency domain this means matching the transfer function of the filter as closely as possible to the signal spectrum, thus eliminating as much noise as possible.

Each of the approaches to establishing a threshold depends on a statistical knowledge of the signal environment, and in the latter two, specifying a cost or penalty. These are not always easy to specify except in the simplest of models. For example, consider the processing of radar returns to detect the presence of a missile. What is the probability of any sample containing an echo return. More profoundly, what is the cost of deciding an echo is not there, when it is; or deciding it is there, when it is not. Yet we have seen that the decision must be made, and that it can only be made based on a comparison to a level, which when set reflects all the assumptions about the statistical environment and the decision costs. In practice a considerable amount of redundancy is built into detection systems to provide increased confidence in a decision, based on some form of averaging.

## 3.2 BASIC CONCEPTS OF ESTIMATION THEORY

### 3.2.0 Introduction

Several important estimation problems are of interest in digital signal processing. From Chapter 1 we are reminded that a digital signal itself is an estimate. More importantly, the statistical averages of signals modelled as random processes, such as the mean, correlation, covariance and power spectral density, are estimates based on finite duration sample sequences.

Another important class of problems is the estimation of a signal, or some parameters of the signal, from noise corrupted data. For example, in a digital communication system it is common to model the received signal as consisting of a source plus an additive noise component, i.e.,

$$r(n) = s(n) + w(n) \tag{3.34}$$

In this example $s(n)$ may be a deterministic signal but $w(n)$ is generally modelled as a random noise process, so the received signal is also a random process. The problem is to estimate the deterministic signal $s(n)$ from the random process $r(n)$. If we wish to determine only the presence or absence of $s(n)$ we have a detection problem. If we wish to measure (estimate) some parameter of $s(n)$ we generally refer to this as an estimation problem.

If $s(n)$ is also considered as a random process we further complicate the problem. We want to estimate some parameters of a random process (itself an estimation problem) from inaccurate sample data. Thus we are first estimating one random process from a related one and then estimating the parameters of the estimated random process. This type of problem is common, for example, in passive sonar where the source signal is considered to be a random process of unknown power spectral density. The requirement is to estimate the power spectrum of the source, given the received signal, that is, a related random process. This problem is further complicated by the fact that the ocean is a complex transmission channel and cannot be accurately modelled as a simple additive noise effect. We shall examine the problems of spectral estimation in more detail in Section 3.4. In the meantime let us consider another aspect of estimation problems.

Consider again the problem of estimating some parameter of either a deterministic or a random signal. The parameter may be related to the signal either by a linear or by a non-linear function. In the absence of noise, and if the relationship is known, the parameter can be computed from the data. However, suppose that the relationship is so complex that an accurate measurement is not practical, such as a complicated non-linear relationship or perhaps that an excessively large amount of data must be examined. In such situations we may wish to estimate a parameter by a simplified model of its relationship. This corresponds to computing the finite time averages of a sequence of data as an estimate of statistical ensemble averages. There are other cases of interest as well.

Such problems can be modelled by expanding an analytic expression in an infinite power series or a Fourier series, and then truncating the series to a tractable number of coefficients to form an estimate. The projection of an infinite dimensional vector onto a finite dimensional space amounts to the same problem. The result is an approximation (or estimate) of a nontractable but deterministic quantity (or signal) in terms of a tractable set of parameters. The general problem is to find the coefficients of this tractable model that estimate the given signal, function or vector, in the most accurate manner possible. In general, the relationship of the parameters to the signal may still be either linear or non-linear. In the non-linear case there is no general method for optimum estimation. However, if we consider only linear models, optimum techniques are available. Optimum linear estimation is the topic of Section 3.3.

In addition, the problem of estimating random parameters also exists. We shall see that for linear relationships between random processes we can optimally estimate one random process or parameter from another.

### 3.2.1 The Quality of an Estimation

In estimating a quantity we carry out a computational procedure based on some observed data, and a model of how the quantity is related to the data. The problem model that specifies how the observed data is related to the quantity to be estimated will dictate to a large extent what computations must be carried out. Thus, we must generally examine estimation problems on a case by case basis. For the case of linear estimation problems a general computational approach can be formulated. We shall examine this approach in detail in the next section. However, given an estimation procedure, we must ask for some assurance that our estimate is reasonable or good, and in what sense is it good. Thus we are interested in error measures.

Two of the common measures of the quality of an estimate are the "bias" of the estimate and the "variance of the error" (assuming that the estimate and the error can be considered as random variables). This assumption is justified when considering estimates from noise corrupted data.

An estimate is said to be "unbiased" if the expected value of the estimate equals the parameter or quantity to be estimated.

$$E\{\hat{p}\} = P \tag{3.35}$$

If $\hat{p}$ is the estimate of a parameter $p$ then the estimate is unbiased. If on the other hand,

$$E\{\hat{p}\} = p + c \tag{3.36}$$

where $c$ is a constant we say the estimate is biased.

If

$$E\{\hat{p}\} = p + f(p) \tag{3.37}$$

where $f(p)$ is some function of the parameter $p$ then the estimate is said to have an unknown bias.

We wish to make unbiased estimates whenever possible. If it can be shown that an estimation procedure produces unbiased estimates we can be reasonably confident in the quality of the estimate, since we are assured that over a large number of trials the average value of the estimates approaches the true value of the quantity being estimated. However, it is often the case that we cannot obtain a large number of estimates. To be confident of the estimate, it is desirable to know how close a single estimate is to the mean. A convenient measure of this closeness is the variance.

The "variance of the estimate" is defined as

$$\text{var}\{\hat{p}\} = E\{[\hat{p} - E\{\hat{p}\}]^2\} = \sigma_{\hat{p}}^2 \tag{3.38}$$

The variance of the error $(\hat{p} - p)$ is equal to $\sigma_{\hat{p}}^2$, i.e.,

$$\begin{aligned} \text{var}\{(\hat{p} - p)^2\} &= E\{[\hat{p} - p - E\{\hat{p} - p\}]^2\} \\ &= \sigma_{(\hat{p}-p)}^2 \end{aligned} \tag{3.39}$$

To have the greatest degree of confidence in an estimate we would like it to be unbiased and have a minimum variance.

If the variance approaches zero as the number of estimation trials approaches infinity, the estimation is said to be "consistent".

### 3.2.2 Optimum Estimation

In some cases we can develop methods for obtaining the optimum estimate for a given optimality criterion. For estimating random parameters from noisy data a statistical approach known as Bayesian estimation can be developed that specifies an optimum estimate in terms of the joint probability density functions of the parameter to be estimated and the observation. The exact specification is dependent on specifying a cost function associated with the error and minimizing the average cost in a manner similar to that discussed in Section 3.1. For a quadratic cost function (see Figure 3.2(a)) the optimum estimate is given by the mean value of the a posteriori density, or the conditional mean, i.e.,

$$\hat{p}_{ms} = \int_{-\infty}^{+\infty} p f_{p|r}(p|r)\, dp \tag{3.40}$$

where $f_{p|r}(p|r)$ is the condition probability density of the parameter $p$ given the observation $r$.

For a uniform cost function (Figure 3.2(b)), that is, all errors beyond a specified amount are equally costly while small errors are of zero cost, the optimum estimate is obtained by taking the maximum of the a posteriori density and is known as the "maximum a posteriori" (MAP) estimate.

For non-random parameters the Bayesian procedure is of little use since the parameter cannot be described by a probability density function. One method of getting around this problem is to assume a uniform distribution for the parameter, that is, to assume all values are equally likely. This results in an optimum estimate being specified as the maximum of the conditional density $f_{r|p}(r|p)$. In words this means we choose as the best estimate that value of $p$ which was most likely to have caused the observation $r$. For obvi-

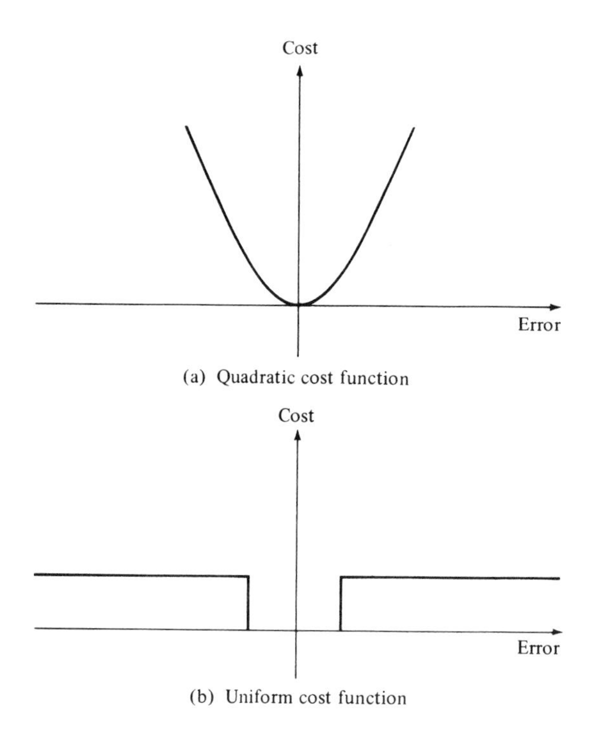


**Figure 3.2:** Cost functions.

ous reasons, this estimate is known as a "maximum likelihood" (ML) estimate.

It can be shown (17) that for any unbiased estimate of $p$ the variance of the estimate satisfies the Cramer-Rao inequality.

$$\text{var}\{\hat{p} - p\} \geqslant (E\{[\frac{\partial \ln f_{r|p}(r|p)}{\partial p}]^2\})^{-1} \tag{3.41}$$

Where equality holds the estimate is called efficient. If an efficient estimate exists, the maximum likelihood estimate gives the efficient estimate (17). The ML estimate is also a consistent estimate.

Clearly, the nature of the probabilistic relationship of the parameter to be estimated to the observation will be dependent on the characteristics of the transmission channel and the relationship of the parameter to the transmitted signal. In the case of a non-linear relationship there is no general procedure for obtaining the optimum estimate. In the case of a linear relationship, however, general procedures can be formulated. We shall look at linear estimation in more detail in Section 3.3.

### 3.2.3 Summary

In this section we have given an overview of the basic concepts of estimation problems and optimum estimation. The concepts of bias and of the variance of an estimator were introduced as measures of the quality of an estimate. The major results of optimum estimation theory were stated without proof. The optimum minimum mean square estimator (i.e., a quadratic cost function) is the mean of the a posteriori density. For a uniform cost function, the optimum estimate is the MAP estimate. For non random parameter estimation a maximum likelihood estimation gives an efficient estimate if one exists, and it is a consistent estimate.

## 3.3 LINEAR MINIMUM MEAN SQUARED ERROR ESTIMATION

### 3.3.0 Introduction

In Section 3.2 we noted that the relationship between the observed data and the quantity to be estimated can be in general either linear or nonlinear. Digital signal processing is concerned mainly with linear systems due to their ease of implementation. For this reason linear estimation problems are of particular interest. In this section we shall consider estimation under the constraint that the relationship between the data and the quantity to be estimated be linear. In addition, we shall consider only quadratic cost function estimates. That is, the optimality criterion is to minimize the total squared error for nonrandom parameter estimation or the mean-squared error for random parameters. We shall see that a powerful tool exists for realizing this criteria in the form of the "projection theorem" or the "orthogonality principle," as it is sometimes called in connection with random variables. Indeed, the projection theorem and the set of linear equations known as the "normal equations" which result from its application form a basis for all linear minimum mean squared error estimation problems. We shall show that while these problems range from the simple estimation of a straight line to the complexities of Kalman filtering they all follow from the application of a common criterion.

This section begins with a presentation of the basic concepts of the projection theorem and the normal equations. Some familiarity with vector spaces is assumed for this discussion. We then proceed to outline a number of general estimation problems that can be solved using the projection theorem. The key result is that a single procedure applies to a wide variety of problems. As we proceed, the models and applications become more complex but the same underlying concepts apply.

### 3.3.1 The Projection Theorem

The basis for all linear minimum mean squared error estimation problems is the "projection theorem." In its simplest form the projection theorem follows from the Pythagorean theorem that assures us that the shortest distance from a point in a plane to a line in the same plane is in a direction that is perpendicular to the line. This can be represented in simple vector form as shown in Figure 3.3. The point $(x,y)$ defines a vector $\underline{v}$ from the origin of the $x - y$ plane. The shortest distance from the point to the line $L$ through the origin is the vector $\underline{v} - \underline{\hat{v}}$ which is perpendicular or orthogonal to the line $L$.

> That is, the dot product* of $(v - \hat{v})$ with any vector along $L$ is zero. The vector $\underline{\hat{v}}$ along $L$ such that $(\underline{v} - \underline{\hat{v}})$ is orthogonal to $L$ is called the projection of $\underline{v}$ on $L$.

The two dimensional case can be extended to an arbitrary number of dimensions. We consider in general "Hilbert spaces" which are abstractions of familiar Euclidean space. The vectors of a Hilbert space may be real or complex sequences, functions or random variables and infinite in number (e.g., a Fourier series). The dot product of Euclidean space is generalized to an inner

*The dot product of two vectors in two-dimensional Euclidean space is defined simply as the scalar sum of the products of the respective coefficients of the two vectors, i.e.,

$$(\underline{x},\underline{y}) = x_1 y_1 + x_2 y_2$$

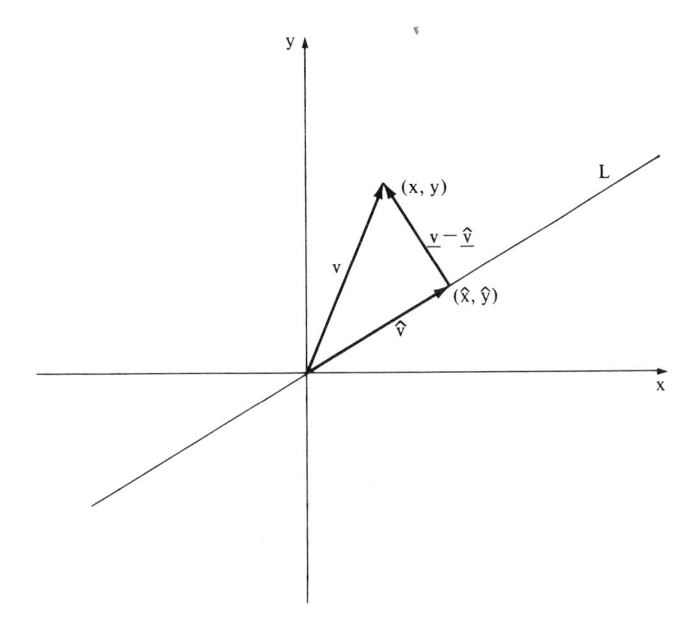


**Figure 3.3:** A simple two-dimensional projection.

product in Hilbert spaces defined as a function that assigns to every pair of vectors $\underline{x}$ and $\underline{y}$ the scalar value $(\underline{x},\underline{y})$ such that:

1. $(\underline{x},\underline{y}) = (\underline{x}*,\underline{y}*)$ where $\underline{x}*$ is the complex conjugate of $\underline{x}$
2. $c\,(\underline{x},\underline{y}) = (c\underline{x},\underline{y}) = (\underline{x},\,,c\underline{y})$; $c$ a scalar
3. $(\underline{x} + \underline{z},\underline{y}) = (\underline{x},\underline{y}) + (\underline{z},\underline{y})$
4. $(\underline{x},\underline{x}) \geqslant 0$
5. $(\underline{x},\underline{x}) = 0$ if and only if $\underline{x} = \emptyset =$ null vector
6. $(\underline{x},\underline{x}) = \| x \|^2 =$ norm of $\underline{x}$

Any two vectors $\underline{x}$ and $\underline{y}$ in a Hilbert space $H$ are orthogonal if their inner product $(\underline{x},\underline{y}) = 0$.

The norm is a direct extension of the concept of the magnitude or length of a vector in Euclidean space.

> The projection theorem assures us that for any vector $\underline{v}$ in a Hilbert space $H$, there is a unique vector $\hat{\underline{v}}$ defined on a complete subspace $S$ of $H$ called the projection of $\underline{v}$ on $S$ such that the norm of $\underline{v} - \hat{\underline{v}}$ is minimum if $\underline{v} - \hat{\underline{v}}$ is orthogonal to $S$.

Given a set of $k$ vectors $\underline{s}_1,\underline{s}_2,\ldots,\underline{s}_k$ each of dimension $N$, they define a subspace $S$ of the $N$-dimensional Hilbert space $H$. We wish to determine the projection $\hat{\underline{v}}$ of an arbitrary vector $\underline{v}$ in $H$ on the subspace $S$. We know from the projection theorem that $\underline{v} - \hat{\underline{v}}$ must be orthogonal to $S$. That is, the inner product of $\underline{v} - \hat{\underline{v}}$ with every vector in $S$ must be zero. Since the set of vectors $\underline{s}_1, \ldots, \underline{s}_k$ define the subspace $S$, it is sufficient to show that $\underline{v} - \hat{\underline{v}}$ is orthogonal to each vector $\underline{s}_i, i = 1, 2, \ldots, k$. Thus we have the condition

$$(v - \underline{\hat{v}}, \underline{s}_i) = (\underline{v}, \underline{s}_i) - (\underline{\hat{v}}, \underline{s}_i) = 0 \tag{3.42}$$

Now $\hat{\underline{v}}$ is a vector in $S$ and can thus be represented as a linear combination of the vectors $\underline{s}_i$, i.e.,

$$\underline{\hat{v}} = \sum_{i=1}^{k} c_i \underline{s}_i = S_{\underline{c}} \tag{3.43}$$

Where $S$ is the $n \times k$ matrix whose columns are the vectors $\underline{s}\, i$ and $c$ is a $k$ element column vector of coefficients.

Therefore, equation (3.42) can be written as

$$(\underline{v}, \underline{s}_i) = \sum_{j=1}^{k} c_j (\underline{s}_j, \underline{s}_i) \tag{3.44}$$

In matrix form this equation can be written as

$$\begin{bmatrix} \underline{v}, \underline{s}_1) \\ \vdots \\ (\underline{v}, \underline{s}_k) \end{bmatrix} = \begin{bmatrix} (\underline{s}_1, \underline{s}_1) & (\underline{s}_1, \underline{s}_2) & \ldots & (\underline{s}_1, \underline{s}_k) \\ \vdots & & & \\ (\underline{s}_k, \underline{s}_1) & (\underline{s}_k, \underline{s}_2) & \ldots & (\underline{s}_k, \underline{s}_k) \end{bmatrix} \begin{bmatrix} c_1 \\ \vdots \\ c_k \end{bmatrix} \tag{3.45}$$

or

$$A = GC \tag{3.46}$$

with the obvious matrix equivalences. The set of linear equations represented by this matrix equation is known as the "normal equations."

It can be shown that the matrix $G$, sometimes called the "Gram matrix," is non-singular if and only if the vectors $\underline{s}_i$ are linearly independent (i.e., the set of vectors $\underline{s}_i$ form a basis set for the subspace $S$). If $G$ is non-singular, then a unique solution to the normal equations exists, i.e.,

$$C = G^{-1}A \tag{3.47}$$

Thus the normal equations provide the mechanism for the application of the projection theorem to linear estimation problems.

It is worth noting that the Gram matrix contains only information about the subspace onto which the projection is being made. All the information regarding the vector to be estimated is contained in the $A$ matrix. Thus, when formulating estimation problems to be solved using the projection theorem, the known or observed data from which the estimation is to be made will define the Gram matrix. Specification of the $A$ matrix may be based on observed data as well in some problems, or it may be derived from a model or an assumption regarding the relationship of the quantity to be estimated to the data from which it is to be estimated.

Equation (3.43) represents the basic model of a linear estimation problem, expressing the desired estimate as a linear combination of the input data. The set of vectors $\underline{s}_i$ correspond to the data. To represent the problem in terms of the normal equations we must define the inner product for the problem and specify the $A$ matrix.

In the remainder of this section we shall outline several linear estimation problems that can be formulated and solved in terms of the projection theorem. We begin by considering two provlems involving deterministic data and continue with problems involving random data.

### 3.3.2 Deterministic Parameter Estimation

As a first example of applying the projection theorem we consider a simple least-squares error approximation problem. Given a set of $M$ predefined $N$-point sequences, $\underline{s}_1, \underline{s}_2, \ldots, \underline{s}_M$; we wish to find the least-squares error approximation of an $N$-point data sequence $x(n)$ that is a linear combination of the sequences $\underline{s}_i, i = 1, \ldots, M$. Clearly, this problem can be viewed as simply finding the projection of $x(n)$ on the space defined by the $M$ sequences $\underline{s}_i$. The approximation of $x(n)$ can be written as the matrix equation

$$\hat{\underline{x}} = S\underline{c} \tag{3.48}$$

where $\hat{\underline{x}}$ is a $N \times 1$ column vector, $S$ is an $N \times M$ matrix whose columns are the sequences $\underline{s}_i$ and $\underline{c}$ is a $M \times 1$ coefficient vector. If we assume that the sequences $\underline{s}_i, i = 1, \ldots, M$ are linearly independent, then we need only define a suitable inner product and we can write the expression for the coefficient vector $\underline{c}$ directly from equation (3.47).

Defining the inner product as

$$(\underline{x}, \underline{y}) = \sum_{i=1}^{N} x_i y_i^* \tag{3.49}$$

we see that the $M \times M$ Gram matrix $G$ is given as

$$G = S^t S \tag{3.50}$$

where $S^t$ represents the conjugate transpose matrix of $S$ and the $M \times 1$ $A$ matrix is

$$A = S^t \underline{x} \tag{3.51}$$

Therefore, the solution can be written from equation (3.47) as

$$\underline{c} = (S^t S)^{-1} S^t \underline{x} \tag{3.52}$$

and

$$\hat{\underline{x}} = S\underline{c} \tag{3.53}$$

is the projection of $\underline{x}$ on the space defined by the columns of $S$. Noting that the error is orthogonal to the estimate, the total squared error is

$$e^2 = (\underline{x} - \hat{\underline{x}})^t(\underline{x} - \hat{\underline{x}}) = \underline{x}^t\underline{x} - \underline{x}^tS(S^tS)^{-1}S^t\underline{x} \tag{3.54}$$

This problem formulation can serve as the basis for several useful processing situations. Consider a situation where it is desired to transmit $N$ point data sequences where $N$ is very large. We can reduce the amount of data transmission significantly if we first calculate the coefficient vector $\underline{c}$ of length $K << N$ (given a suitable set of basis vectors $\underline{s}_i$, $i = 1, 2, \ldots, K$). If the basis vectors $\underline{s}_i$ are known at the receiving end of the transmission then we need only transmit the $K$-element coefficient vector $\underline{c}$ rather than the full $N$ point data sequence. At the receiver the least squared error estimate of the original data sequence $x(n)$, $n = 1, \ldots, N$, can be computed from $\underline{c}$ as $\hat{\underline{x}} = S\underline{c}$. Thus we have a method of implementing a data rate reduction for data transmission based on linear estimation.

We now consider a second problem which may appear to be somewhat different at first glance but which turns out to be identical. We wish to model a real world physical system as a finite impulse response linear filter. We pass a known finite duration input signal through the physical system and observe the output. Given these two known sequences we wish to determine a finite length linear operator (model) that acts on the same input signal to produce a least squared error estimate of the observed output. For obvious reasons this type of problem is referred to as system identification. This type of system modelling is equivalent to assuming that the observed output can be described as a moving average process. The problem then, is to determine the coefficients of a moving average linear operator (an FIR filter) whose output is a least squared error estimate of the observed output of the physical system, given the same input signal. The key point here is that the form of the model is assumed, and within that form the coefficients are then determined in an optimum manner. However, if the initial assumption is not good (i.e., the physical system is not approximated by a moving average model) then large errors may result.

To formulate this problem in terms of the projection theorem we recall that the input and output of a linear time-invariant system are related by the convolution equation

$$y(n) = x(n) * h(n) \tag{3.55}$$

The problem may therefore be stated as follows:

> Given the finite duration input sequence $x(n)$, $n = 0, 1, \ldots N$, and the finite duration output sequence $y(n)$, $n = 0, 1, \ldots N + M$, find the finite impulse response $h(n)$, $n = 0, 1, \ldots M$, such that $x(n) * h(n)$ gives the least squared error estimate of the observed sequence $y(n)$.

The finite convolution operation of the filter can be expressed in matrix form as

$$\begin{bmatrix} y(0) \\ y(1) \\ \\ \\ \\ y(N+M) \end{bmatrix} = \begin{bmatrix} x(0) & 0 \ldots & & 0 \\ x(1) & x(0) \ldots & & 0 \\ \vdots & & & \vdots \\ x(N) & x(N-1) \ldots & & \\ 0 & x(N) \ldots & & x(N-1) \\ 0 & \dot{0} & & x(n) \end{bmatrix} \begin{bmatrix} h(0) \\ h(1) \\ \vdots \\ \\ \\ h(M) \end{bmatrix} \tag{3.56}$$

or

$$\hat{\underline{y}} = X\underline{h} \tag{3.57}$$

Thus, we have expressed the problem in the same form as equations (3.43) or (3.48). If we consider the sequences $x(n)$ and $y(n)$ as deterministic and define the inner product as in equation (3.49) we can write the solution directly as

$$\underline{h} = (X^t X)^{-1} X^t \underline{y} \tag{3.58}$$

Again the total squared error is

$$e^2 = (y - \hat{\underline{y}})(y - \hat{\underline{y}})^t = \underline{y}^t \underline{y} - \underline{y}^t X (X^t X)^{-1} X^t \underline{y} \tag{3.59}$$

If we now consider that the observed output sequence $y(n)$ is corrupted by observation noise we must alter the basic problem model. We now have

$$\underline{y} = X\underline{h} + \underline{w} \tag{3.60}$$

where $\underline{w}$ is considered to be an additive random noise vector. The observed sequence $y(n)$ must also be considered as a random sequence now and the error $(\underline{y} - \hat{\underline{y}})$ must be minimized in the mean square sense. Tretter [16] shows that if we consider $\underline{w}$ to be a zero mean noise process with a positive definite* covariance matrix $C_w$, then the minimum-variance unbiased estimate of $\underline{h}$ is given as

$$\underline{h} = (X^t C_w X)^{-1} X^t C_w \underline{y} \tag{3.61}$$

The mean square errors are the diagonal elements of the error covariance matrix

$$\lceil h = (X^t C_w^{-1} X)^{-1} \tag{3.62}$$

We note that if the noise vector $\underline{w}$ is considered to have uncorrelated

*An $N \times N$ matrix $A$ is said to be positive definite if for every nonzero $N$-dimensional column vector $v$ the following inequality holds: $\underline{v}^t A \underline{v} > 0$.

components, each of which have the same variance $\sigma_w^2$, then equation (3.62) reduces to equation (3.58).

### 3.3.3 The Optimum Linear Filter

In the first two examples of the previous section we assumed the data upon which an estimate was based to be deterministic. In the second example we extended the problem to account for noise corruption in the observed data. We now consider the general case where the estimate is to be based on observed data that is assumed to be a sample sequence of a random process. In such a case the Gram matrix of the normal equations is a matrix of inner products of random variables. The inner product of two random variables $x$ and $y$ is defined as the expectation

$$(x,y) = E\{xy*\} = R_{xy} \tag{3.63}$$

Therefore, when making an estimate based on data sequences of a stationary random process, the Gram matrix is just the auto correlation matrix of the sequence. The $A$ matrix of the normal equations becomes the column vector whose components are the elements of the cross-correlation sequence between the quantity to be estimated and the observed random data sequence. Thus, the normal equations can be written as

$$\begin{bmatrix} E\{vs^*_1\} \\ E\{vs^*_2\} \\ \vdots \\ E\{vs^*_k\} \end{bmatrix} = \begin{bmatrix} R_{ss}(0)\ R_{ss}(-1) & \ldots & R_{ss}(-k) \\ R_{ss}(1)\ R_{ss}(0) & \ldots & R_{ss}(-k+1) \\ & & \\ R_{ss}(k)R_{ss}(k-1) & \ldots & R_{ss}(0) \end{bmatrix} \begin{bmatrix} c_1 \\ c_2 \\ \\ c_k \end{bmatrix} \tag{3.64}$$

or

$$\underline{r}_{vs} = R_s\underline{c} \tag{3.65}$$

where $v$ is the random parameter to be estimated and $s_i$, $i = 1, \ldots, K$, are the individual random variables of the data sequence.

As an alternate method of reaching equation (3.64) consider again the problem of modelling a physical system with a linear filter, but now let us consider $x(n)$ and $y(n)$ to be stationary random sequences. We note that according to equation (1.96) the expression $X^tX$ in equation (3.58) is the estimate of the (M+1) × (M+1) auto correlation matrix of the sample sequence $x(n)$ and similarly the expression $X^t\underline{y}$ is the cross-correlation sequence $R_{xy}(m)$ of length (M+1). Thus, the normal equations can be written as

$$\underline{r}_{xy} = R_x \underline{h} \tag{3.66}$$

where

$$R_x = \begin{bmatrix} R_{xx}(0) & R_{xx}(-1) & \dots & R_{xx}(-M) \\ R_{xx}(1) & R_{xx}(0) & \dots & R_{xx}(-M+1) \\ \\ R_{xx}(M) & R_{xx}(M-1) & \dots & R_{xx}(0) \end{bmatrix}$$

and

$$\underline{r}_{xy} = [R_{xy}(0) \ . \ . \ . \ R_{xy}(M)]^T$$

The matrix multiplication $R_x\underline{h}$ is just the finite convolution of the auto-correlation sequence $R_{xx}(m)$ with $h(n)$. If, for theoretical purposes, we allow the lengths of the sequences $x(n)$, $y(n)$ and $h(n)$ to approach infinity, then we can write equation (3.66) as

$$R_{xx}(m) * h(n) = R_{yx}(m) \tag{3.67}$$

If we now take the Fourier transform of both sides of this equation we get

$$S_{xx}(\omega)H(e^{j\omega}) = S_{yx}(\omega) \tag{3.68}$$

Thus we see that the theoretical optimum linear filter for obtaining the minimum mean square error estimate of $y(n)$ from $x(n)$ has the ideal frequency response

$$H(e^{j\omega}) = \frac{Syx(\omega)}{S_{xx}(\omega)} \tag{3.69}$$

This filter is commonly known as the "Weiner filter." We note two important points regarding this optimum filter:

1. It is nonrealizable since the corresponding impulse response is not causal. A realizable transfer function may be obtained by using one-sided transform techniques.
2. From equation (3.69) it is clear that we do not need to know the exact nature of both sequences $x(n)$ and $y(n)$ to obtain the optimum filter. Given a stationary random process $x(n)$ we need specify only the spectral density $s_{xx}(\omega)$ and the cross spectral density $S_{xy}(\omega)$ to obtain the optimum linear filter for estimating $y(n)$ from $x(n)$.

Consider the case where we have an observed stationary random process $x(n)$ which we assume to be of the form

$$x(n) = y(n) + w(n) \tag{3.70}$$

where $y(n)$ is an unknown stationary random process we wish to estimate, and $w(n)$ is an additive random noise process.

Equation (3.69) gives the optimum non-realizable filter for estimating $y(n)$ from $x(n)$. If we assume that $w(n)$ is uncorrelated with $y(n)$ then

$$\begin{aligned} S_{yw}(\omega) &= 0 \text{ and} \\ S_{xy}(\omega) &= S_{yy}(\omega) \\ S_{xx}(\omega) &= S_{yy}(\omega) + S_{ww}(\omega) \end{aligned} \tag{3.71}$$

and equation (3.69) becomes

$$H(e^{j\omega}) = \frac{S_{yy}(\omega)}{S_{yy}(\omega) + S_{ww}(\omega)} \tag{3.72}$$

This filter has a frequency response near unity at those frequencies where the spectral density of the noise is small compared to that of the signal. The filter response is small where the noise spectral density is large compared to that of the signal. Thus the noise is filtered out while the signal is passed.

If we wish to estimate $y(n + \lambda)$ from $x(n)$, this is equivalent to the case of having

$$R_{xy}(m + \lambda) = R_{xx}(m) * h(m) \tag{3.73}$$

which gives

$$H(e^{j\omega}) = \frac{e^{j\omega\lambda} S_{xy}(\omega)}{S_{xx}(\omega)} \tag{3.74}$$

as the ideal filter. For $\lambda > 0$ this is known as a prediction filter. For $\lambda = 0$ we have the optimum linear filter again and for $\lambda < 0$ it is a smoothing filter.

### 3.3.4 Filtering, Smoothing and Prediction

In our discussion of the optimum linear filter we assumed that the input sequence $x(n)$ was stationary. In many problems encountered in digital signal processing the assumption of stationarity is valid only for a short period. For example, in modelling human speech the waveform can be considered stationary only for about 20 msec. When dealing with this type of signal we are restricted to using short data sequences. Thus, we now consider the general problem of linearly estimating a quantity from the last $N$ received data samples of a random process.

As we have already seen, the general problem can be formatted in terms of the normal equations as given by equation (3.65). We denote the last $N$ received samples of a random process as the vector.

$$\underline{x}_N = [x(n), x(n-1), \ldots, x(n-N+1)]$$

and $p(n)$ as the random quantity to be estimated as $\underline{\hat{p}} = \underline{x}_N \underline{h}$. The normal equations can be written as

$$\begin{bmatrix} E\{p(n).x(n)\} \\ E\{p(n) \times (n-1)\} \\ \vdots \\ E\{p(n) \times (n-N+1)\} \end{bmatrix} = \begin{bmatrix} R_{xx}(0) & R_{xx}(1) & \dots & R_{xx}(N-1) \\ R_{xx}(-1) & R_{xx}(0) & \dots & R_{xx}(N-2) \\ \vdots & & & \\ R_{XX}(-N+1) & & \dots & R_{xx}(0) \end{bmatrix} \begin{bmatrix} h_1 \\ \\ \vdots \\ h_N \end{bmatrix} \tag{3.75}$$

or

$$\underline{r} = R_{xN}\underline{h} \tag{3.76}$$

The coefficient vector $\underline{h}$ defines the optimum linear operator for estimating $p(n)$ from the last $N$ samples of $x(n)$. To complete the specification of the problem we must specify the relationship of $p(n)$ to $x(n)$.

Consider the general case of

$$x(n) = p(n) + w(n) \tag{3.77}$$

where $w(n)$ is zero mean white noise uncorrelated with $p(n)$. We wish to estimate $p(n+\lambda)$ from the last $N$ samples of $x(n)$. The expected value

$$E\{p(n+\lambda)x(n)\} = E\{p(n+\lambda)p(n)\} + E\{p(n+\lambda)w(n)\} \tag{3.78}$$

but

$$E\{p(n+\lambda)w(n)\} = 0$$

since $w(n)$ and $p(n)$ are uncorrelated. Therefore

$$E\{p(n+\lambda)x(n)\} = R_{pp}(\lambda) \tag{3.79}$$

where $R_{pp}(\lambda)$ is the autocorrelation sequence of $P(n)$. We also note

$$\begin{aligned} E\{x(i)x(j)\} &= R_{xx}(i-j) = R_{pp}(i-j), \quad i \neq j \\ &= R_{pp} + \sigma_w^2, \qquad i = j \end{aligned} \tag{3.80}$$

Therefore the matrix $R_{xN}$ of equation (3.76) can be written

$$R_{xN} = R_{pN} + C_w \tag{3.81}$$

where $R_{pN}$ is the $N \times N$ autocorrelation matrix of the random parameter sequence $p(n)$ and $C_w$ is the noise covariance matrix.

Thus, the general problem in normal equation form is expressed as

$$\underline{h} = [R_{pN} + C_w]^{-1}\underline{r}_{pp} \tag{3.82}$$

where $\underline{r}_{pp}$ is a column vector representing the auto-correlation sequence for $P(n)$.

Since we have assumed zero mean noise, then $C_w = R_{wN}$, where $R_{wN}$ is the $N \times N$ autocorrelation matrix of the noise process $w(n)$. Hence, we can write

$$\underline{h} = [R_{pN} + R_{wN}]^{-1}\underline{r}_{pp} = R_{x_N}^{-1}\underline{r}_{pp} \tag{3.83}$$

which we recognize as the finite matrix equivalent to the ideal Weiner filter.

The finite data filtering, prediction and smoothing problems can be formulated in matrix form from this expression. For example, if we wish to predict the next sample from the last $N$ received samples we have the one-step linear prediction problem, i.e.,

$$\hat{x}(n+1) = \sum_{i=0}^{N-1} h(i)x(n-i) \tag{3.84}$$

The corresponding normal equation expression is

$$\underline{r}_{xx} = R_{xN}\underline{h} \tag{3.85}$$

where

$$\underline{r}_{xx} = [R_{xx}(1), R_{xx}(2), \ldots, R_{xx}(N)]^T$$

and

$$R_{xN} = \begin{bmatrix} R_{xx}(0) & R_{xx}(1) & \ldots & R_{xx}(N-1) \\ \vdots & & & \vdots \\ R_{xx}(N-1) & & \ldots & R_{xx}(0) \end{bmatrix}$$

We have

$$\underline{h} = R_{xN}^{-1}\underline{r}_{x\hat{x}} \tag{3.86}$$

and

$$\hat{x}(n+1) = \underline{x}_N\underline{h} \tag{3.87}$$

Thus

$$\hat{x}(n+1) = \underline{x}_N R_{xN}^{-1}\underline{r}_{xx} \tag{3.88}$$

is the minimum mean squared error estimate of $x(n+1)$ from the last $n$ samples of $x(n)$.

The major drawback to using the matrix normal equation formulation for estimating a parameter from the last $N$ data samples is the requirement to set up and solve equation (3.83) for each new data sample received. Clearly this implies a large amount of processing between samples. It is appealing to consider forming a new estimate based on the last one, plus the newly arrived sample. In the next section we shall discuss the basic concepts of Kalman filtering as an example of such a recursive approach.

### 3.3.5 Recursive Estimation (Kalman Filtering)

The presentation given here is not a derivation of the Kalman filter recursion equations but rather a brief explanation of their significance.

The basic assumption in this recursive estimation technique is that the parameter or quantity to be estimated can be modelled as the state of a dynamic system. The measurable signal is modelled as the output of the state representable system. Thus we have a signal source model that can be expressed by the matrix state equations of Chapter 2.

$$\underline{x}(n+1) = \Phi(n+1,n)\underline{x}(n) + \underline{w}(n) \tag{3.89}$$

$$\underline{y}(n) = C(n)\underline{x}(n) + \underline{v}(n) \tag{3.90}$$

These equations represent the system or plant model and the measurement or observation model respectively.

Further, we assume that $w(n)$ and $v(n)$ are independent sample sequences from zero-mean Gaussian white noise sources, i.e.,

$$E\{\underline{w}(n)\} = 0, \; E\{\underline{w}(n)\underline{w}^*(k)\} = 0, \; k \neq n$$

$$E\{\underline{v}(n)\} = 0, \; E\{\underline{v}(n)\underline{v}^*(k)\} = 0, \; k \neq n$$

$$E\{\underline{v}(n)\underline{w}^*(k)\} = 0, \; \text{for all } n,k$$

$$E\{\underline{w}(n)\underline{w}^t(n)\} \underline{\Delta} Q(n) = \text{System State Covariance Matrix}$$

$$E\{\underline{v}(n)\underline{v}^t(n)\} \underline{\Delta} R(n) = \text{Observation noise covariance matrix}$$

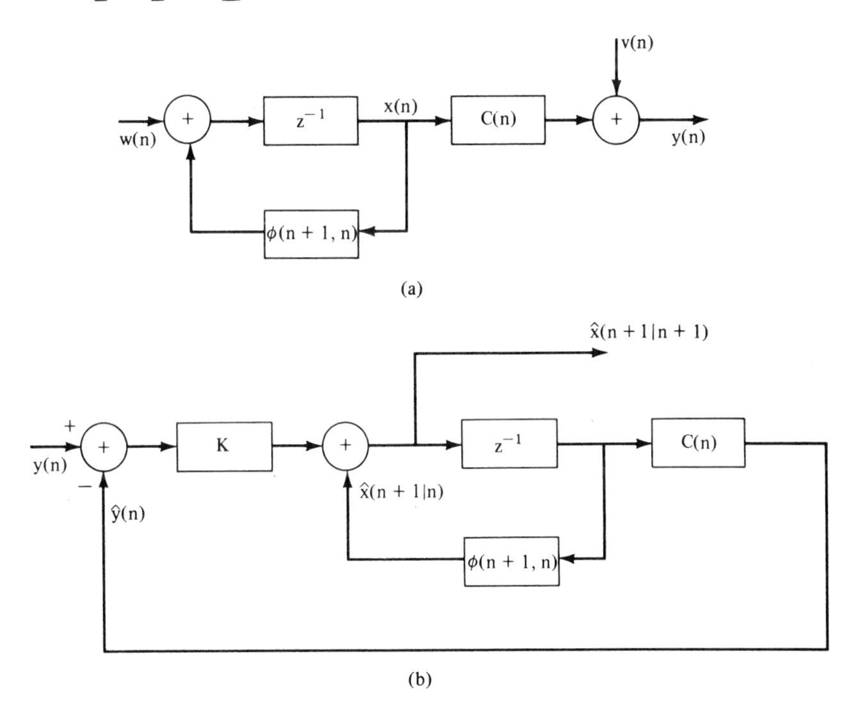


**Figure 3.4:** (a) Source system model. (b) The basic Kalman filter model.

Noise sequences with these properties are often denoted as $N(0,Q(n))$ to indicate a zero-mean normal process with covariance $Q(n)$.

The system equation gives us a recursive definition of the next state and the observation equation relates the observation to the state of the system. Thus, the initial problem is to identify the system and form the state equations.

For the recursive procedure to begin, an initial estimate of the state vector $\underline{x}(n_0)$ is required, as well as an estimate of its mean $(\eta_x)$ and variance $(P(n_0))$.

$$P(n_0) = E\{(\underline{x}(\hat{n}_0) - \underline{\hat{n}}_x(n_0))(\underline{x}(n_0) - \underline{\hat{n}}_x(n_0))'\} \tag{3.91}$$

The first step is to estimate the next state, which we denote

$$\underline{\hat{x}}(n_0 + 1|n_0) = \Phi(n_0 + 1, n_0)\underline{x}(n_0) \tag{3.92}$$

The variance of this estimate is also predicted, i.e.,

$$\hat{P}(n + 1) = \Phi P(n_0)\Phi' + Q(n_0) \tag{3.93}$$

The state estimate is then updated from the new observation $\underline{y}(n + 1)$.

$$\underline{\hat{x}}(n + 1) = \hat{x}(n + 1|n) + K(n + 1)[y(n + 1) - C(n + 1)\underline{\hat{x}}(n + 1|n)] \tag{3.94}$$

where $K(n)$ is the Kalman gain matrix defined as

$$K(n) = \hat{P}(n)C'(n)[C(n)\hat{P}(n)C'(n) + R(n)]^{-1} \tag{3.95}$$

One more equation, the error covariance update, is required to complete the recursion,

$$P(n + 1) = \hat{P}(n) - K(n)C(n)\hat{P}(n) \tag{3.96}$$

From these equations we can summarize the basic concepts of the Kalman filter. The observed data is modelled as the output of a state representable system, and the random quantity to be estimated is modelled as the state of that system. This model is represented by equations (3.89) and (3.90). From the system model the next state is estimated in equation (3.92) and used to predict the next observation. The predicted observation is compared with the actual observation. A weighted portion of the error is added to the state estimate upon which the next estimate will be based (equation (3.94)). The weighting represented by the Kalman gain matrix $K$ is the central issue here. The weighting must be determined such that the mean-square error between the observation and the prediction is minimized. Thus, the gain matrix is dependent upon the relationship of the covariance matrices $Q(n)$ and $R(n)$ to the original covariance matrix $P(n)$, and is recursively computed from equations (3.93), (3.95) and (3.96).

We note that the actual performance of the Kalman filter is ultimately dependent on the ability to derive an accurate linear model of the signal source. We also note that for a good linear model the filter will track the observed input closely and the gain matrix will tend to become very small as the errors diminish. If the source dynamics then vary suddenly from the model, the increased errors will be virtually ignored due to the reduced gain matrix. This can result in the filter diverging from the actual source track.

### 3.3.6 Summary

In this section we have seen that the fundamental basis of linear minimum mean squared error estimation can be viewed geometrically as the projection theorem. The normal equations that result from the projection theorem form the computational basis for a wide variety of signal processing problems ranging from simple least-squared error approximation to Kalman filtering.

The important observation from this review is the requirement to support matrix operations in a general purpose digital signal processor since a large number of signal processing problems can be readily formulated in normal equation format.

## 3.4 SPECTRAL ESTIMATION

### 3.4.0 Introduction

The field of digital signal processing is commonly sub-divided into two major areas—digital filtering and spectral estimation. In this section we discuss spectral estimation as a sub-topic of general estimation theory. We shall discuss three approaches to estimating power spectral densities.

The first approach is based upon a relatively direct interpretation of the definition of power spectral density of a process as the Fourier transform of the autocorrelation function. From this point of view we will discuss first the estimation of autocorrelation functions and introduce the periodogram as the Fourier transform of the autocorrelation estimate. We will find unfortunately (and perhaps surprisingly) that the periodogram does not give a consistent estimate of the spectral density even though the autocorrelation estimate is consistent. We will also discuss two approaches for obtaining consistent spectral density estimates through the averaging or smoothing of periodograms.

In the second approach the data is transformed directly to the frequency domain (without first estimating the auto-correlation function). This results in a signal representation with uncorrelated adjacent frequency bands. The spectrum is then estimated and averaged across consecutive estimates for each frequency band.

The third approach yields a class of autoregressive spectral estimators. Within this class we shall introduce two methods of spectral estimation: the maximum likelihood method (MLM) and the maximum entropy method (MEM). Both of these methods are based on the concepts discussed in Section 3.2.

### 3.4.1 Classical Spectral Estimation

The classical method of spectral estimation is based on the definition of power spectral density as the Fourier transform of the autocorrelation function. Thus, we begin our discussion by considering autocorrelation estimates. In Section 1.3.3 we discussed the use of time averages in place of ensemble averages under an ergodic hypothesis. These time averages were expressed as limits as the number of samples approached infinity. In practice it is always necessary to estimate these time averages from finite duration sample sequences. Section 1.3.3 concluded with the statement that a common estimate of the autocorrelation of a stationary random process is given by equation (1.116), which we repeat here for convenience:

$$\begin{aligned} R_{xx}(m) = R_M(m) &= \langle x(n)x^*(n+m) \rangle \\ &= \frac{1}{2M+1} \sum_{n=-M}^{M} x(n)x^*(n+m) \end{aligned} \tag{3.97}$$

Given a causal sequence of length $N$ denoted $x(n)$, $n = 0, 1, \ldots, n-1$, we can rewrite equation (3.97) as

$$R_N(m) = \frac{1}{N} \sum_{n=0}^{N-|m|-1} x(n)\, x^*(n+m) \qquad |m| < N \tag{3.98}$$

where the upper limit reflects the fact that the number of non zero components of the summation is a function of the lag value $m$. The resulting sequence $R_N(m)$ is of length $2N-1$.

We are concerned with the quality of this estimate in terms of the bias and variance. The expectation of $R_N(m)$ is

$$\begin{aligned} E\{R_N(m)\} &= \frac{1}{N} \sum_{n=0}^{N-|m|-1} E\{x(n)x^*(n+m)\} \\ &= \frac{N-|m|}{N} R_{xx}(m), \ |m| < N \end{aligned} \tag{3.99}$$

Clearly, for any fixed value of $m$ other than $m = 0$ the estimate $R_N(m)$ is a biased estimate. However, the bias approaches zero as $N$ approaches infinity. Thus $R_N(m)$ is said to be "asymptotically unbiased."

The variance of $R_N(m)$ is

$$\operatorname{var}\{R_N(m)\} = E\{R_N^2(m)\} - E^2\{R_N(m)\} \tag{3.100}$$

This expression is not easily evaluated in the general case. It can be shown, however, that for a practical fixed value of $m$ the variance also approaches zero as $N$ approaches infinity [6]. Thus $R_N(m)$ is a consistent estimate.

We note however that for a fixed $N$ the variance of $R_N(m)$ increases as $m$ approaches $N$. This results because fewer observed sample values are used to estimate $R_N(m)$ as $m$ approaches $N$.

Given that $R_N(m)$ is a consistent estimate of the autocorrelation of a signal we might expect that the Fourier transform of $R_N(m)$ would be a good estimate of the spectral density. As it turns out this is not the case.

The Fourier transform of $R_N(m)$ is known as the "periodogram" and is usually denoted as

$$I_N(\omega) = \mathfrak{F}\{R_N(m)\} = \sum_{m=-\infty}^{\infty} R_N(m) e^{-j\omega m} \qquad (3.101)$$

$$= \sum_{m=-(N-1)}^{N-1} R_N(m) e^{-j\omega m}$$

The expectation of $I_N(\omega)$ is

$$E\{I_N(\omega)\} = \sum_{m=-\infty}^{\infty} E\{R_N(m)\} e^{-j\omega m} \qquad (3.102)$$

From equation (3.99) we have

$$E\{I_N(\omega)\} = \sum_{m=-(N-1)}^{N-1} \frac{N - |m|}{N} R_{xx}(m) e^{-j\omega m} \qquad (3.103)$$

Thus, the periodogram is a biased estimate of the true spectrum $S_{xx}(\omega)$. However, $I_N(\omega)$ is asymptotically unbiased for a fixed value of $m$. We note further that reducing the lag parameter $m$ reduces the bias.

Calculation of the variance of the periodogram is not a straightforward matter in the general case. Approximations are generally necessarily based on insight gained from examining special cases such as the zero-mean Gaussian case. For this special case the variance can be shown to be of the order of $\sigma_x^4$ no matter how large $N$ becomes, with additional terms entering for the general case. This non-consistent feature of the periodogram stems from the dependence of the auto-correlation estimate on the lag parameter $m$. We noted that the variance of the autocorrelation estimate increases as $|m|$ approaches $N$. Fourier transforming the autocorrelation estimate gives a representation of that sequence in terms of uncorrelated components based on suitably combining all values of the estimated autocorrelation sequence. Since the variance of these values increases as $m$ approaches $N$ these values introduce increased fluctuation in the individual values of the estimated spectrum.

Increasing the length of the original sequence $N$ does not help because this brings the uncorrelated Fourier coefficients closer together, further increasing the fluctuations about the true spectrum. The variance of the periodogram will be decreased however by reducing the range of $m$; that is, by transforming only those values of the $R_N(m)$ near $m = 0$ and thereby basing the spectral estimate on the most reliable values of the autocorrelation estimate. However, in reducing the number of values of $R_N(m)$ on which the spectrum estimate is based, resolution of the estimated spectrum is also reduced. Thus we have a fundamental tradeoff between spectral resolution and stability.

An alternative method of viewing this tradeoff can be formulated in terms of window functions. The expected value of $R_N(m)$ given by equation (3.99) can be considered as the true autocorrelation of $x(n)$ as seen through a triangular window commonly known as the "Bartlett window." This window function is illustrated in Figure 3.5(a) and is described by the equation

$$w_B(m) = \begin{cases} \dfrac{N - |m|}{N}, & |m| < N \\ 0 & \text{otherwise} \end{cases} \tag{3.104}$$

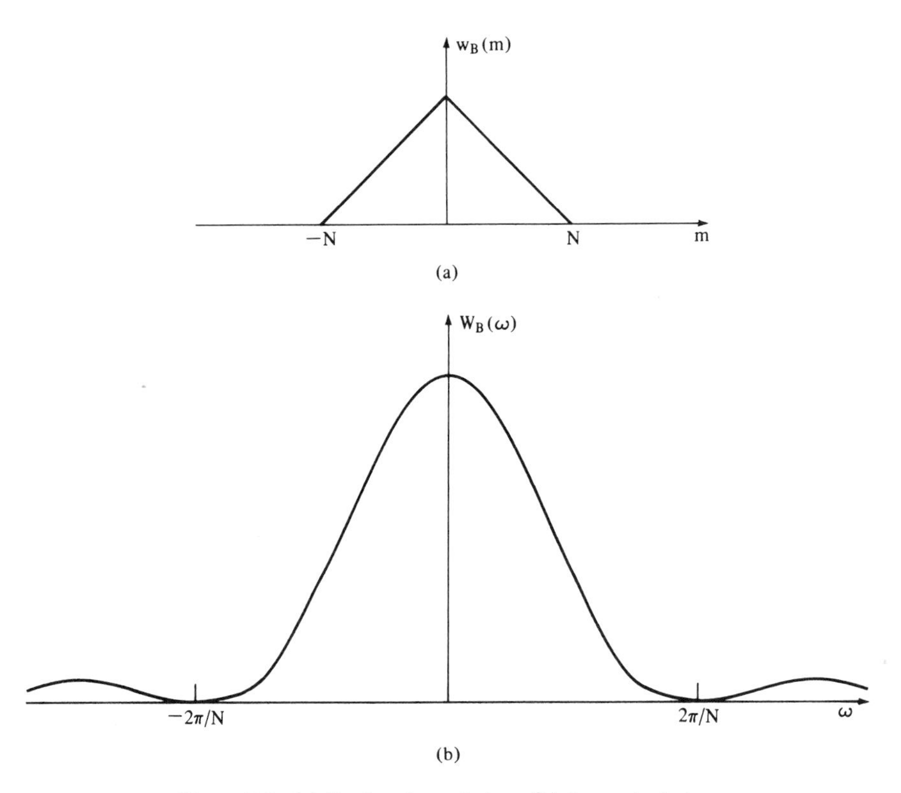


**Figure 3.5:** (a) Bartlett data window. (b) Spectral window.

Thus, equation (3.103) may be viewed as the true spectral density, convolved with the Fourier transform of the window function, i.e.,

$$E\{I_N(\omega)\} = \frac{1}{2\pi}\int_{-\pi}^{\pi} S_{xx}(\Omega)\, W_B(\omega - \Omega)\, d\Omega \tag{3.105}$$

where

$$W_B(\omega) = \mathfrak{F}\,\{w_B(m)\} = \frac{1}{N}\left(\frac{\sin(\omega N/2)}{\sin(\omega/2)}\right)^2 \tag{3.106}$$

is known as the Bartlett spectral window. This window function is illustrated in Figure 3.5(b).

The effect of convolving the true spectrum with a spectral window is a smearing of the spectrum. To maintain high spectral resolution the spectral window must be as narrow as possible. This implies a large data window on the autocorrelation estimate. To minimize the variance of the spectral estimate we want a short data window on the autocorrelation estimate, or equivalently a wide spectral window that spreads the spectral energy and introduces stability.

Consistent autocorrelation estimates other than equation (3.97) can be formed that can be interpreted as applying a data window to the true autocorrelation functions. For example, the autocorrelation estimate

$$R'_N(m) = \frac{1}{N - |m|} \sum_{n=0}^{N-|m|-1} x(n)x^*(n + m) \tag{3.107}$$

is an unbiased estimate with expectation

$$E\{R'_N(m)\} = R_{xx}(m)\,, \quad |m| < N \tag{3.108}$$

Equation (3.108) can be interpreted as the true autocorrelation of $x(n)$ as seen through a rectangular or "boxcar" window

$$W_R(m) = \begin{cases} 1\,, & |m| < N \\ 0 & \text{otherwise} \end{cases} \tag{3.109}$$

which gives a corresponding spectral window

$$W_R(\omega) = \frac{\sin(\omega(2N + 1)/2)}{\sin(\omega/2)} \tag{3.110}$$

as illustrated in Figure 3.6.

Two general approaches are used to improve the periodogram estimate (i.e., to decrease the variance). The first method averages independent periodograms. The second method involves smoothing the periodogram by convolving it with an appropriate spectral window.

The concept of averaging periodograms to reduce the variance is based on the fact that for $M$ uncorrelated random variables each with identical expected value $n$ and variance $\sigma^2$ the sample variance is $\sigma^2/M$. Thus, if we sec-

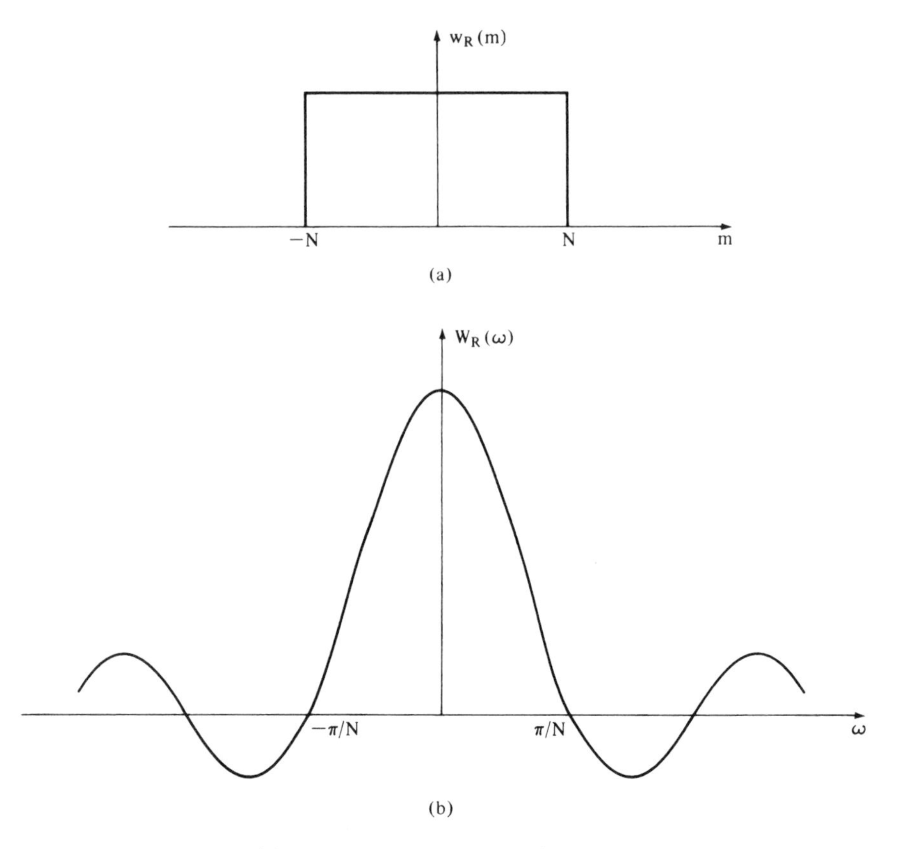


**Figure 3.6:** (a) Rectangular data window. (b) Spectral window.

tion an $N$ point data sequence into $L$-point subsequences such that $N = pL$, $p$ being an integer, and then compute the periodogram for each subsequence

$$I_{L_p}(\omega) = \sum_{m=-(L-1)}^{L-1} R_{L_p}(m)e^{-j\omega m}, \; |m| \leqslant L \tag{3.111}$$

we can reduce the variance of the spectrum by a factor of $p$ by averaging the $p$ values of $I_{Lp}(\omega)$, i.e.,

$$\hat{S}_{xx}(\omega) = \frac{1}{P}\sum_{r=0}^{p-1} I_{L_r}(\omega) \tag{3.112}$$

Thus, $\hat{S}_{xx}(\omega)$ is an "averaged periodogram" estimate of $S_{xx}(\omega)$. If the individual periodograms are uncorrelated then the variance of $\hat{S}_{xx}(\omega)$ is less than that of $I_N(\omega)$ by nearly a factor of $p$. Therefore we would like to break the $N$ point data sequence into as many subsequences as possible to reduce the variance of the spectral estimate. However, as $p$ is increased the length $L$ of the subsequences is decreased resulting in a short data window, i.e.,

$$E\{R_{Lp}(m)\} = \frac{L - |m|}{L} R_{xx}(m), \quad |m| < L \tag{3.113}$$

The short data window decreases spectral resolution due to the smearing of the spectrum by the convolution with the spectral window, which is the Fourier transform of the data window.

Thus we have again the fundamental tradeoff between spectral resolution and variance. For a small variance we require a large number of subsequences but for high spectral resolution we want the subsequences to be as long as possible. Using a very large initial sequence length $N$ and large $L$ leads to a problem with non-stationarity of the data as well as increasing the bias of the estimate, i.e.,

$$\text{Bias} = \sum_{m=-(L-1)}^{L-1} \frac{L-|m|}{L} R_L(m) e^{-j\omega m}, \quad |m| < L \tag{3.114}$$

To increase the number of subsequences and still maintain reasonable spectral resolution without having to increase the initial sequence length $N$, it is common to overlap subsequences. A fifty percent overlap of subsequences has been found to give satisfactory results (6).

The second method of improving the variance of spectral estimates is to smooth a single periodogram. This smoothing is accomplished by applying a suitable lag window to the autocorrelation estimate or equivalently by convolving the periodogram with an appropriate spectral window. This procedure is somewhat analogous to a low pass filtering operation on the periodogram to remove rapid fluctuation.

A number of considerations are necessary in choosing an appropriate lag window:

1. The window must be much shorter than the initial record length. Thus only the most reliable autocorrelation estimates are used.
2. The lag window must be even and peak at zero to maintain the symmetry of $R_N(m)$ and obtain real spectral estimates.
3. The corresponding spectral window should be such that

$$\frac{1}{2\pi} \int_{-\pi}^{\pi} w(\omega)\, d\omega = 1 \tag{3.115}$$

to maintain proper scaling.

Two common windows for spectral smoothing are the "Hamming" and "Hanning" windows. These windows provide improved spectral resolution over the triangular and rectangular windows but they can lead to negative spectral estimates. The Hanning lag window is of the form

$$w_{\text{Han}} = 0.5(1 - \cos(2m/(M - 1)))\, W_R(m) \tag{3.116}$$

where $W_R(m)$ is the rectangular window and $M$ is the width of the lag window. The Hamming window has the form

$$W_H(m) = [0.54 + 0.46 \cos(2\pi m/(M - 1))] W_R(m) \qquad (3.117)$$

These window functions are illustrated in Figure 3.7.

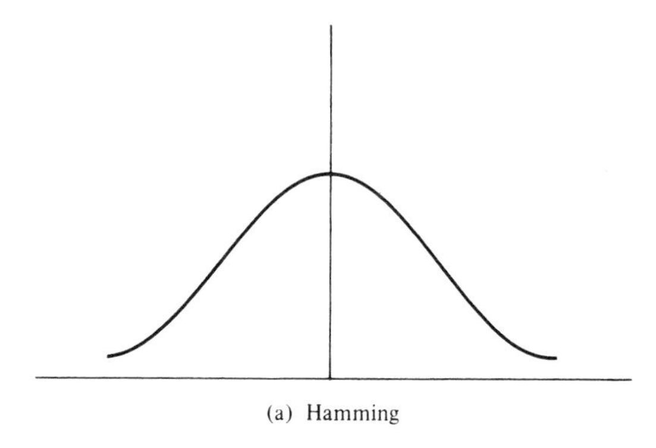

(a) Hamming

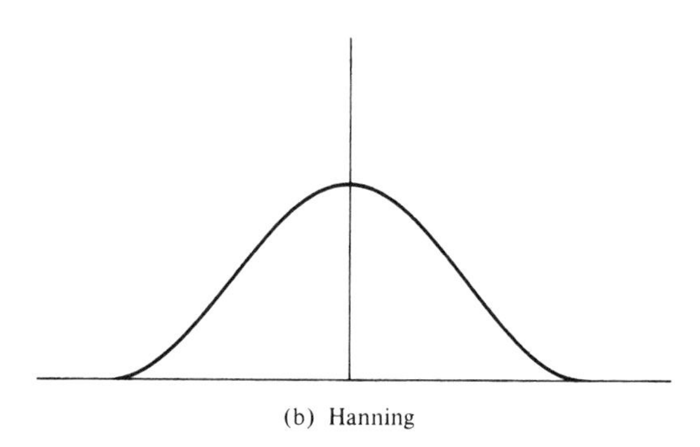

(b) Hanning

**Figure 3.7:** Hamming and Hanning window functions.

### 3.4.2 Direct Transformation Spectral Estimation

The second class of spectral estimator we shall discuss involves the direct transformation of an observed data sequence to the frequency domain without first forming an estimate of the autocorrelation function. This approach is based on the fact that the autocorrelation function may be viewed as the convolution

$$R_{xx}(m) = x(m) * x^*(-m) \qquad (3.118)$$

Thus, the Fourier transform of the autocorrelation can be expressed as

$$\mathcal{F}\{R_{xx}(m)\} = |X(e^{j\omega})|^2 \qquad (3.119)$$

From equation (3.98) we note that $R_N(m)$ may be expressed as

$$R_N(m) = \frac{1}{N}(x(m) * x^*(-m)), \ |m| < N \tag{3.120}$$

Thus, the periodogram $I_N(\omega)$ can be written as

$$I_N(\omega) = \frac{1}{N} \sum_{m=-(n-1)}^{N-1} (x(m) * x^*(-m))e^{j\omega m} \tag{3.121}$$
$$= \frac{1}{N}|X(e^{j\omega})|^2$$

Therefore, the periodogram can be formed by transforming the observed data sequence directly and calculating the squared magnitude of the complex Fourier coefficients. This estimate is still equivalent to a periodogram and as such suffers the same stability versus resolution problem. Averaging and smoothing techniques similar to those discussed in the previous section can be employed, but the lag windows are now applied directly to the observed data.

### 3.4.3 Autoregressive Spectral Estimation

The general class of autoregressive spectral estimators is based on time series modelling of the process that generated the observed data by techniques similar to those of Sections 2.2.1 and 3.2.

We begin our discussion with the "spectral factorization theorem" which states that:

> Any real wide-sense stationary random process $x(n)$ with rational power spectral density can be represented as a ratio (of possibly infinite degree) polynomials, i.e.,

$$S_{xx} = \sigma^2 \frac{A(z)A(z^{-1})}{B(z)B(z^{-1})}, \ \sigma^2 > 0 \tag{3.122}$$

where

$$A(z) = \sum_{k=0}^{M} a_k z^{-k}, \ a_k \text{real} \tag{3.123}$$

and

$$B(z) = \sum_{k=0}^{N} b_k' z^{-k}, \ b_k \text{real} \tag{3.124}$$

A proof for this theorem is given by Tretter [16].

This theorem is equivalent to modelling the random process $x(n)$ as the output of a recursive filter driven by a white noise sequence $w(n)$ with vari-

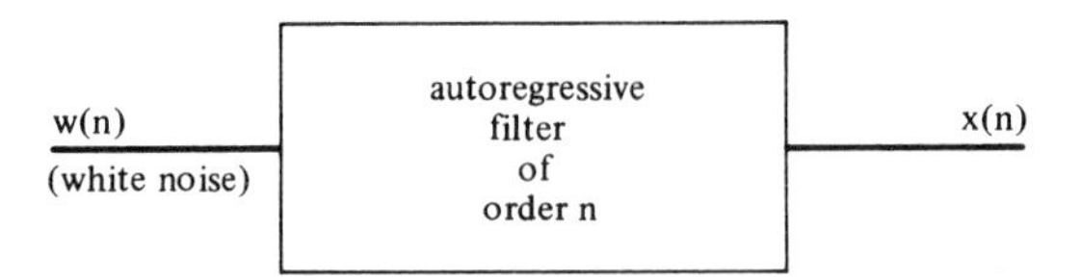


(a) Model for spectral estimation

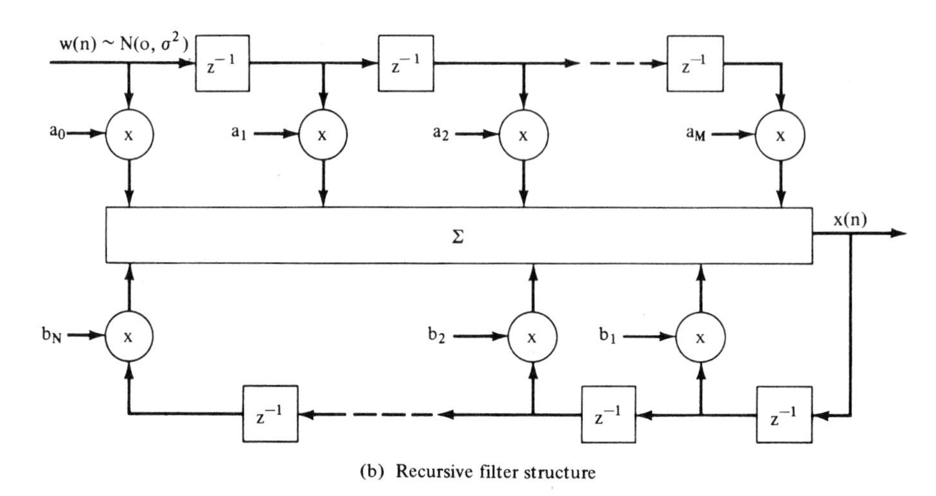


(b) Recursive filter structure

**Figure 3.8:** Signal representations.

ance $\sigma^2$ as illustrated in Figure 3.8. This filter corresponds to the general difference equation

$$\sum_{k=0}^{M} a_k w(n-k) = \sum_{k=0}^{N} b_k x(n-k) \tag{3.125}$$

and has transfer function $A(z)/B(z)$; $b_o = 1$. We recognize this general model as an autoregressive moving average model. If all the $b_k$ are zero except $b_0 = 1$, then we have a moving average model, i.e.,

$$x(n) = \sum_{k=0}^{M} a_k w(n-k), \; a_o = 1 \tag{3.126}$$

The coefficients $a_k$ are the moving average coefficients, and the output spectrum is:

$$S_{xx}(z) = \sigma^2 |A(z)|^2 \tag{3.127}$$

The conventional periodogram spectral estimates correspond to a moving average model where the spectral density is approximated as $|A(z)|^2$.

If all the $a_k$ are zero except $a_o = 1$, then the model is an autoregressive or all-pole model of order $N$, i.e.,

$$x(n) = w(n) - \sum_{k=1}^{N} b_k x(n-k) \tag{3.128}$$

This model has spectral density

$$S_{xx}(z) = \sigma^2/|B(z)|^2 \tag{3.129}$$

and the coefficients $b_k$ are the autoregressive coefficients.

Basically, the estimation of power spectral densities by autoregressive techniques involves estimating the coefficients $b_k$ from the observed data. The spectral density is then proportional to the squared magnitude of the transfer function of the autoregressive model.

If the noise sequence $w(n)$ is considered as a Gaussian process then the maximum likelihood estimate of the parameters $\sigma^2$ and $b_k$ of the autoregressive model can be shown to be given by the solution of a set of normal equations [7].

$$\underline{\hat{b}} = R^{-1}\underline{r} \tag{3.130}$$

where

$$\underline{\hat{b}} = [b_1, \ldots, b_N]^t \tag{3.131}$$

$$\underline{r} = [R_{xx}(0), \ldots, R_{xx}(N)]^t \tag{3.132}$$

$$R = \begin{bmatrix} R_{xx}(0) & \ldots & R_{xx}(N) \\ \cdot & & \cdot \\ \cdot & & \cdot \\ R_{xx}(N) & \ldots & R_{xx}(0) \end{bmatrix} \tag{3.133}$$

and

$$\hat{\sigma}^2 = \sum_{k=0}^{N} \hat{b}_k R_{xx}(k) \tag{3.134}$$

Another interpretation of the autoregressive spectral estimation approach is known as the "maximum entropy method" (MEM). This approach is essentially based on the concept of finding the coefficients of a filter that whitens, or maximizes the entropy of the received time domain data sequence. This is accomplished by implementing a linear prediction filter of order $N + 1$. Again the spectral density is obtained as the squared magnitude of the inverse filter.

In both approaches the properties of white noise play an important role. In the first case, white noise is shaped by a filter to yield a signal with the desired spectrum. The spectral factorization theorem assures us that the spectrum represented by the filter coefficients is the same as the spectral coefficients of the signal. In the second approach, the signal is used as the input to a filter whose coefficients are chosen to whiten the spectrum.

### 3.4.4 Summary

In this section we have examined briefly three major classes of spectral estimators. The classical approach involves a straightforward application of the definition of power spectral density as the Fourier transform of the autocorrelation functions. Thus the processing implied consists of autocorrelation estimate calculations, Fourier transform calculations and windowing operations.

The second method discussed involves the direct Fourier transformation of the observed data, possibly with the application of a data window first. The input data sequences are often broken into overlapping subsequences to add stability to the resulting spectral estimates. The squared magnitude of the Fourier coefficients of each subsequence is proportional to the spectral density of the sequence. Individual frequency samples from successive periodograms can be directly averaged to stabilize further the spectral estimate. The advent of the fast Fourier transform algorithm, which we discuss in the next chapter, has made this approach to spectral estimation quite popular.

The third class of spectral estimator discussed was adaptive spectral estimation. The MLM and MEM spectral estimation techniques were discussed briefly. Both these techniques can be formulated in terms of the matrix normal equations of Section 3.2.

## 3.5 CHAPTER SUMMARY

Chapter 3 has presented an overview of time domain digital filter specification and design from the perspective of detection and estimation theory, followed by a discussion of spectral estimation.

In Section 3.1 we reviewed the basic concepts of detection theory. Four criteria for making a decision with respect to the binary hypothesis testing problems were examined:

1. Maximum a posteriori (MAP) criterion
2. Maximum likelihood (ML) criterion
3. Bayes criterion
4. Neyman Pearson criterion

The common element of each of these criteria is the comparison of a ratio of conditional probabilities with a threshold value as the basis for making an optimum decision. For the case of the last three criteria, the likelihood ratio is the quantity which must be related to measurable signal characteristics. Since the relationship of the measurable characteristics of a signal and the likelihood ratio differ for each situation, depending on the statistical model of the signal, optimum detection processing differs from case to case. For the case of detecting a completely known signal in additive noise, the optimum detector is a matched filter which is equivalent to a replica correlation operation.

Section 3.2 reviewed the basic concepts of optimum estimation and introduced the concepts of the bias and variance as measures of the quality of the estimate. For the general case of non-linear estimation, no general criteria exist for computing an optimum estimate. For linear estimation problems, however, general procedures do exist.

In Section 3.3 we examined the concepts of linear minimum squared error estimation techniques. We saw that the geometric interpretation of this estimation problem in terms of the projection theorem leads to a general solution technique in the form of the normal equations. Several general estimation problems were discussed to show how the projection theorem leads to optimum solutions for both deterministic and random parameter estimation. The optimum linear filter was also shown to result from this approach. The specific filter design problems for smoothing, filters, and prediction were examined and finally the concepts of recursive estimation and the basic Kalman filter were discussed. All of these filtering problems were shown to be based on the same least squared error optimization criteria and all utilized the same normal equation solution technique.

Section 3.4 introduced the two approaches to spectral estimation based on the spectral factorization theorem. Filter coefficients based on shaping the spectrum of white noise to match the signal or on whitening the signal spectrum are equivalent in either case to the unknown spectrum.

It is worth pausing to consider the computational requirements intrinsic to implementing the algorithms discussed in this chapter. First of all, vector arithmetic is an obvious requirement for many approaches. There is an intrinsic parallelism to these operations which would have a strong impact on processor architectures. Matrix inversion is, however, a computation-intensive problem which has had implementations to non-time-critical applications. Conversions from a time to a frequency-domain representation of a signal is also an intrinsic part of many approaches. In this case the FFT with its reduced computational requirements has had a strong impact on implementations. Processor architectures which efficiently handle the basic butterfly of an FFT will emerge as we review example processors in Chapters 6 and 7. The current availability of very high speed multipliers at low cost is an important factor, causing a re-evaluation of both the representation and implementation of signal processing algorithms.

## 3.6 EXERCISES

### Detection and Estimation

**1.** Show that var $\{\hat{p}\} = \text{var}\{[p - E\{P\}]^2\} = \sigma_{\hat{p}}^2$, (i.e., prove equation (3.38)).

**2.** For the situation described in Figure 3.1, suppose the mean of the signal is actually 1.5 rather than the assumed value of unity.

(a) Compute the change in the false alarm and the miss probability assuming the threshold was set, according to an M.L. criterion, for a unity mean situation.

(b) Suppose the samples are obtained from an $n$ bit A/D. Discuss the affects of choosing a specific value of $n$.

**3.** Consider the situation described in Figure 3.1:

(a) For two Gaussian processes of equal variance show that the decision level is ½, if one has zero mean and the other unity.

(b) Suppose one of the processes has a larger variance than the other. Find the decision level as a function of the larger variance.

(c) Compute the probability of false alarms as a function of the larger variance in (b).

**4.** Consider the problem of determining if a coin is unbiased. The problem is to determine if the probability of heads is the same as for tails. Considered as two hypotheses show that a sequence of $N$ tosses in which $N_h$ heads occur is a sufficient statistic for hypothesizing that the coin is unbiased.

**5.** Assume the situation described by equation (3.1). Suppose that both the signal and noise are random variable with exponential densities:

$$f_s(s) = ae^{-as} \text{ for } s \text{ greater than } 0$$

$$f_n(n) = be^{-bn} \text{ for } n \text{ greater than } 0$$

Determine the level $T$ for deciding that the signal is present for the Maximum Likelihood, Bayes, and the Neyman-Pearson test. Assume any further information that is required in each case and retain as a defined literal. (Note: The density of the sum of two independent random variables is given as the convolution of the individual densities [8, p. 189].)

## Spectral Estimation

**6.** A classical technique of detecting a sine wave of unknown frequency is to build a bank of narrow band filters across the frequency range of interest. The output of each filter is monitored and the one with the largest output is assumed to be centered on the unknown signal. Instead of a filter bank assume that an FFT computation is used. Discuss the implementation of such a scheme, and in particular show the improvement in output as successive iterations are computed. What would be the result if the signal fell between the frequency bins of the FFT?

CHAPTER 4

# Digital Signal Processing Algorithms and Techniques

## 4.0 INTRODUCTION

The concepts and the mathematical models developed thus far have proven sufficient, for both the description and the analysis of a wide variety of physical problems. Each application requires a somewhat different utilization of these mechanisms; yet behind the different jargon there is a great similarity from a processing point of view. In Chapter 5, the differences and similarities will be explicitly exposed. In this chapter, the use of the theory in the formulation of algorithms is of interest. One of the driving forces behind the development of these algorithms and techniques has been the attempt to reduce the computational and data transfer requirements to accommodate the performance constraints of available hardware.

Four topics have been chosen as widely representative: the fast Fourier transform (FFT) for spectral estimates, and frequency domain filtering; generalized linear processing (often called homomorphic processing) for multiplicative and convolved signal analysis; data compression; and adaptive processing.

In Section 4.1 the fast Fourier transform (FFT) will be discussed. Because of its lower computational requirement (compared to a direct DFT implementation) this algorithm was one of the major events that shaped the evolution of digital signal processing implementations. Even the development of digital signal processing theory has been affected to some extent by the FFT, in that considerable efforts have been made to formulate signal processing problems in terms of Fourier transforms so that the implementation advantages of the FFT could be exploited. As an example of this, we discuss the use of the FFT for implementing convolution, correlation, and FIR filtering operations in the frequency domain.

In Section 4.2 we introduce the concepts of homomorphic signal processing. Homomorphic processing is based on a generalized linearity or superposition principle and provides a method of separating multiplied or convolved signals. Homomorphic processing of convolved signals also relies on the FFT for efficient implementation.

In Section 4.3 the concepts and techniques of data compression are discussed. These techniques are widely employed to reduce redundancy in signals for storage or for transmission.

Finally, in Section 4.4, adaptive signal processing is discussed. This topic forms the basis for the implementation of a wide variety of processing. Indeed we have already introduced many of the basic concepts under the guise of recursive estimation. Here we shall treat the subject as a separate topic strongly dependent on the concepts from Chapter 3.

## 4.1 THE FAST FOURIER TRANSFORM

### 4.1.0 Introduction

In 1965 Cooley and Tukey published a technique for the efficient digital computation of discrete Fourier transform coefficients [CoTu, 65]. Their results exploited the symmetry properties of complex exponentiation such that the number of multiplications required for an $N$ point DFT was reduced from $N^2$ to $(N/2)\log_2 N$. This provided a dramatic decrease in machine execution time for the computation of DFTs for large $N$, and spawned a large class of computational techniques known as fast Fourier transform (FFT) algorithms.

In this section we discuss the radix two decimation in time, and decimation in frequency algorithms. We then examine some of the major techniques for utilizing the FFT to perform efficient digital signal processing calculations.

### 4.1.1 FFT Algorithms

Many variations of FFT algorithms have been developed for computing a set of $N$ DFT coefficients. In many of these, $N$ is constrained to be an integral power of two, however, algorithms are available for an arbitrary $N$ [10]. In this subsection we shall review two variations of the radix-two FFT algorithms for $N$ a power of two.

#### Decimation in Time

The defining equation for the DFT was given in Chapter 1 as equation (1.21). It is common to denote the complex exponential $e^{-j2\pi/N}$ of that equation as $W_N$. Thus, equation (1.21) can be written as

$$X_k = \sum_{n=0}^{N-1} x(n)\, W_N^{nk}, \quad k=0,1,\ldots,N-1 \tag{4.1}$$

A direct computation of the $N$ DFT coefficients $X_k$ by equation (4.1) requires $N^2$ complex multiplications and $N^2$ complex additions. The object of the FFT algorithm is to reduce the computational complexity. If the time domain sequence $x(n)$ is reordered (decimated) into two subsequences of odd and even indexed samples, then for $N$ a power of 2 equation (4.1) can be expressed as:

$$X_k = \sum_{n=0}^{(N/2)-1} x(2n)\, W_N^{2nk} + \sum_{n=0}^{(N/2)-1} x(2n+1)\, W_N^{(2n+1)k}, \quad k=0,\ldots,N-1 \tag{4.2}$$

$$= \sum_{n=0}^{(N/2)-1} x(2n)\, w_{N/2}^{nk} + W_N^k \sum_{n=0}^{(N/2)-1} x(2n+1)\, W_{N/2}^{nk}, \quad k=0,\ldots,N-1$$

where we have used the identity $W_{N/2}^k = W_N^{2k}$

We note

1. That these last two summations are equivalent to the $N/2$ point DFTs of the two $N/2$ point sequences made up of the even and odd indexed samples of the original $N$ point sample sequence. We shall denote these $N/2$-point DFT sequences as $X_e(k)$ and $X_o(k)$ respectively.
2. Since the DFT is a periodic sequence there are two periods of $X_e(k)$ and $X_o(k)$ represented in equation (4.2) as $K$ takes on values from 0 to $N-1$.

Furthermore, the symmetry of the factor $W_N$, sometimes known as the "twiddle factor," is such that

$$W_N^k = -W^{k+N/2}, k = 0, 1, \ldots, (N/2)-1$$

Based on these observations equation (4.2) can be written as

$$X_k = X_e(k) + W_N^k X_o(k), \quad k = 0, \ldots, (N/2)-1 \tag{4.3a}$$

$$X_{k+N/2} = X_e(k) - W_N^k X_o(k), \quad k = 0, \ldots, (N/2)-1 \tag{4.3b}$$

These equations represent the basic computational requirements for the FFT algorithm and are commonly known as the "decimation in time butterfly" operation due to the shape of the flow graph often used to represent this operation, as illustrated in Figure 4.1. The butterfly is a 2-point FFT. It represents a basic computational macro which must be designed in software or hardware. It requires one complex multiplication and two complex additions. Using the butterfly with the appropriate $W_N$, any $N$ point FFT can be designed.

This method of computing an $N$ point DFT from 2 $N/2$-point DFT sequences is called decimation in time because the original $N$-point time

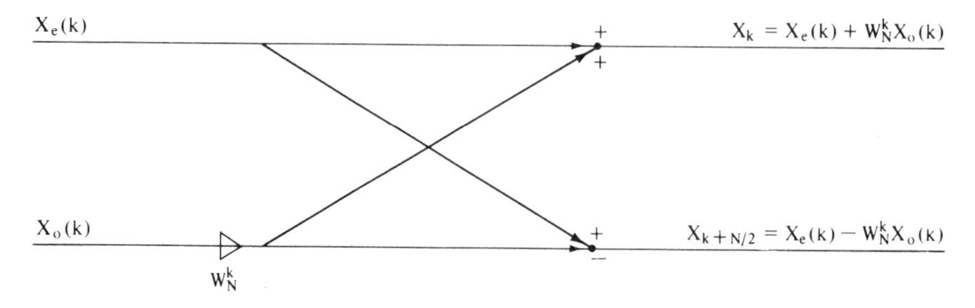


**Figure 4.1:** Basic butterfly operation for a decimation in time FFT.

domain sequence is decimated or reordered into two $N/2$ point sequences from which the two $N/2$ point DFT sequences $X_e(k)$ and $X_o(k)$ are computed. The computation of $X_e(k)$ and $X_o(k)$ can be accomplished in a similar manner by further decimating the time domain sequence into $N/4$ point sequences

$$x_{ee}(n) = x(4n),\ x_{eo}(n) = x(4n+2),\ n = 0, 1, \ldots, (N/4) - 1$$

$$x_{oe}(n) = x(4n+1),\ x_{oo}(n) = x(4n+3),\ n = 0, 1, \ldots, (N/4) - 1$$

from which four $N/4$ point DFT sequences are computed and combined to form $X_e(k)$ and $X_o(k)$.

The formation of $X_e(k)$ and $X_o(k)$ will be of the same form as equation (4.3), i.e.,

$$X_e(k) = X_{ee}(k) + W_N^{2k} X_{eo}(k),\ k = 0, 1, \ldots, \frac{N}{4} - 1 \qquad (4.4a)$$

$$X_e(k + N/4) = X_{ee}(k) - W_N^{2k} X_{eo}(k),\ k = 0, 1, \ldots, \frac{N}{4} - 1 \qquad (4.4b)$$

where $W_N^k/2 = W_N^{2k} \cdot X_{ee}(k)$ is the $N/4$ point DFT of $X_{ee}(n) = X(4n)$ and $X_{eo}(k)$ is the $N/4$ point of DFT of $X_{eo}(n) = X(4n+2)$. $X_o(k)$ is computed in a similar manner from $X_{oe}(k)$ and $X_{oo}(k)$.

This decomposition process is continued until the $N/2$ 2-point DFTs are computed directly from $N/2$ 2-point sequences, each consisting of two samples of the original sequence spaced $N/2$ samples apart. We note that the computation of a two-point DFT from equation (4.1) also has the form of a butterfly operation.

A complete flow graph for an 8-point radix-two decimation in time FFT is illustrated in Figure 4.2. The large arrowheads represent multiplications by twiddle factors and the nodes are addition/subtraction points. The small arrowheads indicate data flow directions.

From this graph we note two interesting points about the algorithm:

1. If all butterflies of pass 1 are computed before starting pass 2, then output data samples from every butterfly computation can be placed

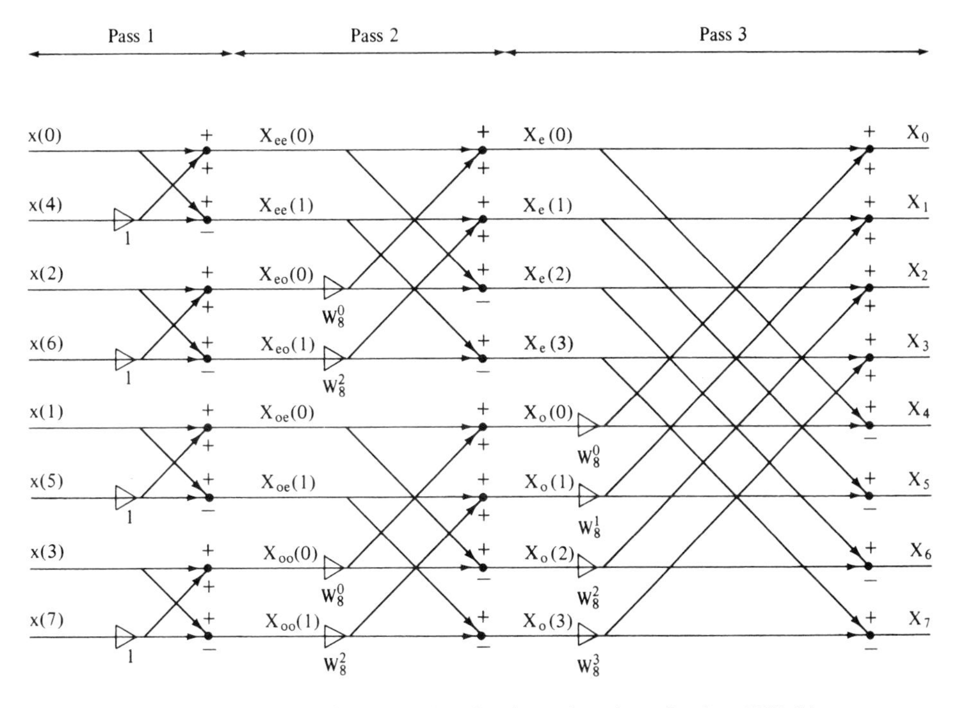


**Figure 4.2:** Radix 2 decimation in time, in place 8-point FFT bit reversed input; normal output.

back in the same memory locations from which the inputs were taken, since these input data points are not required for further computations. Thus, the computation of an $N$ point FFT can be accomplished using only $N$ memory locations. This is known as "in place" computation.

2. The order of the input data of Figure 4.2 is shuffled into what is known as "bit reversed order." That is, if the binary digit representation of the sample indices is complemented, then the resulting organization of the input data is as shown in Figure 4.2. Figure 4.3 illustrates bit reversal.

| Normal Index Order | | Bit Reversed Index Order | |
|---|---|---|---|
| Decimal | Binary | Binary | Decimal |
| 0 | 000 | 000 | 0 |
| 1 | 001 | 100 | 4 |
| 2 | 010 | 010 | 2 |
| 3 | 011 | 110 | 6 |
| 4 | 100 | 001 | 1 |
| 5 | 101 | 101 | 5 |
| 6 | 110 | 011 | 3 |
| 7 | 111 | 111 | 7 |

**Figure 4.3:** Bit reversal.

Figure 4.4 illustrates an alternate flow graph for an 8-point in place decimation in time algorithm. This flow graph shows the inputs in normal order and the outputs in bit reversed order. This graph can be constructed from that of Figure 4.2 by following each step in sequence.

The radix 2 decimation in time algorithm is based on the observation that $N$ complex DFT coefficients for an $N$ point sample sequence can be computed from two $N/2$ point DFT sequences with only $N/2$ complex multiplications and $N$ complex additions. Similarly the $N/2$-point sequences can each be computed from two $N/4$ point DFT sequences with only $N/4$ complex multiplications and $N/2$ complex additions. Assuming $N$ is an integral power of 2 this decomposition can be repeated until $N/4$ 4-point DFTs are to be derived from $N/2$ 2-point DFTs. The initial $N/2$ 2-point DFTs can be computed directly from equation (4.1).

For each stage or pass of the algorithm $k$ DFT sequences of length $N/k$ are computed from 2 DFT sequences of length $N/2k$. The computation of each $N/2k$ point DFT requires $N/2k$ complex multiplication and $N/k$ complex additions. Thus the total computational requirement for each pass is $N/2$ complex multiplication and $N$ complex additions. Since $N$ is a power of 2 a total of $\log_2 N$ passes are required to compute the final $N$ point DFT. Thus the complete algorithm requires only $(N/2)\log_2 N$ complex multiplications and $N\log_2 N$ complex additions.

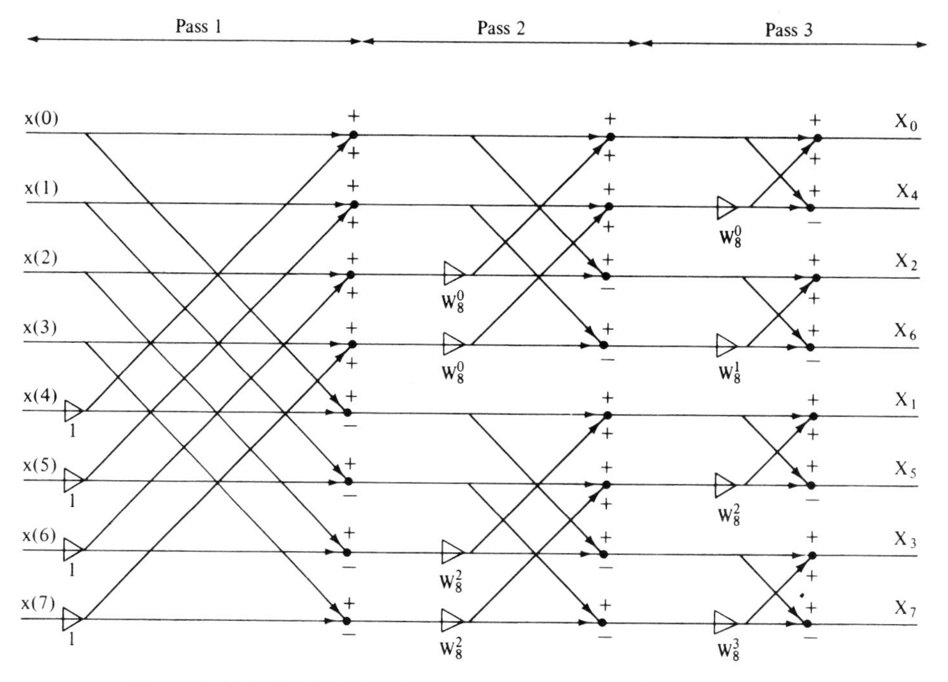


**Figure 4.4:** Radix 2 decimation in time, in place 8-point FFT normal ordered input; bit reversed output.

Since the multiplications and additions of the FFT algorithm are complex, the algorithm is ideally suited to handle complex input sequences. In transforming real data sequences an additional computational saving can be realized by combining two real $N$-point sequences $x(n)$ and $y(n)$ into a single $N$-point complex sequence.

$$z(n) = x(n) + jy(n) \tag{4.5}$$

Since the DFT is a linear operation we note that

$$DFT\{z(n)\} = Z_k = X_k + jY_k \tag{4.6}$$

and from the properties of the DFT listed in Chapter 1 we have

$$DFT\{z^*(n)\} = Z_k^* = X_k - jY_k \tag{4.7}$$

Thus we can obtain the sequences $X_k$ and $Y_k$ as

$$X_k = \frac{Z_k + Z_k^*}{2} \tag{4.8}$$

and

$$Y_k = \frac{Z_k - Z_k^*}{2} \tag{4.9}$$

This ability to combine two real sequences into a single complex sequence allows the computation of two $N$-point DFT sequences with a single $N$-point FFT operation plus the added $2N$ complex additions and divisions (shifts) to "unravel" the results.

The same technique can be used to compute the DFT of an $N$ point real sequence $x(n)$ with an $N/2$-point FFT by forming the $N/2$-point complex sequence $z(n) = x(2n) + jx(2n + 1)$. The result of unravelling the sequence $Z_k$ is two $N/2$ point sequences $X_e(k)$ and $X_o(k)$, $k = 0, 1, \ldots, N/2 - 1$. These two sequences can be combined according to equation (4.3) to produce the final $N$-point sequence $X_k$, $k = 0, \ldots, N - 1$.

We note finally that the similarity of the IDFT to the DFT allows the same algorithm to be used to compute both forward and reverse FFTs simply by altering the twiddle factors.

### Decimation in Frequency

A second common FFT algorithm involves the radix 2 decimation in frequency. This algorithm is very similar to the decimation in time version in that it also requires $N$ to be a power of 2 and is based on reducing an $N$-point DFT to the combination of two $N/2$-point DFTs. The term decimation in frequency is applied because it is the frequency domain samples that are reordered into even and odd subsequences of the index $k$ in equation (4.1). These sequences correspond to the DFTs of the $N/2$-point data sequences representing the sum of the first and second halves of the original $N$-point input sequence and the difference of the first and second halves multiplied by $W_N$.

The combination of the two $N/2$-point DFTs to form an $N$-point DFT has the form

$$X_k = X_1(k) + X_2(k), \quad k=0, 1, \ldots, (N/2) - 1 \tag{4.10a}$$

$$X_{k+N/2} = (X_1(k) - X_2(k))\, W_N^k, \quad k=0, 1, \ldots, (N/2) - 1 \tag{4.10b}$$

where

$$X_1(k) = DFT\{x(n) + x(2n + N/2)\}, \quad n = 0, \ldots, (N/2) - 1$$

and

$$X_2(k) = DFT\{x(n) - x(2n + N/2))\, W_N^2\}, \quad n = 0, \ldots, (N/2) - 1$$

These equations define the decimation-in-frequency butterfly operations as illustrated in Figure 4.5. Figures 4.6 and 4.7 illustrate the complete flow graphs for an 8-point radix-2 decimation-in-frequency FFT for normal ordered inputs and bit reversed inputs respectively.

### 4.1.2 Fast Convolution

The computational efficiency of the FFT can be exploited in the implementation of convolution and correlation operations by a technique known as fast convolution. This technique takes advantage of the fact that the product of two $N$-point DFT sequences $X_k$ and $H_k$ is equal to the DFT of the periodic convolution of the two $N$-point sequences $x(n)$ and $h(n)$, i.e.,

$$X_k H_k = Y_k \tag{4.11}$$

where

$$Y_k = DFT\{y(n)\}, \quad n = 0, \ldots, N - 1 \tag{4.12}$$

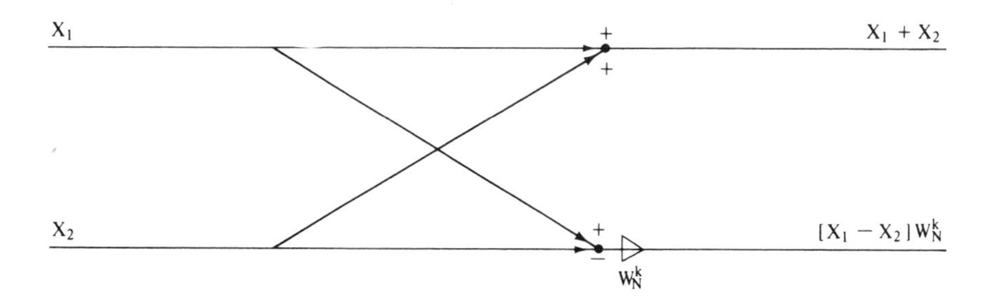


Figure 4.5: Basic butterfly operation for a decimation in frequency FFT.

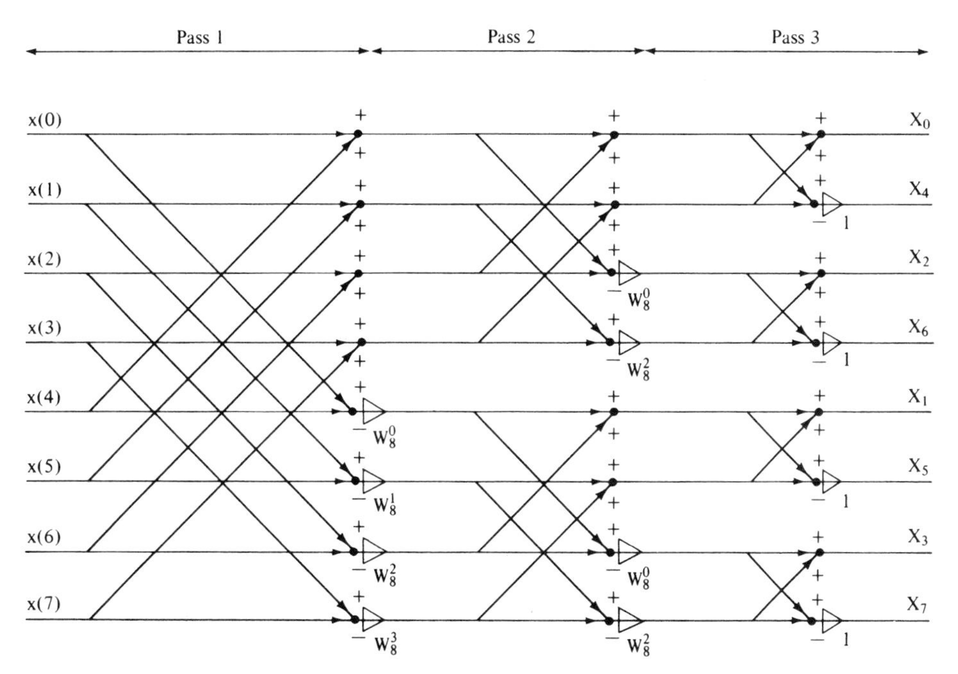


**Figure 4.6:** Radix 2 decimation in frequency, in place, 8-point FFT normal ordered input; bit reversed output.

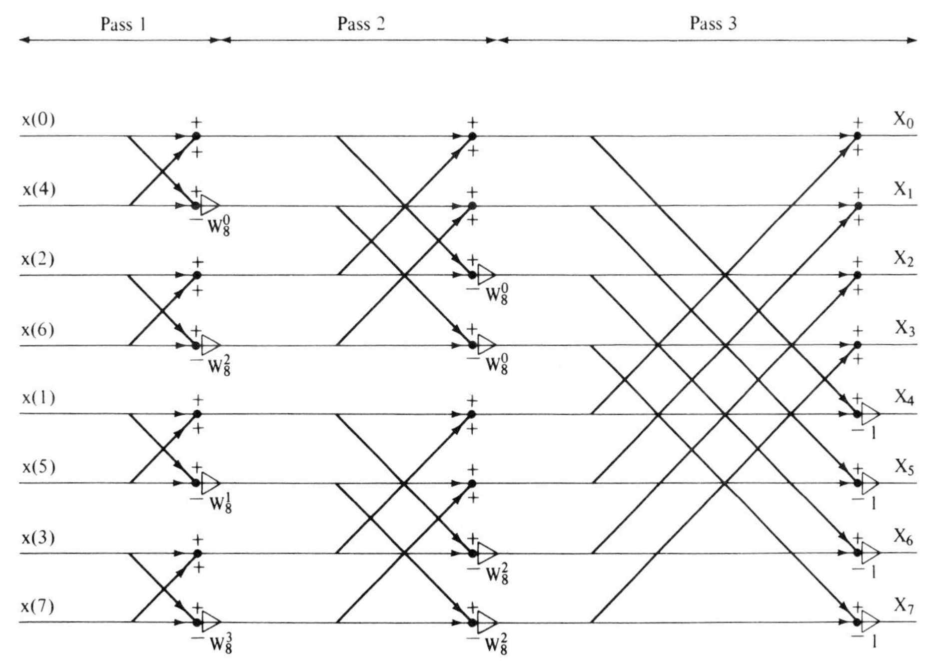


**Figure 4.7:** Radix 2 decimation in frequency, in place, 8-point FFT normal ordered output; bit reversed output.

and
$$y(n) = x(n) \circledast h(n) = \sum_{i=0}^{N-1} x_p(i) h_p(n-i)$$

where $x_p(n)$ and $h_p(n)$ are periodic extensions of $x(n)$ and $h(n)$, i.e.,

$$x_p(n + kN) = x(n), \; n = 0, \ldots, N-1, \; k \text{ any integer}$$

$$h_p(n + kN) = h(n), \; n = 0, \ldots, N-1, \; k \text{ any integer}$$

Thus, $y(n)$ is a periodic sequence of period $N$ such that

$$y(n + kN) = y(n), \; n = 0, \ldots, N-1, \; k \text{ any integer} \quad (4.13)$$

Obtaining the periodic convolution of two finite sequences of length $N$ using the FFT requires two $N$-point forward FFTs, the forming of the product of the resulting DFT sequences, and an $N$-point reverse FFT operation. This is known as the "fast convolution technique."

To obtain the linear convolution of two finite sequences the same basic technique is applied with a slight variation. Given two finite sequences $x(n)$, $n = 0, \ldots, N_1 - 1$, and $h(n)$, $n = 0, \ldots N_2 - 1$, their linear convolution

$$y(n) = x(n) * h(n) = \sum_{k=0}^{N-1} x(k) h(n-k), \; N \geqslant (N_1 + N_2 - 1) \quad (4.14)$$

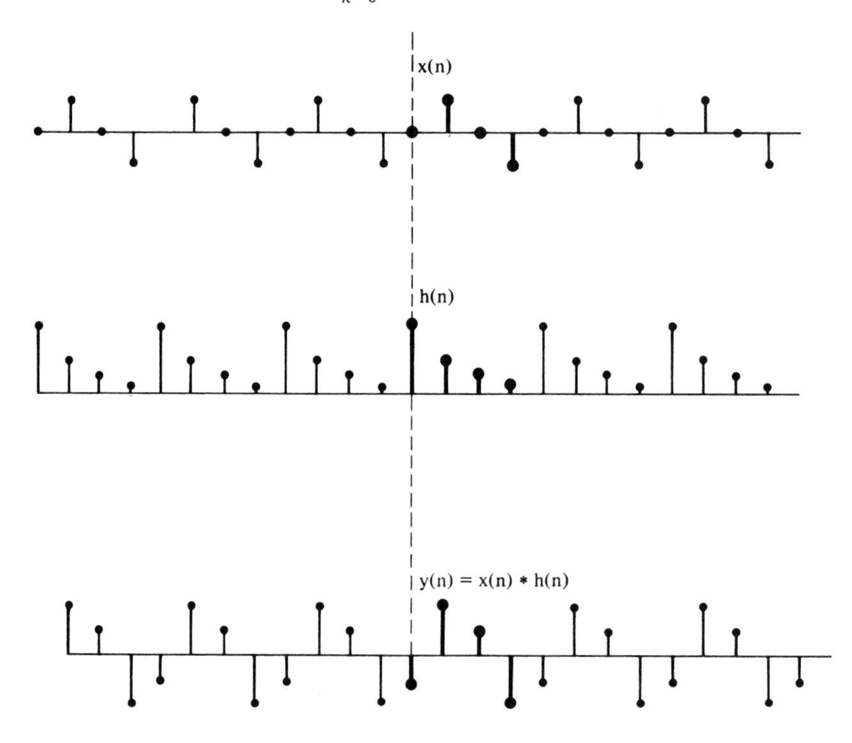


**Figure 4.8:** Periodic convolution.

is an $(N_1 + N_2 - 1)$-point sequence. To compute $y(n)$ using the radix two FFT we zero extend both $x(n)$ and $h(n)$ to a length $N$ that is the smallest power of two such that $N \geqslant N_1 + N_2 - 1$. The periodic convolution of these two extended sequences can then be computed by the fast convolution technique as described above. The resulting sequence will consist of $(N_1 + N_2 - 1)$ non-zero values that are the linear convolution of $x(n)$ and $h(n)$ followed by $N - (N_1 + N_2 - 1)$ zeros. If an FFT algorithm is used

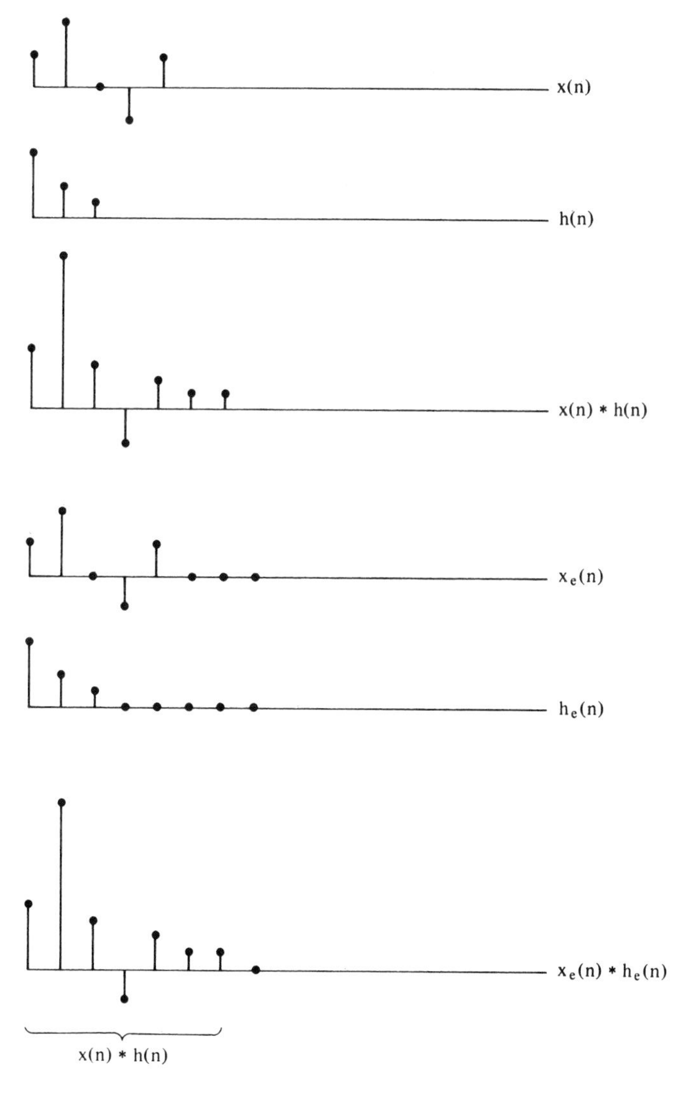


**Figure 4.9:** Equivalence of periodic convolution of zero extended sequences to linear convolution.

that is not constrained to have $N$ a power of two then the sequences need only be extended to a length of $(N_1 + N_2 - 1)$ and the result of the fast convolution will be the desired linear convolution.

To see that this technique is an efficient method of implementing a convolution consider the direct application of equation (4.14) which requires $N_1 N_2$ multiplications and $(N_1 - 1)(N_2 - 1)$ additions. The fast convolution techniques using a radix two FFT requires three $N$-point FFTs, two forward and one reverse. These three FFT operations require $(3N/2)\log_2 N$ complex multiplications and $3N\log_2 N$ complex additions. In addition $N$ complex multiplications are required to form the product, giving a total of $(3N/2)\log_2 N + N$ complex multiplications and $3N\log_2 N$ complex additions.

For sequences of the order of one thousand points the direct convolution method required roughly one million multiplications and additions. The fast convolution technique requires only about 35 thousand complex multiplications and 68 thousand complex additions. If the data sequences are real they can be combined as described in the previous subsection to further reduce the number of operations required for the fast convolution.

The technique of fast convolution can be employed to perform fast correlation as well, since correlation is equivalent to convolution with one sequence time reversed.

### 4.1.3 Frequency Domain Filtering

In some situations such as the filtering of very long sequences the direct application of the FFT to perform convolution as outlined in the previous subsection is not practical. The transform size and storage requirements may get unreasonably large. For real-time applications the allowable processing delay may also limit the length of input signal that can be handled. There are two common approaches to the use of the FFT for performing continuous convolution of very long sequences. Both are based on the segmentation of the input sequence into sections of manageable length, performing finite sequence fast convolution and suitably combining outputs.

#### Overlap Add Method

Assume $h(n)$ is a finite impulse response of length $N_1$ and $x(n)$ is a very long input sequence which is segmented into non-overlapping segments of length $N_2 > N_1$. Denoting the $i$th section of $x(n)$ as $x_i(n)$, i.e.,

$$x_i(n) = \begin{cases} x(n), & iN_2 \leqslant n \leqslant (i+1)N_2 - 1 \\ 0 & \text{otherwise} \end{cases} \tag{4.15}$$

we can write
$$x(n) = \sum_{i=-\infty}^{\infty} x_i(n) \tag{4.16}$$

To complete the filter output sequence

$$y(n) = \sum_{i=0}^{N_1-1} h(i)x(n-i) \tag{4.17}$$

we first compute the output segments

$$\begin{aligned} y_i(n) &= x_i(n) * h(n) \\ &= \sum_{k=0}^{N_1-1} h(k)x_i(n-k) \end{aligned} \tag{4.18}$$

The outcome segments $y_i(n)$ can be computed using the fast convolution technique of the previous subsection. The output from this technique will contain $(N_1 + N_2 - 1)$ non zero samples. The final output sequence $y(n)$ is created by concatenating the segments $y_i(n)$ [each of length $N > (N_1 + N_2 - 1)$] such that the last $N_1 - 1$ non-zero points of segment $y_i(n)$ are summed (overlapped and added) with the first $N_1 - 1$ points of the segment $y_{i+1}(n)$. This process is illustrated in Figure 4.10.

**Overlap Save Method**

A second method known as the overlap save method involves performing a periodic convolution of the impulse response $h(n)$, of length $N_1$, with the overlapped segments of the input sequence $x(n)$. The impulse response $h(n)$, of length $N_1$, is zero extended to length $N$ for periodic convolution with input segments $x_i(n)$ of length $N > N_1$. The first $(N_1 - 1)$ points of the periodic convolution are incorrect but the remaining $N_2$ points $(N_2 = N - N_1 + 1)$ are the same as would be obtained from a normal linear convolution. Thus, we segment the input such that successive segments overlap by $N_1 - 1$ points. That is, the last $(N_1 - 1)$ points of section $x_i(n)$ are the same as the first $(N_1 - 1)$ points of section $x_{i+1}(n)$. The first $(N_1 - 1)$ points of each output section $y_i(n)$ are discarded and the remaining $N_2$ points are concatenated with the previous outputs to form the desired sequence $y(n)$. This method is illustrated in Figure 4.11.

### 4.1.4 Finite Word Length Effects in FFT

In Chapter 1 we introduced the concept of quantization errors in digital signal representation which are due to the finite number of bits used to represent sample values. In addition to this initial error introduced when the signal is converted to digital form, errors are introduced during processing. The quantization of coefficient values in both digital filtering operations and

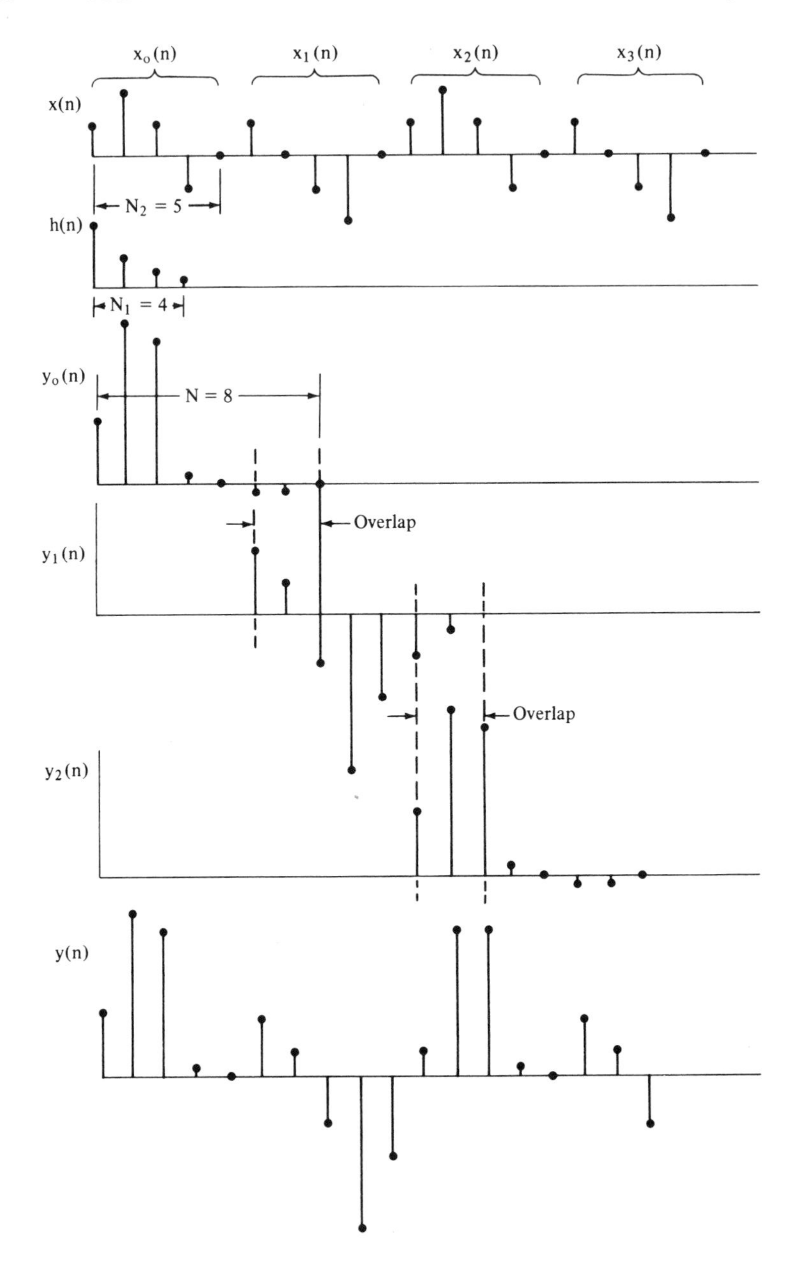


**Figure 4.10:** Overlap add.

FFT computations introduces further errors in the results. However, coefficient quantization errors are generally not the major error source in processing digital signals. Depending on the application requirements, coefficient word lengths can usually be chosen, which will provide sufficient accuracy,

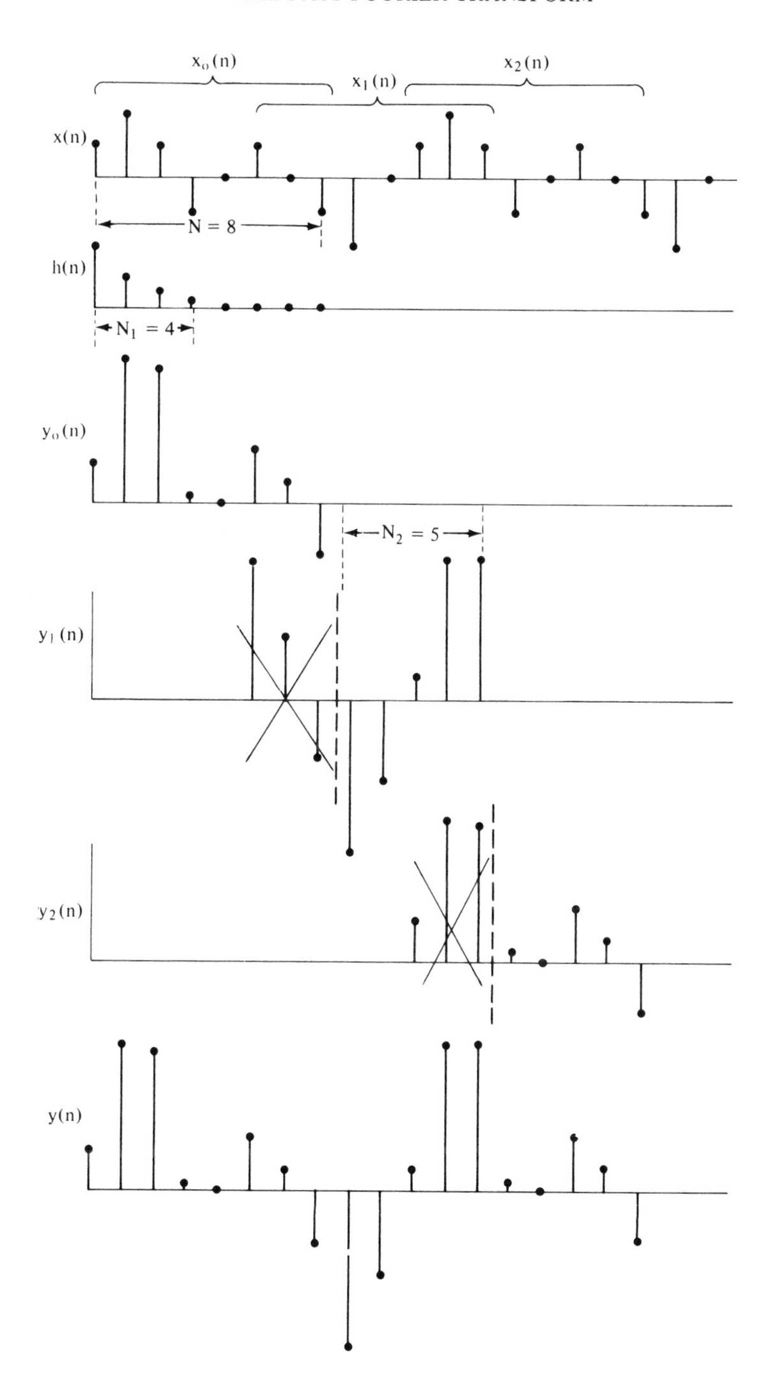


**Figure 4.11:** Overlap save method.

and which are equal to or less than the sample word lengths. The major source of processing errors is in the round-off or truncation of arithmetic operation results, particularly products. The effects of finite word length are well documented in the literature and the reader is referred to the references for details. Here we present only a brief overview of some major concerns with respect to the FFT.

There are two major areas of concern with finite word lengths in the computation of DFT coefficients using the FFT: processing gain, and quantization errors. By processing gain, we mean the general increase in the magnitude of array values throughout the processing which may cause overflow for fixed point number representation.

From Parseval's theorem we have

$$\sum_{n=0}^{N-1} |x(n)|^2 = \frac{1}{N} \sum_{k=0}^{N-1} |X_k|^2 \tag{4.19}$$

which implies that in general the magnitude of the DFT coefficients will be considerably larger than the magnitude of input samples. To accommodate this increase in dynamic range of numeric values some form of scaling mechanism must be provided for fixed point implementations, or floating point arithmetic must be used.

It can be shown [10] that the maximum value of any intermediate result at pass $i + 1$ of a radix 2 FFT is greater than the maximum value at pass $i$ and less than twice the maximum value at pass $i$. Thus, scaling the results down by a factor of two (right shift) after each pass guarantees that overflow will not occur. However, this approach reduces the accuracy of the output since scaling is always performed even if overflow would not have occurred at the next pass. Various schemes for scaling only when necessary have been devised [10]. If floating point arithmetic is used the overflow problem does not generally arise.

As noted above there are two basic sources of error due to quantization, the quantization of coefficients and the rounding or truncation of products. The major source of errors in the FFT is due to the rounding of products. In general the effects of rounding and truncation are analyzed in a statistical manner that results in a theoretical upper bound on the errors expressed in terms of a root mean squared error to signal ratio. For fixed point FFT implementation this upper bound is [10]:

$$\frac{\text{rms error}}{\text{rms output signal}} = \frac{\sqrt{N}\, 2^{-b}(0.3)\sqrt{8}}{\text{rms input}} \tag{4.20}$$

where $N$ is the transform size and $b$ is the word length in bits. Thus, the upper bound increases as $\sqrt{N}$, or 1/2 bit per pass. From this expression a suitable word length $b$ can be determined for a given transform size $N$ to provide the desired rms noise to rms signal output.

The effects of coefficient quantization have been shown to be relatively insensitive to transform size with the error variance increasing very slowly with $N$ [9].

### 4.1.5 Summary

In this section we have examined the basic concepts of the efficient computation of DFT coefficients by the FFT algorithm. The advent of the FFT algorithm has been one of the major influencing factors in the evolution of digital signal processing over the past 15 years. We note, however, that this reduction in computational complexity is accompanied by an increase in data handling and control complexity compared to the simple sum of products form of the original DFT-equation.

We have also examined how the FFT can be utilized to perform fast convolution and implement FIR filters in the frequency domain. Finally, we looked at the effect of finite word lengths on the FFT.

In the next section we shall examine a processing technique that exploits the efficiency of the FFT to implement the separation of convolved signals.

## 4.2 GENERALIZED LINEAR PROCESSING

### 4.2.0 Introduction

In general, nonlinear systems do not yield tractable models upon which to base processor design. A certain class of nonlinear signals do however yield to a model which is both successful and tractable. Signals with amplitude modulation are widely used and can be represented as a multiplication of a carrier and the modulation signal. The resulting signal can be modelled and analysed by a generalized linear model. These concepts and techniques are now introduced.

### 4.2.1 Generalized Linearity

To a large extent the digital processing of signals revolves around linear systems theory. As we saw in Chapter 2, a linear system obeys a superposition principle such that

$$L[x_1(n) + x_2(n)] = L[x_1(n)] + L[x_2(n)] \tag{4.21}$$

and

$$L[cx(n)] = cL[x(n)] \tag{4.22}$$

where $x_1(n)$ and $x_2(n)$ are any two inputs to the linear system $L[\ ]$ and $c$ is a scalar constant.

A generalized principle of superposition can be derived that is fulfilled by certain non-linear systems[6]. We denote $\square$ as an operation for combining inputs and $\bigcirc$ as an operation for combining outputs from a system. Define the operation $:$ to combine inputs with scalars and the operation $\rceil\!\lfloor$ to combine outputs with scalars. A generalized principle of superposition can then be said to hold if

$$T[x_1(n)\square x_2(n)] = T[x_1(n)]\bigcirc T[x_2(n)] \tag{4.23}$$

and

$$T[c:x_1(n)] = c\rceil\!\lfloor T[x_1(n)] \tag{4.24}$$

If $T$ is a linear operation then both $\square$ and $\bigcirc$ represent addition, and both $:$ and $\rceil\!\lfloor$ represent multiplication. However, if $T$ represents a non-linear transformation, e.g.,

$$T[x_1(n)] = \log x_2(n) \tag{4.25}$$

then where $\square$ representing multiplication, $\bigcirc$ must represent addition, i.e.,

$$\begin{aligned} T[x_1(n)\ \square\ x_2(n)] &= \log[x_1(n)\ \square\ x_2(n)] \\ &= \log[x_1(n)] + \log[x_2(n)] \\ &= T[x_1(n)]\ \bigcirc\ T[x_2(n)] \end{aligned} \tag{4.26}$$

Similarly, if $:$ is exponentiation, i.e.,

$$c:x_1(n) = [x_1(n)]^c \tag{4.27}$$

then $\rceil\!\lfloor$ must represent multiplication, i.e.,

$$\begin{aligned} T[c:x_1(n)] &= \log[x_1(n)] = c\log x_1(n) \\ &= cT[x_1(n)] \end{aligned} \tag{4.28}$$

Figure 4.12 illustrates a typical homomorphic system for performing generalized linear processing with the system $T_\square[\ ]$ representing a logarithmic transformation while the system $T_\bigcirc^{-1}[\ ]$ represents exponentiation. Such systems can be usefully employed to separate multiplied and convolved signals such that the individual signal components can be processed separately.

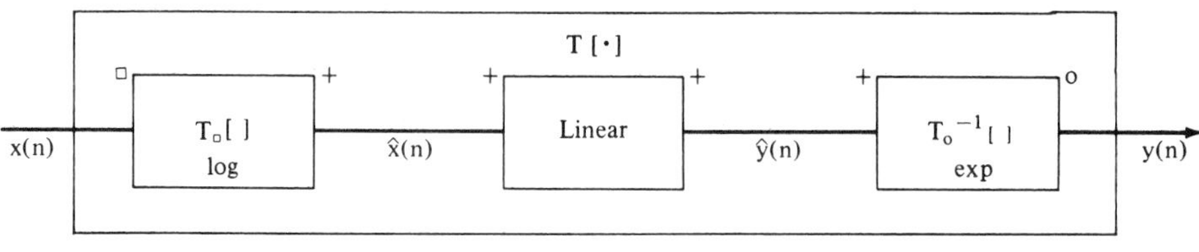


$x(n) = e(n)\ s(n)$
$x(n) = \log[e(n)] + \log[s(n)]$
$y(n) = \log[e(n)]$ or $\log[s(n)]$
$y(n) = e(n)$ or $s(n)$

**Figure 4.12:** Typical representation of homomorphic systems.

### 4.2.2 Multiplicative Signals

Consider the simple signal model for a multiplicative signal

$$s(n) = e(n)v(n) \tag{4.29}$$

where $e(n)$ and $v(n)$ are both constrained to be non-negative. By taking the logarithm of this signal, a signal representation is obtained that expresses the components of $s(n)$ in an additive form, i.e.,

$$\log[s(n)] = \log[e(n)] + \log[v(n)] \tag{4.30}$$

If the signal components $e(n)$ and $v(n)$ have minimal overlap of their respective spectra, such as in the case of a low frequency envelope modulating a higher frequency carrier, then the two spectra can be separated by linear filtering of $\log[s(n)]$. The filtered signal can then be exponentiated to return it to its original domain. (This represents the essence of homomorphic processing). Figure 4.12 illustrates this basic model.

In general the signal components $e(n)$ and $v(n)$ will not be strictly non-negative. In order to deal with non-negative signals it is necessary to use the complex logarithm.

### 4.2.3 The Complex Logarithm

If we denote the general complex sequence as

$$s(n) = |s(n)|e^{jarg[s(n)]} \tag{4.31}$$

then the complex logarithm of $s(n)$ is defined as

$$\log[s(n)] \leqslant \log|s(n)| + jarg[s(n)] \tag{4.32}$$

and the inverse of $\log[s(n)]$ is the complex exponential

$$\epsilon^{\log[s(n)]} = \epsilon^{\log|s(n)|}\epsilon^{jarg[s(n)]} \tag{4.33}$$

We note the existence of an ambiguity in this definition of the complex logarithm in that any integral multiple of $2\pi$ may be added to the imaginary part without changing the result of the inverse of $\log[s(n)]$. Thus, the inverse operation is not unique, as it must be to satisfy an overall system definition in which generalized linearity or superposition holds. That is, the general system of Figure 4.13 must provide a unique output.

For $\log[s(n)]$ to be defined such that generalized superposition holds we require that if

$$s(n) = e(n)v(n) \tag{4.34}$$

then

$$\log[s(n)] = \log[e(n)] + \log[v(n)] \tag{4.35}$$

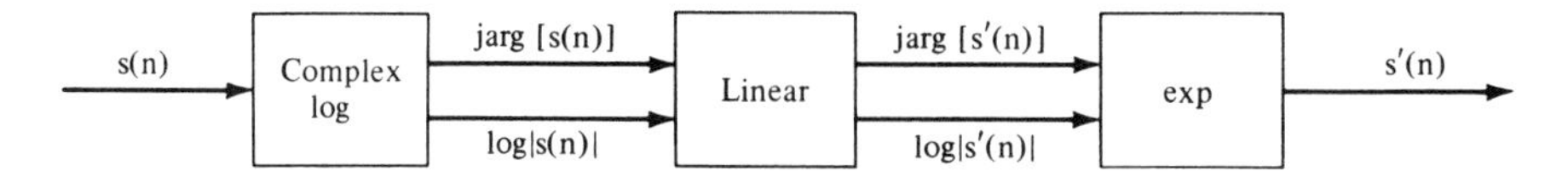


**Figure 4.13:** Basic homomorphic processing of multiplicative signals.

This implies that

$$\log|s(n)| = \log|e(n)| + \log|v(n)| \tag{4.36}$$

and

$$arg[s(n)] = arg[e(n)] + arg[v(n)] \tag{4.37}$$

A common method of resolving the ambiguity is to replace $arg[s(n)]$ by its principal value (i.e., its value modulo 2). However, for such a definition, equation (4.37) does not generally hold. Thus, a definition of $arg[x]$ is required such that $arg[x]$ is a continuous function of $x$. An approach to this problem is to assume that the continuous complex logarithm is obtained by integrating its derivative(6).

For real signals the imaginary part of the complex logarithm will be either 0 or $\pi$ and the ambiguity is not generally a problem.

### 4.2.4 Convolved Signals

Consider the processing of a convolved signal

$$s(n) = e(n) * v(n) \tag{4.38}$$

by homomorphic means. We first compute the Fourier transform of $s(n)$.

$$\mathfrak{F}\{s(n)\} = S(\epsilon^{j\omega}) = E(\epsilon^{j\omega})V(\epsilon^{j\omega}) \tag{4.39}$$

In the frequency domain we have a multiplicative relationship that can be dealt with by multiplicative homomorphic processing techniques, i.e.,

$$\log|S(\epsilon^{j\omega})| = \log|E(\epsilon^{j\omega})| + \log|V(\epsilon^{j\omega})| \tag{4.40}$$

We noted in Section 4.2.2 that if the spectra of $e(n)$ and $v(n)$ did not overlap they could be separated by conventional linear filtering of $\log[s(n)]$. For the present situation we are concerned with the spectrum of the log magnitude spectrum of $s(n)$, which is known as the "cepstrum", i.e.,

$$\mathfrak{F}\{\log|S(\epsilon^{j\omega})|\} = \mathfrak{F}\{\log|E(\epsilon^{j\omega})|\} + \mathfrak{F}\{\log|V(\epsilon^{j\omega})|\} \tag{4.41}$$

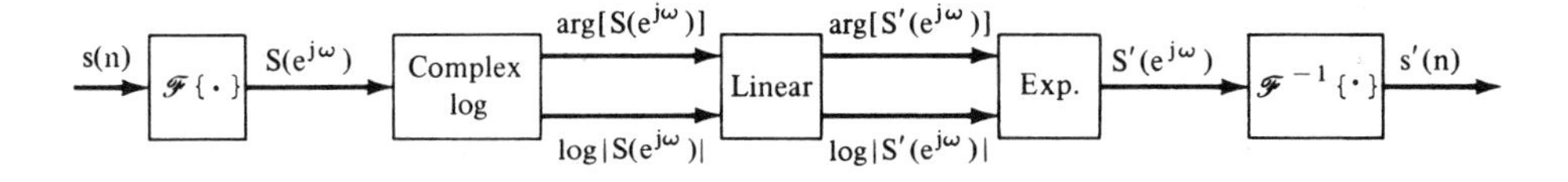


**Figure 4.14:** Basic homomorphic processing of convolved signals.

The cepstrum may be considered as a measure of the periodicity of the spectrum of $s(n)$. If $s(n)$ is a signal containing echoes or reverberation, etc., the cepstrum will exhibit peaks at the repetition rate. Thus, the cepstrum is a kind of "event time" measure that shows the repetition time of events in the signal. These repetition peaks can be identified from the cepstrum and the echoes or reverberation can then be removed by filtering.

If the signal $s(n) = e(n) * v(n)$ is considered as the output of a linear system with frequency response $E(e^{j\omega})$, then the signal may be deconvolved by subtracting $\log|E(e^{j\omega})|$ from $\log|S(e^{j\omega})|$ if we can identify the frequency response $E(e^{j\omega})$.

### 4.2.5 Summary

In this section we have briefly introduced the elementary concepts of generalized linear processing or homomorphic processing. We have shown how this general technique can be utilized to separately process the components of multiplicative and convolved signals.

The homomorphic processing of convolved signals led to the definition of the cepstrum of a signal as the Fourier transform of the log magnitude spectrum of the signal. The concept of the cepstrum is useful in such applications as echo removal and deconvolution processing.

## 4.3 DATA COMPRESSION TECHNIQUES

### 4.3.0 Introduction

The transmission of digital signals is an important activity in several areas of processing such as speech, telecommunications and image processing. Associated with the transmission of digital data is a class of signal processing techniques for the reduction of data rates. In general these techniques involve the modulation, multiplexing and encoding of digital signals in a manner that attempts to maximize the information content per bit of transmitted data.

In Chapter 1 we discussed the general principles of analog to digital conversion. In that discussion we considered the representations of signals as a sequence of finite length digital numbers representing the signal amplitude at equally spaced points in time. This form of signal encoding is generally known as "pulse code modulation" (PCM) in connection with data transmission. To transmit such a signal as a serial bit stream requires a transmission bit rate of at least the number of bits per sample times the sample rate of the signal. When this bit rate is sufficiently below the capacity of the transmission channel it is common to interleave the samples of several signals onto a single channel in a fixed order. This technique is known as "time division multiplexing."

The bit rate required to transmit a specific signal is also affected by the requirement to transmit synchronization, error detection and error correction information to ensure proper reception of the signal. The details of the techniques for synchronization, error detection and correction encoding are major considerations in digital data communications. We shall not pursue these topics further but simply note their effect is to increase the required transmission bit rate. Further information on these topics has been provided in the references.

If a PCM-encoded signal is represented by fixed-point binary numbers there is a fundamental trade-off between the required transmission bit rate, the dynamic range of the signal and the signal to quantization noise ratio. If a given signal to noise ratio or quantization granularity is to be maintained then the sample word length must be increased if the dynamic range of the signal is to be increased. This in turn implies an increased bit rate to transmit the signal at the same sample rate. In this section we shall review a number of techniques for achieving lower bit rates while attempting to maintain desired dynamic range and SNR.

### 4.3.1 Companding

The first technique we shall examine to accomplish data rate reduction is known as "companding." Essentially this technique involves the compression of the dynamic range of the signal by passing the signal through a non-linear device prior to analog to digital conversion. These devices usually have a logarithmic characteristic for high signal amplitudes and a linear characteristic for small signal amplitudes. The reason for this type of characteristic is to minimize the loss in signal to quantization noise ratio for small amplitude signals. Upon reception of the transmitted signal it is passed through another non-linear device with an inverse transfer characteristic to restore the original dynamic range of the signal.

An example of a companding characteristic is the $\mu$ law.

$$C(x) = Sgn(x)\frac{1n(1 + \mu|x|)}{1n(1 + \mu)}, \quad 1 \geqslant x \geqslant -1 \tag{4.42}$$

where $C(x)$ is the companded signal and $x$ is the input signal. The compression factor $\mu$ controls the amount of compression by maintaining a constant but lower value signal to noise ratio over a larger dynamic range as $\mu$ is increased. The overall characteristic is linear for small signal values but logarithmic for larger values, with the transition level determined by the factor $\mu$. Figure 4.15 illustrates such a linear-logarithmic companding characteristic.

Companding may be done on a sample by sample basis or on a block floating point basis, sometimes called "syllabic companding."

The homomorphic processing techniques of the previous section can be employed to perform companding digitally. Consider a signal that is

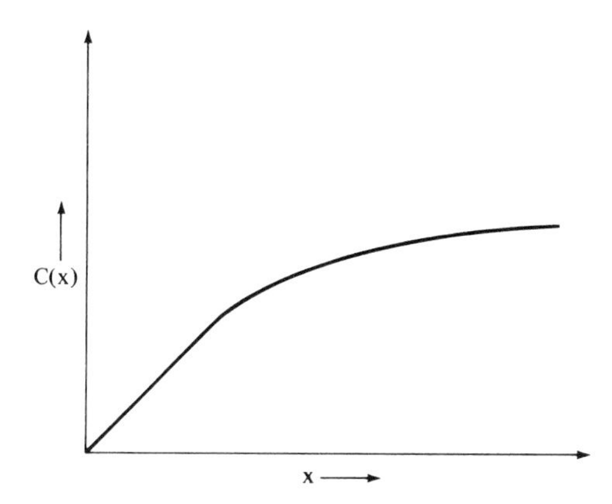


Figure 4.15: Linear-logarithmic companding characteristic.

modelled as a slowly varying envelope modulating a higher frequency carrier, i.e.,

$$s(n) = e(n)v(n) \tag{4.43}$$

Taking the logarithm of this signal we can write

$$\log|s(n)| = \log|e(n)| + \log|v(n)| \tag{4.44}$$

The typical log-magnitude spectrum for an audio band signal due to [Stoc 68] is illustrated in Figure 4.16. The wide dynamic range of the signal is represented by the low frequency portion of the log spectrum corresponding to $e(n)$. A linear filtering operation on the signal log $|s(n)|$ that has a unity gain above 16 Hz and a gain of G below 16 hz will affect only the log $|e(n)|$ portion of the signal. If G is less than one, the envelope amplitude is reduced and the dynamic range is compressed. A block diagram of this system is illustrated in Figure 4.17. We note also that the same system can be used to expand the dynamic range by setting G greater than one. If G is zero then the system operates as an automatic gain control.

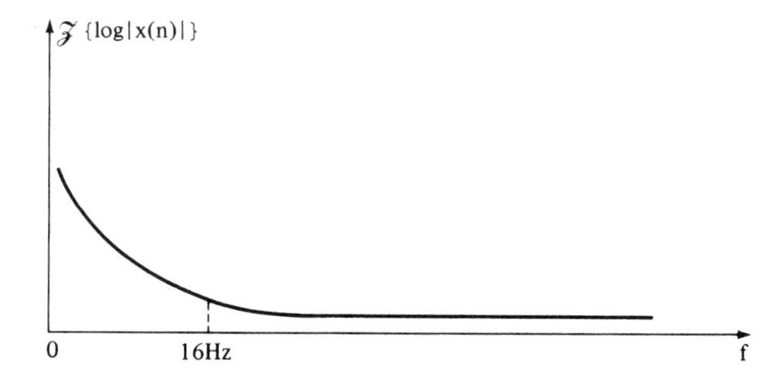


Figure 4.16: Typical audio log spectrum.

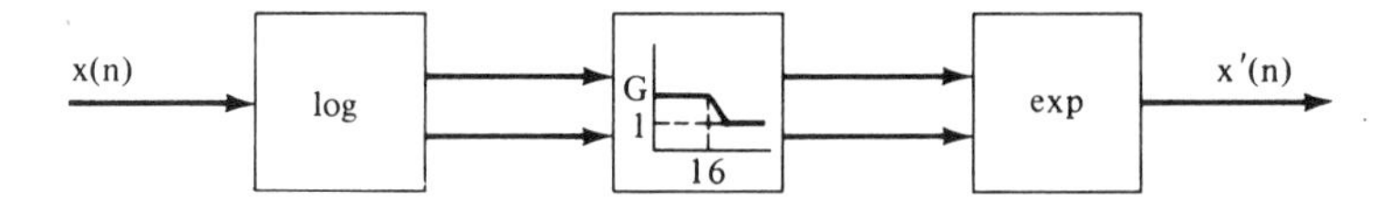


**Figure 4.17:** Homomorphic companding. © Alan V. Oppenheim, *Applications of Digital Signal Processing*, 1978, pp. 100-101. Reprinted by permission of Prentice-Hall, Inc., Englewood Cliffs, N.J.

## 4.3.2 Floating Point PCM

A simple method of increasing the dynamic range of a PCM signal, without increasing quantization granularity for small signals, or dramatically increasing bit rate, is to use a floating point encoding scheme. The basic concepts of a floating point PCM encoder are illustrated in Figure 4.18.

The conversion of a wide dynamic range input signal to floating point PCM is accomplished by passing each analog sample through a prescaler that attempts to maintain the sample voltage at the input to the ADC in the upper half of its input voltage range. This ensures that the maximum number of bits will be used to represent the sample value. The prescale control compares the input signal level with fixed references for the maximum and 1/2 scale values. If the sample value is below half scale the sample is successively prescaled by a factor of 2 until the sample value lies in the upper half of the converter's input voltage range or until the maximum prescale factor is reached without overload. The exponent bits are determined from the prescaling used. Thus, the resulting floating point samples always have a full n bit mantissa except for small signal values that fall below half scale with maximum prescaling. The floating point PCM sample sequence may be digitally companded to further reduce the bit rate.

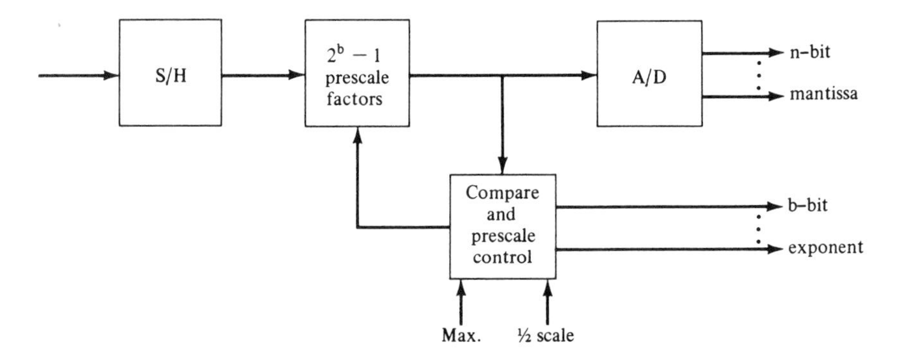


**Figure 4.18:** Floating point PCM. © Alan V. Oppenheim, *Applications of Digital Signal Processing*, 1978, p. 40. Reprinted by permission of Prentice-Hall, Inc., Englewood Cliffs, N.J.

### 4.3.3 Adaptive Differential Modulation

The next method of data compression we shall consider is "adaptive differential PCM." Recall from Chapter 1 that we introduced the concept of differential signal encoding whereby the transmitted sample represented not the actual sample value but the difference between the current sample and the previous sample. This concept of data compression is based on the assumption that the difference between successive samples can be represented by fewer bits than the actual sample value.

The simplest differential modulation scheme is delta modulation (DM) where an analog signal is converted to a binary bit stream with each bit representing a unit step change in amplitude of the signal in either a positive or negative direction. Figure 4.19 illustrates the concept of DM. The step size is fixed and if the signal is changing rapidly the DM representation may become "slew-rate limited" as indicated by a repeated sequence of ones or zeros. This is also known as "slope overload." To avoid slew-rate limiting a higher sample rate is needed for simple DM. Another approach to avoid slew-rate limiting without increasing the bit rate is to make the step size variable. Since the receiver must know when the step size is altered, the criteria for altering the step size must be based on the transmitted bit stream, this being the only information the receiver has available from which to reconstruct the signal. Figure 4.20 illustrates the concept of adaptive DM. The

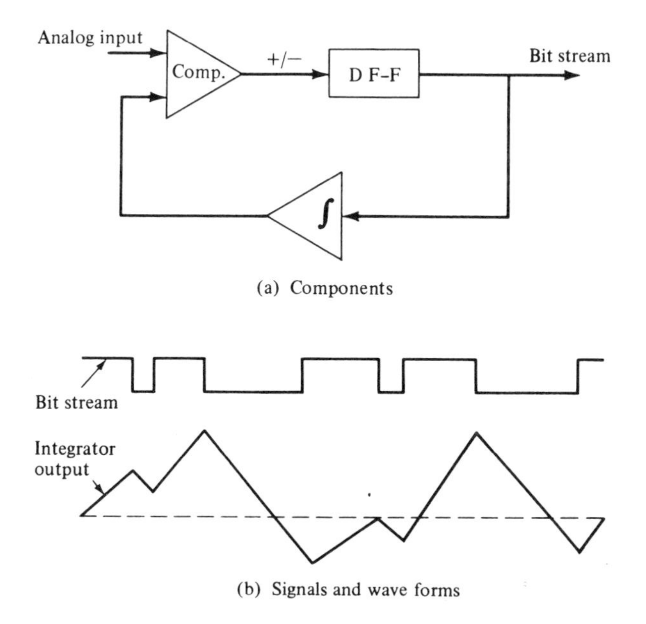


**Figure 4.19:** Delta modulation.

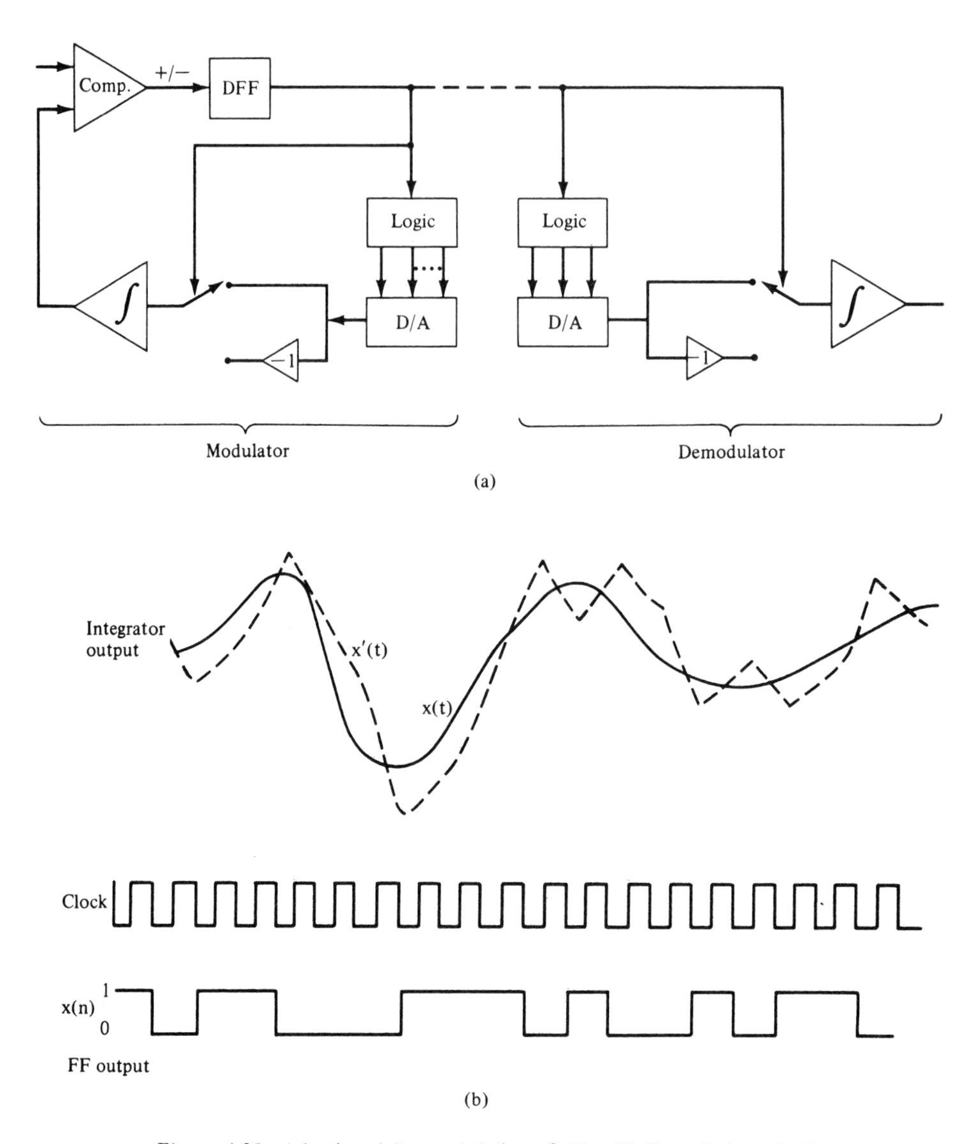


**Figure 4.20:** Adaptive delta modulation. © Alan V. Oppenheim, *Applications of Digital Signal Processing*, 1978. Reprinted by permission of Prentice-Hall, Inc., Englewood Cliffs, N.J.

block marked "logic" in Figure 4.20 compares the current bit with the previous bit; if they are the same the step size is increased, if they are different the basic step size is used. As with simple delta modulation the direction of change (i.e., the sign of the voltage level into the integrator) is determined by the current output of the flip-flop.

The adaptive differential PCM technique extends the concept of adaptive

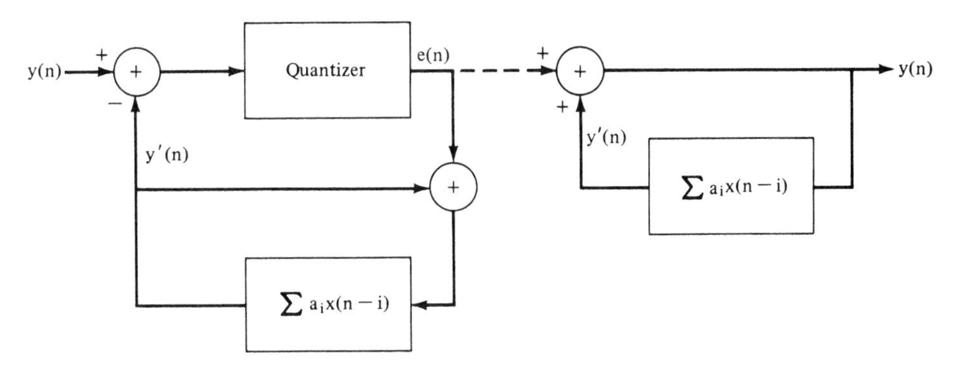


**Figure 4.21:** Adaptive DPCM. © Alan V. Oppenheim, *Applications of Digital Signal Processing*, 1978. Reprinted by permission of Prentice-Hall, Inc., Englewood Cliffs, N.J.

delta modulation to include a multi bit difference signal. Thus, the flip flop becomes an $n$ bit A/D converter and the integrator becomes a linear predictor; that is, the next sample value is predicted as a linear combination of the past $N$ samples and the difference between the predicted value and the actual value or the prediction error is the quantity transmitted.

At the receiver, an identical set of prediction coefficients is used to linearly combine the last $N$ received error samples. The predicted signal sample is added to the current received error signal to reconstruct the original signal. This compression technique is also known as "linear predictive coding," shown in Figure 4.21.

### 4.3.4 Transform Compression

Another data compression technique that has found application in image data processing is based on transform techniques. The data is transformed into another domain that represents an essentially uncorrelated sequence. If the majority of the information of interest is concentrated in a particular region of the transform domain then the transform coefficients having small values in other regions are not transmitted. This truncated sequence is zero-filled after reception to restore it to its original length and then inverse transformed to recreate the original data. The accuracy quality of the reconstruction will depend on how much information was lost due to the truncation of the transmitted transform domain sequence.

Transform compression techniques can also lead to noise reduction applications in data transmission. Consider using the Fourier transform in a data compression sense. A sequence of signal samples is converted to the frequency domain using the FFT. The FFT size is chosen (in conjunction with the sample rate) to cover the full frequency range of the transmission channel. If the signal contains significant spectral components only in the lower end of the channel band width then the high frequency spectral coefficients can be truncated and not transmitted. When the signal is reconstructed at the receiver by zero-filling the transmitted sequence and performing an inverse FFT, the high frequency components of the time domain signal are no longer present. Since these higher frequency components essentially represented noise in the original signal an overall signal to noise ratio improvement has been achieved.

### 4.3.5 Summary

The field of digital data communications has been growing steadily over the past decade as a major application area for digital signal processing techniques. In this section we have examined briefly some digital approaches to the problem of data rate compression. Historically, analog companding techniques provided a solution to the data rate compression problem. However, as with all analog processing systems, realizing an exact inverse process leads to problems of component matching and fine tuning of analog circuits. Several digital signal processing approaches have been developed that provide exact repeatability (and invertibility). Of these, we have looked at using homomorphic processing techniques for digital companding. We also examined differential signal encoding approaches to data rate reduction. Finally the concepts of transform compression were introduced as another method of using digital signal processing to achieve data rate compression (as well as noise suppression).

In our discussion of differential signal encoding we looked at adaptive approaches that lead to better performance. In the next sections we shall examine the basic concepts of adaptive signal processing techniques in a more general manner as they apply to a wide range of applications.

## 4.4 ADAPTIVE SIGNAL PROCESSING

### 4.4.0 Introduction

The term adaptive processing is applied to a wide variety of apparently differing problems. In general, the adaption mechanisms are concerned with identifying certain characteristic parameters of the observed data. Based on these parameters an error measure is computed which is used to adjust the signal processing so as to minimize the error. The sequence of steps required to minimize the error measure is called "adaptation," and the resulting system is therefore adaptive. Usually, the sequence of error measurements, and the algorithm for adapting the processing, must occur within the sample period if the results are to be relevant. This real time constraint forms a pragmatic boundary which forces many compromises on mathematically ideal solutions to adaptation.

As an example, recall the discussion of the Kalman filter (Section 3.3.5). The Kalman filter recursion process assumes that the transition matrix $\phi$ and the observation matrix $C$ are known. In addition, the initial covariances $P$, $Q$ and $R$ were assumed to be known. If these quantities are not known, then various empirical estimates must be tried until satisfactory performance is achieved (the $\phi$ and $C$ matrices are part of the assumed source model).

An alternative to this off-line tuning of the filter is to incorporate an adaptive procedure in the filter equations that attempts to optimize these parameters in real-time based on available information within the filter. If the filter is performing well (e.g., the source system has been accurately modeled) then the error between the predicted and observed data will be an uncorrelated zero mean random process with variance $\sigma^2$, known as the "innovations process," $a(n)$. If this signal is observed within the filter and found to exhibit non-random or non-zero mean characteristics, then the filter is not behaving in an optimum manner. By the use of various estimation techniques, the parameters of the filter can be adaptively estimated for use in the recursion equations such that the variation of $a(n)$ from its desired zero mean, random characteristic is minimized.

As another example of the basic concept of adaptive processing consider the problem of removing echoes from a transmitted signal (usually on a transmission line). In Section 4.2 we observed that echoes appear as peaks in the cepstrum of a signal. To adaptively filter out echoes the cepstrum of the signal may initially be processed to identify the location of echo peaks. The location of these peaks can be used as input data to an algorithm that alters the frequency response of a linear filter applied to the cepstrum to remove the peaks, thus filtering out the echoes when the signal is returned to its original domain.

It is our purpose in this section to develop a general model for adaptive filters, and to discuss the main issues in their implementation.

### 4.4.1 An Adaptation Model

An adaptive digital filter is implemented so that its coefficients can be changed between sample periods (or between blocks of sample values). Two possibilities are shown in Figure 4.22. In Figure 4.22(a) the input signal $x(n)$ is processed to obtain an output $d(n)$, which is compared to $x(n)$ to obtain an error measure. This measure is used to adjust the coefficients of the adaptive processor to produce the desired output $y(n)$. This represents a generalized "feed forward" adaption process. In Figure 4.22(b) an input stream $x(n)$ is processed in a recursive filter and the filter output $y(n)$ subtracted from an input stream $d(n)$. The object is to minimize some measure of the difference $e(n)$ representing a general "feedback" adaption. In both cases the difference is usually a function of the mean square. In the following we derive first the basic expression to be minimized, then the issues

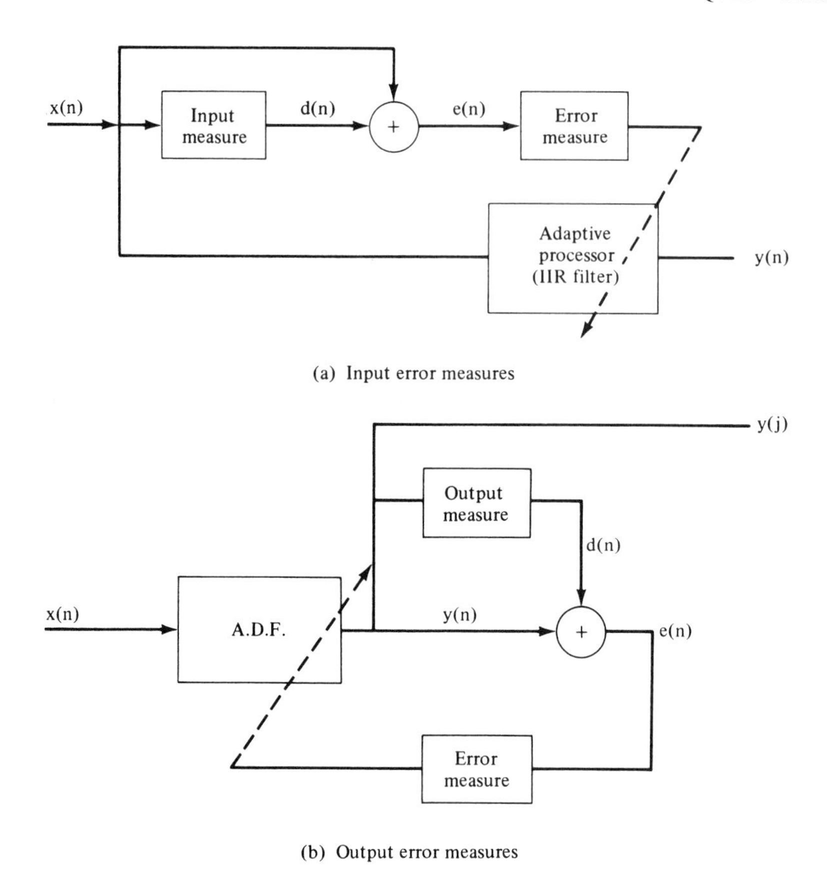


**Figure 4.22:** General adaptive architectures.

involved in choosing a particular approach and finally illustrate several alternatives.

### The General Model

The input-output relationship of a general IIR filter with time varying coefficients is an extension of equation (2.59), i.e.,

$$y(n) = \sum_{k=0}^{N-1} a_k(n)x(n-k) + \sum_{k=1}^{M-1} b_k(n)y(n-k) \tag{4.45}$$

This can be conveniently represented in vector notation by defining

$$W = Col[a_o(n), a_1(n), \ldots a_{N-1}(n), b(n), \ldots, b_{M-1}(n)]$$

$$(4.46)$$

$$Z(n) = Col[x(n), x(n-1), \ldots, x(n-N+1), y(n-1), \ldots, y(n-M+1)]$$

Thus equation (4.45) can be written

$$y(n) = W^T z(n) \tag{4.47}$$

The column matrix $W$ represents the coefficients of the filter and it is these coefficients which must be adjusted to minimize the difference between $y$ and $d$.

**An Error Function**

Consider the error on the $n$-th sample for the model in Figure 4.22(b):

$$e(n) = d(n) - y(n) \tag{4.48}$$

It is a function of this error which must be minimized by adjusting $W$. The mean square error is the most often chosen, because of computational tractability and its general utility. Thus, let the criterion be to minimize $J(w)$ where

$$J(W) = E\{e^2(n)\} \tag{4.49}$$

By direct substitution

$$\begin{aligned} e^2(n) &= d^2(n) - 2d(n)y(n) + y^2(n) \\ &= d^2(n) - 2d(n)Z^T(n)W + Z^T(n)WW^TZ(n) \end{aligned} \tag{4.50}$$

from which

$$E\{e^2(n)\} = E\{d^2(n)\} - 2E\{d(n)Z^T(n)\}W + W^TE\{Z^T(n)Z(n)\}W \tag{4.51}$$

thus

$$J(W) = \sigma_d^2 - 2R_{dZ}W + W^TR_{ZZ}W \tag{4.52}$$

where

$$\sigma_d^2 = E\{d^2(n)\}, \; R_{dZ} = \begin{bmatrix} E\{d(n)x(n)\} \\ E\{d(n)x(n-1)\} \\ \vdots \\ \{E\{d(n)y(n-M+1)\}\} \end{bmatrix} \tag{4.53}$$

$$R_{ZZ} = E\{Z^T(n)Z(n)\}$$

Assuming for a moment that these can be calculated, the second problem is to adopt a strategy for adjusting the coefficients to minimize the error. The major difficulties are based on the pragmatic facts that:

1. There is a limitation on the computational time available if an adaptation is to be in effect before the next sample (or block of samples) arrives; and
2. There is a limited amount of information that is available about the success of previous attempts or on the character of the input.

***Example*** *(An FIR, Two-Tap Adaptive Filter):*

To illustrate the issues consider an FIR filter with two taps, i.e.,

$$y(n) = a_o(n)x(n) + a_1(n)x(n-1) \tag{4.54}$$

For this filter

$$W = \begin{bmatrix} a_o(n) \\ a_1(n) \end{bmatrix} \quad \text{and} \quad Z(n) = \begin{bmatrix} x(n) \\ x(n-1) \end{bmatrix} \tag{4.55}$$

From equation (4.51)

$$\begin{aligned} J(W) = \sigma_d^2(n) - 2 &\begin{bmatrix} E\{d(n)x(n)\} \\ E\{d(n)x(n-1)\} \end{bmatrix} \begin{bmatrix} a_o(n) \\ a_1(n) \end{bmatrix} \\ &+ [a_o(n)a_1(n)]E\left\{[x(n)x(n-1)]\begin{bmatrix} x(n) \\ x(n-1) \end{bmatrix}\right\} \begin{bmatrix} a_o(n) \\ a_1(n) \end{bmatrix} \end{aligned} \tag{4.56}$$

which upon simplification becomes

$$\begin{aligned} J(W) = \sigma_d^2(n) &- 2a_o(n)E\{d(n)x(n)\} - 2a_1(n)E\{d(n)x(n-1)\} \\ &+ a_o^2(n)E\{x^2(n)\} + a_1^2(n)E\{x^2(n-1)\} + 2a_o(n)a_1(n)E\{x(n)x(n-1)\} \end{aligned} \tag{4.57}$$

This equation describes an error surface as shown in Figure (4.23). The purpose of the error calculation is to determine the position on the bowl. The adaptation algorithm must adjust the coefficients $a_o$ and $a_1$ so that the next error calculation is closer to the minimum. We note that the gradient (i.e., the partial differentials of $J(W)$ with respect to $a_i$) is zero at the minimum point. Before considering this we note further that unless the process is stationary and ergodic, the minimum point will move with time. In this case the adaptive process must also track this movement. This aspect of adaptive processing will not be pursued here.

### The Adaptation Algorithm

The overall purpose of the adaptation algorithm is to generate a new set of coefficients $W_{t_n+1}$ after computing the error at time $t_n$. In the general case this requires a two step procedure; first, given the error, compute the gradient vector which points toward the minimum, and second compute a new set of tap coefficients to achieve the minimum. In principle this is straightforward; however, the real-time constraints force many compromises.

Consider first the gradient of the error surface. From equation (4.51) the gradient vector of $J(W)$ is:

$$\nabla_n\{J(W)\} = -2R_{dZ} + 2R_{ZZ}W \tag{4.58}$$

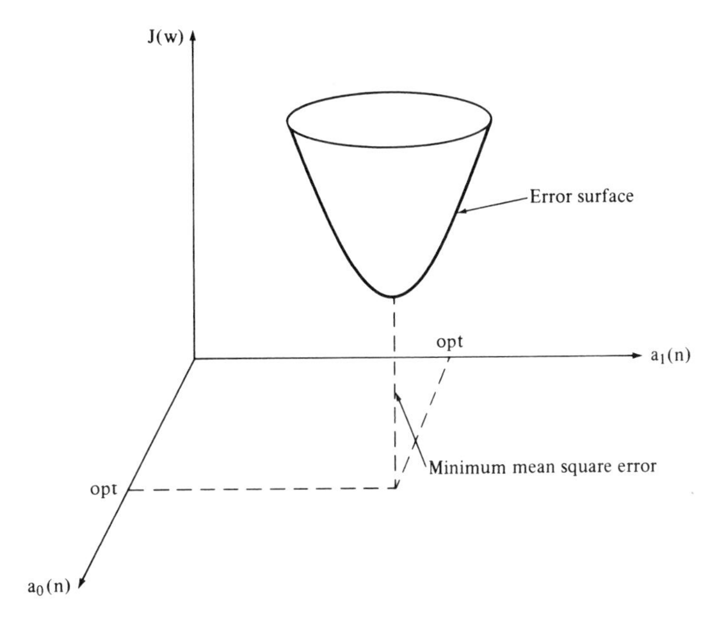


**Figure 4.23:** Error surface for a two-tap FIR filter.

Setting this equal to zero and solving for $W$ yields

$$W_{OPT} = R_{ZZ}^{-1} R_{dZ} \tag{4.59}$$

(this is the Weiner-Hopf equation in vector form). Substituting this into equation (4.51) gives the minimum error

$$J(W)_{MIN} = d^2(n) - R_{dZ}^T W_{OPT} \tag{4.60}$$

The mean square error computed at any time in terms of the minimum mean square error is

$$J(W) = J(W)_{MIN} - W_{OPT}^T R_{ZZ} W + W^T R_{ZZ} W \tag{4.61}$$

Several approaches exist for computing new coefficients which will reduce the computed error to its minimum value. They are all based on some function of the gradient of the error with respect to the filter coefficients. In practice there is always a trade-off with the speed of convergence and the time available between samples.

Assume that the tap weights are changed as follows:

$$W_{n+1} = W + \hat{S}[-\nabla J(W)] \tag{4.62}$$

where $\hat{S} = \text{diag}[S_o, \cdots, S_{n-1}, \cdots, S_{M-1}]$ is a diagonal matrix whose elements are a weighting of the relative importance of the parameters (i.e., a proportionality constant). The generality of this expression is appealing; however, the amount of computation involved requires long itnervals between samples. As a first simplification the terms in the matrix are considered constant, i.e.,

$$S_o = S_1 = \cdots S_{N-1} = \mu/2;\ S_{n'} = \cdots S_{M-1} = \nu/2 \tag{4.63}$$

***Example*** *(An FIR Adaptive Filter):*

Consider once again the FIR filter (i.e., $b_i = 0$).

$$y(n) = \sum_{k=0}^{M-1} \alpha_k x(n-k) \tag{4.64}$$

In this case

$$\begin{aligned} W_{n+1} &= W_n + \mu/2\left[-\frac{\partial J(W)}{\partial W}\right] \\ &= W_n + \mu/2E\left\{-\frac{\partial}{\partial W}(d^2(n) + y^2(n) - 2d(n)y(n))\right\} \\ &= W_n + \mu/2\left[2E\{d(n) - y(n)\}\frac{\partial y(n)}{\partial W}\right] \\ &= W_n + \mu E\{e(n)x(n-k)\} \end{aligned} \tag{4.65}$$

The last step follows from the definition of $y(n)$. This equation allows an iterative approach which always descends the error surface toward the minimum error. The second term should become zero when the minimum mean square error is found. In terms of each coefficient

$$a_k(n+1) = a_k(n) + \mu E\{e(n)x(n-k)\} \tag{4.66}$$

It is the cross correlation of the computed error and the past samples which provides some difficulty, since the statistical properties are usually unknown.

It is easy to show, following this example, that the general IIR filter coefficients can be obtained in a similar manner.

Because of the difficulty in obtaining a model which yields the cross correlation of the error and the stored sample, a more pragmatic approach has proven useful. Each new coefficient is computed without the cross correlation, i.e.,

$$W_{n+1}\ W_n + \hat{S}\ e(n)x(n-k) \tag{4.67}$$

This is known as the LMS adaptation algorithm. The major question is, "How fast does it converge?"

**Convergence of the LMS Algorithm**

The convergence properties of the LMS algorithm depend to a large extent on the statistical properties of the samples and the error signal. These are independent variables, thus the only computational variables are $\mu$ and $\nu$. It has been shown [Widr 76] that the LMS algorithm will converge to the optimum values provided.

$$o < \mu < 2/(N-1)\sigma_x^2 \qquad (4.68)$$

$$o < \nu < 2/M\sigma_y^2$$

The speed of convergence depends on the choice of $\mu$ and as well as on $\sigma_x^2$ and $\sigma_y^2$. The time constant has a lower bound (in an FIR filter) of

$$\tau > \frac{\mathrm{T}}{\mu\sigma_x^2} \qquad (4.69)$$

where T is the sample period. This can be fairly long; for example, at an 8KHz sampling rate (T=125 microseconds) of a 1 VRMS signal, $\tau$ is 12.5 milliseconds, or 100 sample periods.

The LMS algorithm represents a compromise between an exact but lengthy calculation with speedy convergence and an inexact but shorter calculation with a more lengthy convergence. Many other approximations are possible. In general any function of $J(W)$ could be used before the gradient is computed provided it were computationally feasible (in the time available) and it were to converge quickly enough for the application. The function should be a positive even function of the error.

Consider merely clipping the error signal and adjusting the tap values by a fixed amount depending on the sign of the error. Such a scheme might converge slowly in the general case, but it is simple to implement and once convergence is achieved, the system would be stable. Such a scheme is often referred to as bang-bang control, a term adopted from early control theory. A simple extension of this would account for a range of error signals and adjust the weight depending on the error, e.g., for an error between 0.5 and 1, adjust $W_n$ by a given amount and for larger values increase the adjustment. The convergence properties of these systems depend on the signals and will not be pursued here. We note only that the functional requirements of an adaptive system are the same; it is only in the details that variations are discovered.

### 4.4.2 Summary

Adaptive procedures require in general a large computational load between each sample. The extremes in implementation range from almost "blind" iterative trial-and-error searches to a complete computational determination of the optimum receiver response. The trade-offs which influence the final choice are:

- real time response
- accuracy
- stability
- equipment simplicity

Examples of search algorithms include

- Steepest descent: the local gradient is followed to a minimum
- Uniform step search: steps of uniform size toward zero error (viz., delta modulation)
- Variable step search: step size is chosen depending on the error in a direction to minimize the error along each coordinate (viz., adaptive delta modulation)

Each procedure requires knowledge or an estimate of the gradient of the error surface. The mean square estimator has a particularly well behaved surface which accounts for its wide use.

## 4.5 CHAPTER SUMMARY

The overall purpose of this chapter was to introduce a range of processing techniques and algorithms that are of wide practical importance. A second purpose was to provide evidence that the concepts introduced in the preceding chapters form an adequate basis for understanding these applications.

Four areas were chosen to illustrate a range of algorithms: the FFT, generalized linear processing, data compression, and adaptive processing.

The FFT is a fast technique for computing the DFT. It trades data flow complexity for a reduced computational requirement. In general this trade-off is advantageous. The FFT has had a profound effect on signal processing, and it is intrinsically useful for obtaining spectral coefficients in a real time environment. On the other hand, it has forced designers to contort processing algorithms in this direction rather than toward time-domain solutions.

Our two examples of the FFT were fast convolution, and frequency domain filtering. Convolution in the time domain is accomplished easily by multiplication in the $Z$ domain. Obtaining the frequency coefficients through the $Z$-plane was a natural application of the FFT. Spectral shaping in the frequency domain is another application dependent on the fast determination of the required coefficients.

The problem of finite word length effects is always present in digital processing. In the FFT (or in filtering generally), the system specification must account for anticipated errors due to limitations imposed by finite length registers.

The concepts of generalized linear processing allow us to separate and to individually process the components of multiplicative and convolved signals. This technique greatly extends the range of signal types to which digital signal processing can be usefully applied when the operations of taking logarithms and performing exponentiation are added to the capabilities of the processor.

Data compression is widely used in both storage and transmission. Several approaches were discussed, including companding, differential modulation techniques, and transform compression.

Adaptive processing was originally introduced in Chapter 3 and illustrated with the Kalman filter. In this chapter we have summarized the general requirements from a broader point of view. The key issues were the error criteria and the mechanisms (or strategy) for reducing this error in an adaptive manner. The point to emerge in this presentation was the necessity of finding reduction strategies which could be computed in the time available, and which reduced the error at a satisfactory rate. The resolution of these two antagonistic requirements is not always straightforward.

We have now completed our review of basic theory and techniques of digital signal processing. This review serves primarily as an indication of the range of general background theory upon which the implementation of practical digital signal processing functions is based. In the next chapter, applications will be discussed and a range of processing operations identified which completely specify the requirements of signal processing algorithms.

## 4.6 EXERCISES AND PROBLEMS

### FFT

**1.** Derive the decimation in frequency, radix 2 FFT algorithm for $N = 8$.

**2.** For $N = 16$, consider the radix 2 decimation-in-time algorithm. Draw the flow graph (similiar to Figure 4.2). Structure your figure carefully so that each butterfly and higher order section is identifiable.

**3.** For $N = 16$, develop the radix 2 decimation-in-frequency algorithm. Draw the flow graph (similiar to Figure 4.16). Structure your diagram carefully so that each butterfly and higher order sections are identifiable.

**4.** Using as an example $N = 16$, show that if the input is addressed in bit reversed order then the twiddle factors must be addressed in normal order. Is this true if the situation is reversed?

**5.** Prove that the twiddle factors can be computed recursively as

$$W_N^{ij} = W_N^i W_N^{i(j-1)}$$

Write a HOL program to compute successive factors. Comment on the error accumulation as the iterations proceed. Suggest ways of limiting the errors at intermediate points.

**6.** An input signal is sampled at 40KHz. Suppose the spectrum of the sample point is to be analyzed using a DFT algorithm. Determine the frequency spacing between the resulting spectral samples.

## Convolution

**7.** Develop an example to compare the number of multiplications required when doing direct convolution as opposed to fast convolution (using overlap and add).
Suppose an FIR filter is to be implemented having 64 taps. Choose a segment length which will minimize the number of multiplications per output sample when using fast convolution.

**8.** Assume the following requirement:
It is required to measure the signal strength of a broad band source. Assume there exists a background narrow-band noise which varies slowly in frequency. The power of this noise is sufficient to invalidate the desired measurement.
It is proposed to proceed as follows:

(i) Build an adaptive notch filter to block the noise.

(ii) The center frequency of the notch is to be computed by continuously computing an $N$ point FFT. The frequency bins of the FFT are to be maintained and if the power rises above $\sigma^2/N$ it will be assumed that the undesired signal is present at that frequency.

(iii) The notch filter coefficients will be adjusted to eliminate the noise.

Compute the coefficients of the filter.
Comment on the suggestion that the notched filter is unnecessary.

## Adaptive Processing

**9.** Consider a two-tap FIR adaptive filter. Compare the number of calculations required for an exact implementation of the LMS algorithm, and for the bang-bang approximation. Base your estimate on a unit of time for multiplication, and one for addition, the real time constraints for each. Comment on the speed of convergence if the signal is 1 volt RMS.

CHAPTER 5

# From Processing to Processors

## 5.0 INTRODUCTION

This chapter forms the bridge between the concepts of signal processing developed in previous chapters and the subject of signal processors to be discussed in Part B. Our concern thus far has been with both the underlying theoretical structure of digital signals and with some of the common algorithms used in processing. The architecture of processors which actually implement these algorithms and techniques for measuring and comparing performance will be studied in Chapters 6 and 7.

Two related factors seem to have constrained earlier processor designs; technology and perspective. The high cost/performance ratio of available components severely limited what was affordable and therefore what could be considered. It is the continuous lowering of this ratio which is responsible for the increasing pervasiveness of digital processing. And of course it is the mechanisms for lowering this ratio (i.e., VLSI chips) which lead to the requirement for new perspectives and design methodologies which are the subjects of Volume II. It is apparent, however, that early architects were gradually evolving mechanisms for resolving the often conflicting requirements for data flow and for procedural control of a system. The outstanding contribution of von Neumann was the concept of a program counter which provided for centralized control of the instruction sequence and the data flow of a computer. This mechanism, however, became a bottleneck for input/output data flow, even in early computers, and resulted in such ameliorative techniques as interrupts, direct memory access (DMA), I/O channels, etc. Strict sequentiality, enforced by the program counter, prevented any form of parallelism in execution. High performance systems of the future must address these problems and others to be exposed later, in a unified way. This chapter provides the focusing mechanism for the preceding material and begins to highlight the issues which must be resolved in a processor design methodology as we proceed toward the design of processors in Volume II.

To provide this focus, we require two steps. The first is to examine application areas and to abstract from them the character and structure of processing requirements. In the second, it is shown that the algorithmic structure of all signal processing can be represented by a set of basic computations. These form a powerful means of moving processing onto processors.

Section 5.1 examines a range of typical applications and examines the algorithmic requirements of the associated processing, and in addition quantizes where possible the signal models as well as the performance requirements and constraints on the processors. All the required processing theory discussed thus far can be reduced to a set of basis operations, which in various combinations form the processing components of any application. It is these combinations of signal models, the processing algorithms, and the performance requirements which drive processor designs.

In Section 5.2, a unifying perspective of signal processing is presented in which the apparent diversity of various application areas is drawn together in terms of the partitioning of processing requirements as basis operations. This partitioning of the computational requirements forms the basis of many algorithms and explains also many of the architectural features of VLSI chips designed for signal processing. However, the architectural issues that arise when creating processors for conglomerates of such chips are of overall interest.

The two sections of this chapter yield two messages. First, that the application areas of signal processing form a context in which the processing requirements can be represented by a relatively small set of mathematical operations. Second, that these basis operations, coupled with the signal characteristics and the performance requirements, completely characterize an application and form the input to a processor design methodology which exploits current technology. That the latter assertion is true still remains to be demonstrated. The demonstration requires many more concepts which will be the subject of Volume II.

## 5.1 APPLICATION OF DIGITAL SIGNAL PROCESSING

### 5.1.0 Introduction

This section establishes the first of our two objectives for this chapter. Perhaps the overriding concern is to present convincing evidence that all signal processing applications can be systematically quantified by the answers to a small set of questions. The design of processors will be shown later to follow from this same set; a most important attribute from our point of view. In order to do this we first review a selection of application areas to provide a common introduction. Then in each of the next three subsections the key attributes are illustrated; signal modeling, processing requirements, and performance requirements. The last subsection discusses some implementation trends to illustrate how these three characteristics have been accommodated in the past.

### 5.1.1 Application Areas

The concepts and techniques of digital signal processing discussed in the previous chapters have been, and continue to be, applied to a wide range of applications. The exact nature of individual signal processing problems tends to be highly dependent on the particular area of interest. Our interest in this section is to expose those common issues and attributes of any signal processing problem or application area that will directly impact the processing and hence the processor requirements.

The approach taken to achieve this aim is to list and discuss a number of major application areas in which digital signal processing techniques have been successfully used. The primary purpose of this approach is to indicate the range of applications that exist and to note the general nature of the problems involved in each area. No attempt is made to discuss signal processing within the context of any one application area since this would require a volume in itself, even for a brief examination. An excellent overview of applications edited by Oppenheim [7] does exist, and the reader is directed to this for a more extensive survey.

#### Speech Processing

One of the earliest fields of research to employ digital signal processing techniques was that of speech processing. Two major problems exist in this field. First is the analysis of human speech signals for such applications as automatic speech recognition systems, voice or speaker identification or verification and speech waveform parameterization, encoding and compression for efficient transmission and storage of speech signals. The second general problem area is that of speech synthesis. It is in this area that the greatest advances are currently being made. Digital speech synthesis systems are becoming available for a host of applications such as automatic reading machines for the blind, speech synthesizers for the speech handicapped, voice response computer terminals, and even talking toys and appliances. A variety of speech synthesis systems for the reconstruction of intelligible speech from compressed and encoded speech signals are currently available, several consisting of only two or three integrated circuits. A good overview is contained in [7:Ch.3]; for a complete discussion on the digital processing of speech signals the text by Rabiner and Schafer [11] is recommended.

### Music Processing

There are several signal processing problems in the area of music processing to which digital techniques have been applied. These include the problems of editing and mixing multiple music signals into a single performance, and the enhancement of music signals by the addition of special effects such as reverberation and chorus effects. Digital techniques have also been employed for the composition, synthesis, recording and transmission of music as well as in the restoration of old recordings. An introductory overview of digital signal processing for music applications is contained in Oppenheim [7:Ch.2].

### Geophysics

Another area that has a considerable history of utilizing digital processing techniques is geophysics. The major processing problems are concerned with the analysis of seismic signals to aid in the modelling of the structure and properties of the earth's interior, the study of earthquakes and volcanic activity and of course in the exploration for oil [7:Ch.7]. Digital processing techniques have also been utilized more recently in connection with the analysis of radiowaves reflected from the upper atmosphere for determining atmospheric properties such as electron content, as outlined by Bernhardt [18:p.705].

### Radar

Radar systems are an example of how the increased speed and performance capabilities of new digital integrated circuits have fostered the move toward digital signal processing techniques in high performance applications. The major signal processing functions of a modern radar system include signal generation, matched filtering, threshold comparisons and the estimation of target parameters such as range, bearing and velocity. Due to the high data rates associated with the wide bandwidth of radar signals (10 to 100 MHz), the primary concern in radar is the reduction of data volume and rate to levels that can be suitably handled by the data processing portions of the radar system. The application of modern high speed digital components to the problems of designing and implementing modern radar systems forms a highly active area of current research and development in high performance digital signal processing [7:Ch.5].

### Sonar

Sonar signal processing may be conveniently considered with respect to two major subcategories, active sonar and passive sonar. Active sonar systems share many common signal processing concepts with radar. In both cases signal pulses are generated and transmitted and the primary processing is associated with the detection and analysis of echo returns for such applications as target detection and localization, navigation and mapping. As with radar, the functions of matched filtering, thresholding and target metric generation are major signal processing activities. Passive sonar systems differ from active systems in that they do not generate and transmit acoustic signals but simply listen to the acoustic environment. The major signal processing activities are high resolution spectral analysis, and transducer array processing such as beamforming techniques for selective directional listening. For a summary of sonar signal processing issues and further references, Baggeroer [7:Ch6] is recommended.

**Image Processing**

The application of digital signal processing techniques to the processing of images has been strongly influenced by the recent advances in integrated circuit technology. With a single digital image signal representing on the order of a million sample values the requirement for high performance processors and high density data storage is obvious. The major categories of image processing problems to which digital processing techniques have been successfully applied include data compression, image restoration, deblurring and enhancement as well as the creation of visual images from x-ray projections, radar, sonar, ultrasonic and infrared signals [7:Ch.4].

**Communications**

There is scarcely any branch of the overall field of communications where digital signal processing techniques have not had some influence. Digital techniques have been applied to the problems of signal modulation, multiplexing, encoding and data rate compression. In the area of telecommunications digital processing techniques have been developed for tone detection, echo cancellation and digital switching networks. Many audio band communication signal processing functions have been implemented as single integrated circuit components as illustrated by almost any current catalogue of communications products from a semiconductor manufacturer (e.g., Intel Corporation, American Microsystems Inc., Texas Instruments, etc.).

**Biomedical Signal Processing**

The use of digital signal processing techniques is becoming widespread in medical applications such as the analysis of EEG and ECG signals, and computer aided tomography (the creation of two and three dimensional images of the interior of the body from x-ray projections and ultrasonics) [18:p.697; 7:Ch.4, Sec.4.5]. In addition to these application areas, some recent applications have been reported in such diverse areas as power distribution planning, environmental studies on air pollution and intrusion detection systems [18]. Also missing from the above list is a host of military applications such as navigation systems, guidance and control, electronic counter measures and magnetic anomaly detection. The message here is clear, however. The use of digital techniques is rapidly spreading to every application area where a signal processing problem can be identified.

### 5.1.2 Signal Modelling

The first obvious application dependent factor to affect the processing and the processor requirements is the characterization of the signals. In general there are two application dependent aspects of signal characteristics:

1. The first is concerned with the basic signal parameters such as frequency content, dynamic range and signal to noise ratio. These considerations will affect both the sample rate and sample quantization requirements. In terms of these signal characteristics, many individual application areas will have similar requirements. For example, of the application areas listed in the previous subsection, many are concerned with signals that fall within the audio band.
2. A second aspect of signal characterization is concerned with how the information content of the signal is modeled. This signal modelling determines how the signal is interpreted to obtain information. It is often concerned with modelling the signal source and the effects of the transmission medium. Signal modeling is of prime importance in the determination of the actual signal processing requirements. As an example, consider the areas of speech processing and telecommunications. Both of these areas are concerned with low audio band signals that can be adequately handled with similar sample rates and sample quantization. However, the signal models differ greatly. For telecommunications, the signal modelling is chiefly concerned with the various modulation, multiplexing and signal encoding schemes employed to achieve efficient transmission. In addition these signals are often modeled with respect to the transmission channel characteristics. For speech processing the signal model is based on the mechanisms for human speech production. In general the speech waveform is modeled as the response of a slowly varying linear system to either a periodic or noiselike excitation signal. As a result of these differences in the information content and format of signal models, the overall processing requirements in the two areas differ considerably.

### 5.1.3 Processing Requirements

Based on the signal models and the specific problems or goals of the various applications, the required structure of the processing operations must be formulated. The specification of processing requirements is carried out in terms of the basic mathematical tools as discussed in earlier chapters. In general, this amounts to delineating the order in which the signal is to be manipulated to get the information of interest into a desired form or representation. The actual signal manipulations tend to be based on a relatively small set of basic signal processing operations such as convolution, correlation or difference equation calculations for filtering operations, DFT coefficient calculations, vector or matrix arithmetic operations, etc. The actual problem at hand simply specifies the appropriate combination of these operations to accomplish the required processing and the minimum performance requirements.

This view of processing requirements specification is of central importance since it provides a common set of processing function types needed for processing. These common processing function types, which we designate as the "basis operations" of digital signal processing, are summarized in the following list.

**1. Difference Equation Calculations**

$$y(n) = \sum_{k=0}^{N} a_k x(n-k) - \sum_{k=1}^{M} b_k y(n-k) \tag{5.1}$$

This equation represents the general computational requirement for a recursive or infinite impulse response (IIR) filtering operation; where $x(n)$ is the input sequence and $y(n)$ is the filtered output sequence. The parameters $N$, $M$, $a_k$ and $b_k$ define the actual transfer function or equivalently the phase and amplitude response of the filter. As noted in previous chapters the limiting of the parameters $N$ and $M$ to a value of 2 leads to the standard second order filter section from which arbitrary filters can be created in parallel and/or cascade realizations.

If the coefficients $b_k$ are all zero then equation (5.1) reduces to a finite convolution sum representing a non-recursive or finite impulse response (FIR) digital filtering operation.

$$y(n) = \sum_{k=0}^{N} a_k x(n-k) \tag{5.2}$$

In this case the $a_k$'s of equation (5.2) can be interpreted as the impulse response of the filter.

**2. Correlation Coefficient Calculations**

The correlation operation is another frequently required operation:

$$\phi(n) = \sum_{k=-N}^{N} y^*(k)x(n+k) \tag{5.3}$$

where the interval $-N$ to $N$ represents the window over which the two signals are correlated and $y^*(k)$ represents the complex conjugate of $y(k)$.

**3. Complex Frequency Translation**

The translation of a digital signal to a different frequency band can be represented as a complex multiplication

$$y(n) = e^{j2\pi nk}x(n) \tag{5.4}$$

where the parameter $k$ represents the ratio of the translation frequency to the sample frequency of the input sequence $x(n)$.

**4. Discrete Fourier Transform Calculation**

One of the most widely required calculations of DSP is the computation of discrete Fourier transform coefficients, represented by the equation

$$X_k = \sum_{n=0}^{N-1} x(n) e^{-j(2\pi/N)nk}, \quad k = 0, 1, \ldots, N-1 \tag{5.5a}$$

and the inverse relation

$$x(n) = \sum_{k=0}^{N-1} X_k e^{j(2\pi/N)nk}, \quad n = 0, 1, \ldots, N-1 \tag{5.5b}$$

As we have seen in Chapter 4, these computations are most often carried out in the form of an FFT algorithm based on the butterfly computation;

$$X_k = A +/- BW_N^k \tag{5.6}$$

where $A$ and $B$ are in general complex and $W_N^k = e^{j(2\pi/N)k}$.

**5. Power Spectral Density or Magnitude Squared Calculation**

The calculation of squared magnitude values of a complex sequence represents another basic computational requirement of digital signal processing.

$$|X_k|^2 = X_k X_k^* \tag{5.7}$$

**6. Matrix Arithmetic and Matrix Manipulation**

The formulation of a great many signal processing problems in vector/matrix format leads to the requirement for carrying out the fundamental operations of matrix arithmetic and matrix manipulation. The arithmetic requirements are the addition, subtraction and multiplication of variable dimension matrices, as well as scalar and cross-vector multiplication. The major manipulation functions required are matrix inversion, transposition and complex conjugation.

**7. Logarithms and Exponentiation**

To carry out the operations of homomorphic signal processing the ability to perform both logarithmic and exponential transforms is required.

**8. ADC and DAC**

Finally, the functions of analog to digital and digital to analog conversion are fundamental requirements for digital signal processing.

We note that these basis operations all require the specification of some parameters to identify their exact computational requirements or functional 'personality.' The specification of these parameters and a suitable combination of operations allows the processing requirements of any processing problem to be defined.

A mechanism is required to represent these basis operations and to structure their interaction as a processor. Such concepts and mechanisms will be developed in Volume II.

The important concept to emerge from this section is that of the basis operations. These operations, which can be derived from the requirements of signal processing, form a set of primitive operations which when given their personality (i.e., parameterized) and connected, become a logical description of a signal processor. This concept drives the search for a suitable representation mechanism which will be presented in Volume II. Clearly, it is irrelevant at the logical level whether the primitive is implemented by special hardware devices or as a subroutine in a computer program. What is important is the response requirements and the resultant physical mechanisms for implemention and interconnection.

### 5.1.4 Performance Requirements and Constraints

In order to specify the architecture of a processor (or a processing system) it is necessary to consider both the algorithmic requirements of the application area as well as the implementation details.

The specific application area will dictate not only what sequences of processing operations are required, but also the performance requirements in terms of throughput. The performance requirements will be determined by such factors as the sample rate and word size requirements, the need for real-time and multichannel processing, and the exact parameterization of the basis operations required.

In addition to the specification of functional performance requirements, the individual application will determine the non-functional requirements and constraints such as "physical characteristics of the hardware" (size, weight and power consumption considerations), software and programming goals, cost constraints, reliability, and maintainability, growth potential and expandability, etc.

### 5.1.5 Implementation Trends

The actual implementation of the various signal processing techniques and functions in a specific application area may range from the use of dedicated fixed function hardware to the use of software on a general purpose digital computer. The choice of implementation method tends to be driven by cost and performance considerations and processing flexibility requirements. The following list indicates the general trends in implementation hardware for several application areas.

Speech: single chip processors to mainframes
Music: minicomputer and special purpose processors
Radar: special purpose processors
Sonar: special purpose processors/array processors
Seismic: general purpose mainframes/array processors
Image: general purpose mainframes/array processors
Biomedical: special purpose hardware/micro and mini computers
Telecommunications: special purpose systems/general purpose processors dedicated functional IC's.

The general trend for implementing very high performance digital signal processing has been through the use of special purpose hardware, especially where real time performance requirements exist. For applications where processing flexibility is a major consideration, the trend has been away from large mainframe scientific processors toward the specialized attached processors that are optimized for high speed arithmetic operations on arrays of data. This represents a first step toward multiple processor solutions to flexible real time digital signal processing.

Advances in integrated circuit technology have lead to the implementation of many signal processing functions as single chip special purpose processors, particularly in the area of audio band communications. More recently, the development of single-chip general purpose signal processors for real time audio signal processing has begun to replace the fixed function approach. Clearly, the continued advances of integrated circuit technology will be a major factor in shaping the general trends in the implementation of digital signal processing.

### 5.1.6 Summary

In this section the wide range of problems and many diverse fields to which digital signal processing techniques have been applied has been explored. There are several attributes of the intended application that impact implementation decisions and processor design. In general, these attributes can be categorized in terms of signal models, the processing functions required to manipulate the signals, and the performance requirements and constraints imposed. In discussing these attributes we have stressed three points. First, there are considerable similarities in the basic parameters of the signals in a number of major application areas, but the models of the information content of the signals differ widely from one area to the next. It is this information content modeling that generally determines the detailed processing requirements of an application. Second, we noted that the specific processing requirements dictated by the signal models are always some combination of a relatively small number of basis operations. Using these, the processing requirements of virtually all applications can be described by specifying the appropriate combinations and parameterizations. The third point we stress is that the intended application will not only dictate the characteristics of the signals and what processing is required, but also processing throughput and other performance requirements.

## 5.2 A PERSPECTIVE OF DIGITAL SIGNAL PROCESSING

### 5.2.0 Introduction

This section establishes the first perspective of a methodology required to specify processors from the requirements of a problem. Its purpose is twofold. First, the material presented thus far must be shown to be part of this methodology. Second, the ongoing requirements for new concepts or structure must be postulated. The methodology is in a sense tentative at this point since our acquisition of these new concepts is not complete. These will be mentioned here and presented in detail in Volume II.

The design of a processor can be viewed as consisting of two major areas:

1. Theoretical problem modelling
2. Implementation of processing models as processors

These two areas can be further partitioned into five logical levels. The general structure of this partitioning is illustrated in Figure 5.1. A detailed description of these levels forms the subject matter of this section.

Also indicated in Figure 5.1 are two major activities associated with the theory and implementation of digital signal processing:

1. Algorithm development
2. Hardware architecture development

Subsection 5.2.3 discusses the changing importance of these activities as a result of advances in component technology.

### 5.2.1 Problem Modelling and Theory

In this subsection we shall deal with the first three layers of the perspective. The first two layers are concerned with the application areas and with the basic theory. Both of these have already been discussed. The key issue to be presented here is that digital signal processing can be reduced to a set of basis operations. These operations can each be characterized by a set of parameters and interconnected to perform over-all signal processing functions. This is an important point which becomes embedded in both the

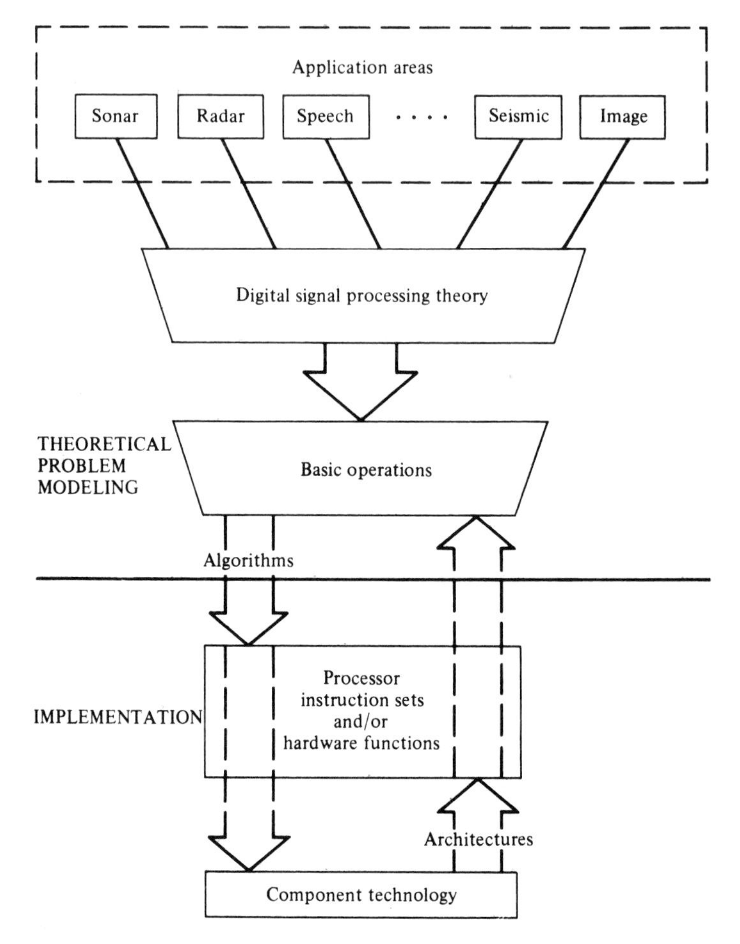


**Figure 5.1:** A perspective of the digital signal processing problem.

design approach to be considered later and in the analysis of digital processors. Indeed this concept is implicit in the examination of classical processors presented in Chapter 6.

### Level 1—Application Areas

At the highest level of this perspective are the major areas of application for DSP techniques. Within each of these areas are specific signal processing problems, the parameters of which are defined by environmental factors of the particular field of interest. Several of these areas, such as control systems, geophysics, and speech provided the earliest motivations for the development of DSP theory. As the theory and associated techniques became well established and the performance of digital systems increased, the range of application areas widened. With the continuing advances in DSP system capabilities, at decreasing costs, the number of major application areas can be expected to continue to grow.

#### Level 2—DSP Theory

This second logical level represents the collective body of theory concerned with the discrete-time or digital processing of signals. The content of this area has been discussed in Chapters 1, 2, 3 and 4.

#### Level 3—The Basis Functions

At the third level are the common computational requirements of DSP as outlined in the previous section. Virtually all DSP problems can be expressed as some linear combination of these "basis operations." These form the interface between DSP theory and implementation. The modelling of a signal processing problem in any particular application area determines the combination of operations to be performed, their parameters, and the system response. The basis operations define precisely the operations that must be implemented, either in software or hardware, or some combination of both.

### 5.2.2 Implementation

In Section 5.1.3 it was proposed that a set of basis operations could be used to represent the computational algorithm of any processor. Concepts for describing and creating such algorithms will be presented in Volume II. Beyond this description there is the final necessity of allocating the basis operations to a processor architecture (possibly multiple processors) and accommodating the interconnection requirements. We now introduce the overall requirements for this phase of the design.

#### Level 4—Hardware Functions and Instruction Sets

The fourth logical level of our perspective represents the hardware systems which must realize the functions of level 3. These range from special purpose systems that realize the required operations in dedicated hardware to general purpose computers in which the operations are implemented in software. These systems can therefore be characterized by the hardware functions and/or instruction sets they provide for implementing and interconnecting the basis operations.

For design purposes it would be convenient to have a synthesis procedure for creating the required hardware architecture from a logical design. This will be discussed in Volume II.

For the present, we note that in implementing the basis operations there is a fundamental division in the requirements for mapping processing to processor. First, we must implement the individual functions according to the required parameterizations and interconnections. Second, we must ensure the correct sequencing and synchronization of the basic processing operations. These two aspects of implementation may be considered as the data flow and the control aspects of processing. They form powerful concepts with which to analyze both processing and processors as we shall see in Chapters 6 and 7. In Volume II, we shall see that they also provide a basis for the systematic design of processors.

**Level 5—Component Technology**

The basic building blocks of both dedicated hardware signal processors and general purpose processors are determined by the current state of component technology. In 25 years component technology has evolved from single logic gate components to a single component capable of high speed multiplication and addition of complex numbers. Complete processors on a single chip are now widely available. Yet the cost increase for these components has been almost insignificant in comparison to the performance increases. Clearly the cost-performance constraints of past processor implementations no longer apply. Projections for continued component technology advances indicate that cost-performance characteristics are unlikely to stabilize in the near future. The net result of this rapidly advancing component technology is a drastic alteration in hardware cost constraints for digital processor design.

A review of technology and a forcast will be presented in Chapter 7. A designer in this field must maintain currency in components and capabilities before beginning a design. An implicit requirement for the designer is a wide ranging knowledge of alternative architectures and interconnection mechanisms. Such a presentation is beyond our scope although several examples will be discussed in Volume II. The rapid advances in VLSI technology are not only dramatically altering the processing throughput and price of components but also creating new interconnection capability. Thus, it becomes clear that designers must track technology not only for capability but also for new perspectives on what is possible. An explicit discussion of this is reserved for Section 5.2.3.

The thrust of this subsection has been the role of technology in driving new capabilities in hardware architectures which become the targets for processor implementation. Many such architectures have evolved and are reasonably well understood. Technology is, however, creating new capabilities and these directly impact the ability to create viable architectures. In the next section we shall deal with the overall twin thrusts of processing algorithms and architectures.

### 5.2.3 Algorithms and Architecture

In this section we shall deal explicitly with the abstract features of Figure 5.1. On the one hand, the thrust from application requirements, through basic theory to algorithm. On the other hand the flow from components to architectures to processor configurations. The aim is not to quantify these concepts but to create this overview; for it is the one shared by the authors and it is the motivation for the approaches taken to design in Volume II; and has biased our approach to the study of theory.

**Algorithm Development**

The implementation of DSP techniques for any particular application has traditionally involved either the development of efficient algorithms and/or the development of efficient system architectures. In both cases hardware cost constraints and the real-time performance requirements of the application have been major considerations. In addition, the two must be related by design algorithms which yield acceptable systems within the myriad constraints imposed from all sides.

The activity of efficient algorithm development is indicated in Figure 5.1 as a downward mapping from the required functions of level 3 to the operations of level 4. The term "efficient" has usually been translated to mean reducing the number of multiplications required since multiplication has traditionally required either increased hardware costs or increased execution time if implemented in software. Efforts to develop such efficient algorithms have not been without considerable success, the fast Fourier transform (FFT) algorithm being a notable example. However, the reduction in multiplication requirements has generally been at the expense of more complex data handling requirements. This translates directly into more complex control logic for hardware implementations and generally increased algorithm complexity for software implementations.

Advances in component technology are removing the hardware cost and execution time constraints previously associated with multiplication operations. Single components are now available that carry out multiplication and addition operations with equal speed. Clearly, the traditional approach to assessing algorithm time-complexity strictly in terms of the number of multiplications required is no longer valid for systems employing these new components.

If the rapid advances in component technology continue, the advantages to be gained from continued efforts in algorithm development are likely to be of more interest at the level of component design than at the systems implementation level.

**Architecture Development**

Architectural development is represented in Figure 5.1 as an upward mapping from the available component technology of level 5 to the basic system operations of level 4.

The computational requirements of DSP techniques, especially in real-time applications, have generally been too severe to be met by any but the largest and most expensive general purpose computing systems. As a result, many applications of DSP techniques could only be realized by the design of special purpose processors. In the past the basis operations presented in Section 5.1.3 represented relatively complex operations for implementation in hardware. It is not uncommon to find a special purpose processor devoted entirely to the performance of a single function such as digital filtering, correlation or the calculation of discrete Fourier transform coefficients. In general a special purpose processing system could be designed that would meet any given performance requirement by employing large enough quantities of dedicated hardware. However, the cost of such systems, both in terms of hardware and design complexity, was often prohibitive in the past. The lack of flexibility of these systems meant also that a new design was required for each new application as well as new component technology.

Attempts to reduce recurring design costs and to improve flexibility lead to the development of programmable special purpose processors. Typically these processors were limited to the performance of one type of function such as digital filtering or FFT calculations but they provided programmable control of parameters. These processors could be used as modules within a variety of systems having different requirements, thus reducing overall system design costs as well as introducing some degree of flexibility into the system.

For applications with less severe real-time performance requirements (i.e., the analysis of recorded seismic data in geophysics) the available general purpose computing systems could be programmed to perform a variety of DSP functions. General purpose systems reduced hardware costs and provided flexibility but were generally unable to meet even modest real-time performance requirements. The instruction sets of general purpose machines rarely provided basic assembly level instructions that were more complicated than simple arithmetic and logic functions. Clearly the implementation of even the most elementary DSP functions involved considerable programming and were severely limited by realtime processing capabilities, particularly on low cost systems with average instruction cycles of a few microseconds. High performance general purpose systems capable of meeting some real-time performance requirements (i.e., Illiac IV, Cray I) tended to be more expensive than designing a dedicated system in many cases and they were generally difficult to program.

Efforts to develop general purpose architectures specifically suited to DSP requirements began to gain momentum in the early 1970s with the design of several processors. These processors all made use of some form of parallelism or pipelining to increase throughput and the extra hardware required for these designs kept their prices high. Several of these processors will be discussed in Chapter 7.

Recent advances in component technology are removing the hardware cost constraints previously associated with multiprocessor architectures. The use of multiple processors to provide the real-time performance required in many DSP applications is rapidly becoming a cost effective approach to the design of general purpose signal processing hardware. The development of efficient multiprocessor architectures will undoubtedly become a major concern in DSP implementations in the future. The key issues to be resolved concern the partitioning of overall system processing requirements, the allocation of tasks to processors, and the synchronization and control of multiple processors.

### 5.2.4 Summary

In this section we have presented a perspective of the interaction of processing algorithms and the availability of compatible components for implementation. The overriding thesis is simple: the one influences the other. The search for algorithms from the basic theory is influenced by the perception of target implementation mechanisms. This is a most important concept. The fast Fourier transform was developed under the constraint of limited execution capability of general purpose processors. It traded complex data flow for faster execution. Many new components may alter this balance in favour of different algorithms. Thus an interesting quandary for the designer; "Is my algorithm intrinsically hostile to my target components?" There is no easy answer to this question and good designers must be aware of the interaction and continually question their assumptions.

## 5.3 CHAPTER SUMMARY

Perhaps the overriding result from this chapter is that the processing requirements of a wide range of applications can be represented by a set of basis operations. It is this set of operations in various interconnected structures which characterizes the processing algorithm and eventually drive the design of processors.

These basic operations all involve multiplications and additions which could be programmed on any general purpose computer; however; the performance requirements of most applications are such that this is usually unacceptable. It is the structure of the computations in the operations which must be supported by the architecture of the host processor.

We have seen in previous chapters that a difference equation can be represented by an equivalent set of matrix equations. Yet there is an enormous difference in processor architectures to support either representation. In the state space approach the intrinsic parallelism of vector operations leads to designs with an equivalent data path parallelism. However, an intrinsic requirement for vector manipulations is to invert a matrix. No fast algorithm for this operation has appeared; and processors which reflect the state space representation of a system have also not emerged.

On the other hand, high speed multipliers and adders with supporting data paths have been sufficient to yield processors which implement algorithms represented by difference equations. In addition, the FFT provides an expedient means of moving in and out of the frequency domain, so that the appropriate approach could be exploited. Many processors reflect an architecture which exploits this approach to signal processing. Some of these will be explored in Chapters 6 and 7.

General purpose processors must provide a capability for implementing any of the operations within a wide range of parameterizations and interconnections. These operations become important mechanisms for comparing processor architectures and performance. Application areas are differentiable by the basis operations they use and by the time constraints imposed on their execution.

PART **B**

# SIGNAL PROCESSORS

The design of this part of the book was based on two global requirements: first, to build a self contained conclusion which would relate processing theory to processors; second, to point onward to the problems of designing both processors and processing systems in a VLSI environment.

There are perhaps several ways of accomplishing this; for example, one massive case study of a large scale processor exposing the performance enhancement mechanisms and design issues. This did not seem to satisfy our requirements and an alternate route was chosen.

Chapter 6 is a blended stream of tutorial material with real examples. In the first section mechanisms for enhancing and analyzing performance are introduced. Then two older processors are examined. These two are fundamentally different in concept and during the examination many new issues are uncovered and discussed. In the last section a whole range of design issues is examined.

In Chapter 7 new processors are examined based on the concepts and experience gained in the previous chapter. Here the examination is more finely focussed and more structured. In addition, the impact of technology is introduced and examined.

At the conclusion of this volume the reader should possess sufficient background in both theory and processors, to appreciate in depth how to relate a new processor to his processing requirements. We hope the reader will also proceed onward to Volume II in which the concepts and methodologies of processor design are presented.

Some background in computers is required to appreciate the material. The general concepts of a simple processor architecture (any micro or mini) and the concepts of assembly language programming would be sufficient. Ideally, a course in a mini or micro computer with some experience in programming in both assembler and a high order language would serve as prerequisites. An introductory course in computer architecture would provide more than adequate preparation.

CHAPTER 6

# Performance Measures and Limitations

## 6.0 INTRODUCTION

This chapter continues the process begun in Chapter 5 of relating the theoretical concepts of signal processing to the range of often conflicting requirements which must be resolved in creating signal processors. To accomplish this it is first necessary to agree on some measures of performance, and to have some historical context in which to view attempts at enhancing these measures. Many issues, in addition to raw processing power, must be satisfied in the design process; cost is a notable concern for example. The concept of cost extends beyond the purchase price of a system to the total costs of creating software, maintenance, and even less quantifiable considerations such as growth modularity. Discussions of these broader issues are contained in Volumes II and III.

In this chapter we examine the fundamental concepts of high performance processing in terms of two early super computers. In Chapter 7 the concepts developed here will be extended to more recent processors. It may come as a surprise that the mechanisms for achieving high performance were well thought out by the mid-1960s, even though the technology was not yet able to support the concepts.

At the most fundamental level in any processor, operands must be moved from memory to processing units for computation, and the results returned. The throughput of the processing units (e.g., a multiplier) seems an overwhelming concern, for they are a potential bottleneck; however, the arrangements for the flow of data and the control mechanisms are ultimately of equal importance. Our aims in this chapter are therefore threefold:

1. to introduce mechanisms for enhancing processing performance and to quantify their potential;

2. to examine two processors in order to provide some realism to the performance models and at the same time to appreciate the other factors which limit performance and/or the scope of a processor architecture; and
3. to assemble and discuss other performance limitations which are real, yet difficult to model or quantify in general.

In order to accomplish these aims there are a host of concepts and implementation mechanisms that come under the general field of computer architecture, which must be introduced. This is done in Section 6.1. It is our task to choose selectively a path which reveals the concepts relative to high performance signal processors and which places the key issues of high performance in perspective. In order to accomplish this we will use von Neumann's original architecture as an example, not only to develop the concepts he introduced to computing, but also to highlight the issues involved in increasing the computing power of this type of machine. When these are implemented, we must search for some form of multiplicity if further performance enhancement is required, which leads us to the concepts of pipelines and to arrays of processors.

In Section 6.2 two classical processors, the ILLIAC-IV, and the Advanced Scientific Processor (ASC) built by Texas Instruments, are examined as examples of early attempts to achieve high performance and to illustrate the concepts developed in Section 6.1. The ILLIAC-IV was designed at the University of Illinois during the late 1960s and manufactured by Burroughs Corporation. The original design contained four quadrants of 64 parallel processing units each and other performance enhancement mechanisms. The ASC contains an array of pipelines for high speed arithmetic operations. It will serve to illustrate the throughput potential of such arrays.

In Section 6.3, the previous sections are drawn together in a summary of performance limitations which often seem peripheral to the designer, but which can, if ignored, negate the usefulness of an otherwise elegant design. These limitations are often difficult to model and must be faced as issues, the resolution of which is based on principles unique to the particular environment. They are, however, real and should be explicitly acknowledged when designing or when evaluating an existing or proposed processor architecture. In particular, the impact of technology in creating a demand for innovative architectures is introduced. Influences of a rapidly changing technology will be further discussed in Chapter 7.

## 6.1 PERFORMANCE MEASURES

### 6.1.0 Introduction

This section provides the first discussion of processor architectures. Unfortunately the word architecture has a wide variety of connotations, some of which are apparently quite different. From our point of view it implies a block diagram of the major functional components of a processor and their interconnection for data flow. That part of the processor which controls the operations of the units and/or the data paths is normally omitted. The granularity of the block diagram in terms of the operational detail of each block varies, depending on the purpose of the discussion which it supports.

An architectural block diagram corresponds in some ways to pseudo code in programming. It is a logical view of the required operations which hides (or makes transparent) the complexity of the implementation details.

We begin in Section 6.1.1 by establishing a historical context. The initial concepts for a programmable computing machine were established over a century and a half ago. The architecture of modern machines however was strongly influenced by the concepts introduced by John von Neumann during the early 1940s. His contributions form a comparison benchmark and will be examined in some detail. In particular we are interested in the mechanisms for achieving and controlling the flow of data within the machine and in the intrinsic performance limitations which result.

In Section 6.1.2 two mechanisms for enhancing throughput are discussed and analyzed. The requirements for control mechanisms to support them are tabulated. In Section 6.1.3 it is shown that the control of data flow leads to an architectural classification scheme which has proven useful for many years.

Enhancing processor performance within a myriad of constraints is the ultimate goal of a design team. There is still a long path to travel before processor design can be fully understood, but this section is the first step.

### 6.1.1 A Historical Perspective

This section introduces the preliminary concepts of data flow and control which will gradually occupy a central theme of both a perspective by which processors may be examined and as part of a methodology to design them. Establishing the flow of data through a computing machine and the mechanisms for controlling this flow are perhaps the most crucial tasks an architect faces.

The first well documented plans for a programmable computing machine were formulated by Charles Babbage in 1834. His concept called for data and for instructions to be punched on two separate sets of cards which would be fed into his machine by separate card readers. A clear separation existed in his mind between the flow of data and its control (including manipulation). The motivation seems to have been that a mathematical algorithm is essentially data independent; and thus a set of cards which faithfully executed an algorithm became a resource which could be used on any suitably coded set of data cards. Despite the logical advantages, numerous synchronization problems existed, including those introduced by the vagaries of card readers (it is perhaps safe to assume Babbage's card readers were no better than their present day counterparts).

From Babbage's time until about the mid-1930s, the realizations of computing machines used mechanical components. In these machines the control was an intrinsic part of the connections between the components, which is a way of saying that the machines were special purpose. This approach solved the control problem by eliminating it, but the life cycle cost of a machine which could execute only one algorithm was prohibitive. The problem of achieving a general purpose machine was conceptionally simple enough; given a set of processing resources, how do you interconnect them to execute a variety of algorithms? And how do you control this interconnection and the flow of data (i.e., operands and results)? The solution was published by Burk, Goldstine, and von Neumann in 1946. The architecture is shown in Figure 6.1. The concept was so deceptively simple it is difficult to believe there was ever a problem.

In order to appreciate the development of signal processors, it is necessary to understand the concepts and limitations of the von Neumann architecture, for his work has dominated the design of computing systems over four decades. In the following we present an overview, then a more detailed examination.

The operation of the von Neumann machine depends on a word addressable memory. Such memories are called random access memories or RAMs. The bits in the word are either data or instructions. Consider for the moment instructions; each instruction can be logically divided into two fields. These two fields are coded to represent an operation and an address in memory containing the operand. A controller and a processor are designed to perform as follows: first an instruction is "fetched" from memory; the operation is decoded; the operand is then fetched from memory and the operation "executed." The algorithm is performed in a step-by-step sequence of fetch/execute cycles. To expand this, consider the instruction set shown in Figure 6.3, the architecture in Figure 6.1, and the sequence of control operations in Figure 6.2. These represent a machine implemented at the Institute for Advanced Studies (IAS) at Princeton.

Before proceeding, note that in this implementation two instructions were fetched at a time. This was an early attempt to increase the performance of slow memories. This feature was dropped from later computers and replaced by cache memories and/or instruction prefetching.

The control sequence of Figure 6.2 contains the key to the operation of the computer. The sequence of steps determine the required transfers of information, including arithmetic and logical operations. The registers in the architecture shown in Figure 6.1 are required to support the fetch/execute sequence and to perform the data manipulations.

The break with conventional wisdom was based on the following premises and conclusions.

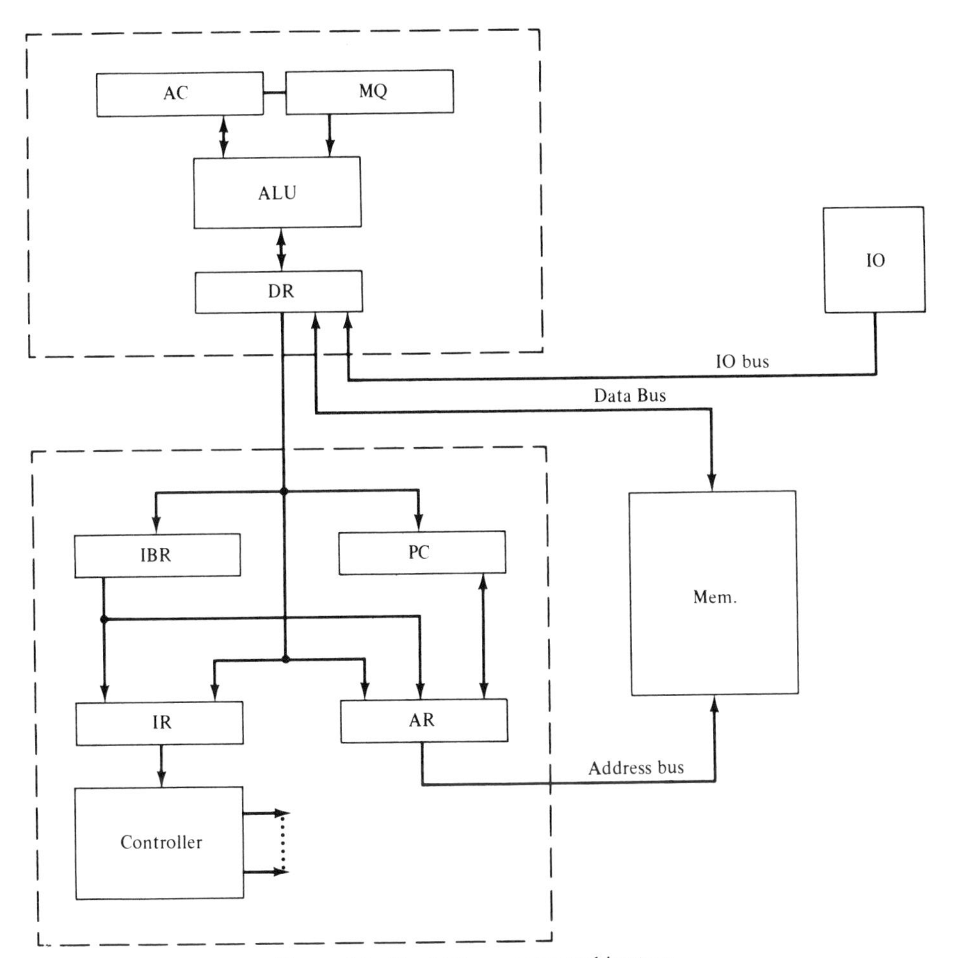


**Figure 6.1:** The IAS computer architecture.

1. *Premise*: There is no intrinsic difference between instructions and data. Each can be represented in the same physical media by the same mechanisms.
   *Conclusion*: There is no need for a separate memory (e.g., punched cards) for data and for instructions.
   *Requirement*: Memory must be organized in such a way so that addresses can be associated with individual instructions and/or operands. And the processor must address memory, not merely read the next set of characters.

2. *Premise*: Instructions can be partitioned into two major fields containing, in addition to the operation, the address of the operand.
   *Conclusion*: An instruction can be fetched from memory, the address field interpreted, the operand fetched, and the operation performed.
   *Requirement*: The processor architecture for fetching, interpretation, and execution.

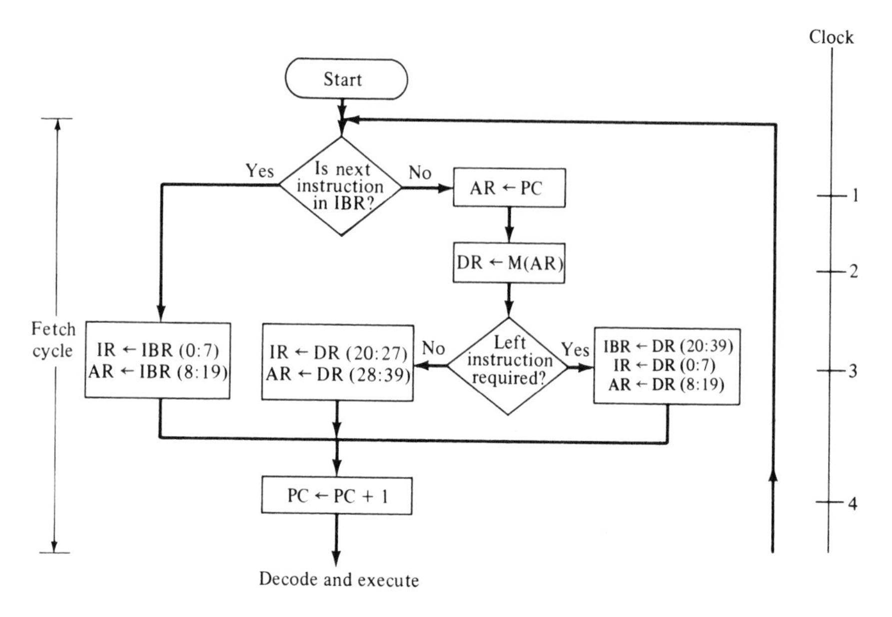


**Figure 6.2:** Instruction fetch; control microcode.

The architecture shown in Figure 6.1 fulfills all these requirements. Based on coding an operation and address in separate fields of an instruction, control and data flow were inextricably tied together; synchronization problems vanished.

The elegance of the von Neumann solution forms the basis of all general purpose maxi, mini and micro computers since his time. The irony of this is that he was attempting to design a special purpose machine to compute artillery trajectories.

Several criticisms of this machine can be made at this point although these are almost universally accommodated in later implementations:

- Limited arithmetic capability.
- No subroutine linking facilities.
- No index registers. (The address modification instructions were intended for iterative loops—a programming nightmare in practice.)
- Two word fetch, complicated the control (later replaced with cache memories and instruction queues).
- Input/output was primitive.
- Instruction set design was not suited to logical operations.

In the intervening years every aspect of this architecture has been subjected to extensive scrutiny in order to enhance both the raw instruction

Data Transfer

| Action | Description |
|---|---|
| $AC \leftarrow MQ$ | Transfer contents of register MQ to the accumulator AC |
| $MQ \leftarrow M(X)$ | Transfer contents of memory location X to MQ |
| $M(X) \leftarrow AC$ | Transfer contents of accumulator to memory location X |
| $AC \leftarrow M(X)$ | Transfer M(X) to the accumulator |
| $AC \leftarrow -M(X)$ | Transfer −M(X) to the accumulator |
| $AC \leftarrow \lvert M(X) \rvert$ | Transfer absolute value of M(X) to the accumulator |
| $AC \leftarrow -\lvert M(X) \rvert$ | Transfer the negative of the absolute value to the accumulator |
| $AC \leftarrow AC + M(X)$ | Add M(X) to AC; put the result in AC |
| $AC \leftarrow AC + \lvert M(X) \rvert$ | Add $\lvert M(X) \rvert$ to AC; put the result in AC |
| $AC \leftarrow AC - M(X)$ | Subtract M(X) from AC; put the result in AC |
| $AC \leftarrow AC - \lvert M(X) \rvert$ | Multiply M(X) by MQ; put most significant bits of result in AC, put least significant bits in MQ |
| $AC, MQ \leftarrow MQ \times M(X)$ | Multiply M(X) by MQ; put most significant bits of result in AC, put least significant bits in MQ |
| $MQ, AC \leftarrow AC \div M(X)$ | Divide AC by M(X); put the quotient in MQ and the remainder in AC |
| $AC \leftarrow AC \times 2$ | Multiply accumulator by 2, i.e., shift left one bit position |
| $AC \leftarrow AC \div 2$ | Divide accumulator by 2, i.e., shift right one bit position |

Control

| Action | Description |
|---|---|
| go to M(X, 0:19) | Take next instruction from left half of M(X) |
| go to M(X, 20:39) | Take next instruction from left half of M(X) |
| if $AC \geqslant 0$ then go to M(X, 0:19) | If number in the accumulator is nonnegative, take next instruction from left half of M(X) |
| if $AC \geqslant 0$ then go to M(X, 20:39) | If number in the accumulator is nonnegative, take next instruction from right half of M(X) |

**Figure 6.3:** IAS instruction set.

execution rate and the suitability of the instruction set to a wide range of applications. Three considerations interact to produce a machine: instruction set, processor architecture, and technology.

In earlier days technology tended to drive architectural design or at least to limit the range of feasible realizations. Today the application can exert a predominant influence, with the architect choosing from a wide range of components or indeed even creating his own for special purposes. The three factors interact as follows:

Instruction Set: The instruction set reflects the basic operations and data transfers required to implement the algorithm. The design is highly influenced by the requirements for compilers. Of particular interest is the flow of data; and memory addressing for data structures. Typically, several dozen addressing modes are implemented.

Processor Architectures: The register structure must be devised to fulfill two requirements: first, the implementation of the instruction set, and second, the implementation of the fetch/execute cycle. In the latter case, the architecture can be designed to overlap these two functions by utilizing instruction queues and control mechanisms for prefetching blocks of instructions.

Technology: The basic clock cycle depends ultimately on the speed of the components. In addition, technology has provided higher functional densities at lower costs so that operations such as multiplication can now be implemented with single clock period delays.

In a broader examination of the flow of data and control, it becomes apparent that the von Neumann machine is heavily biased in favor of control. The strict sequentiality of instruction execution forms an intrinsic bottleneck to performance, for data can never be processed faster than its accompanying control. This is illustrated in Figure 6.4(a). Note the strict sequentiality of instruction flow in a computer and that the address of and the operation on each item of data is tied to this control flow. The overall problem is to maintain control while enhancing the flow of data. We now address this problem.

In the earliest days of programming it was found convenient to group segments of instruction into logical entities called "subroutines" or "procedures." A program could be regarded as a sequence of procedures rather than instructions as shown in Figure 6.4(b). Extensions to this concept will prove most useful for logically describing concurrent processing and will be discussed in Volume II. Here we note only that each procedure is executed sequentially based on the control maintained by the component instructions.

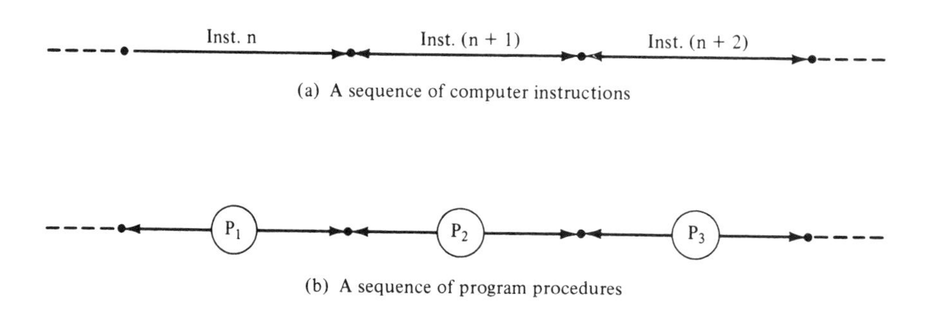


**Figure 6.4:** Computer instruction and procedure flow.

Three options emerge at this point if higher performance (i.e., faster execution time) is required:

1. the execution speed of each instruction could be increased; this requires faster hardware
2. the design of the instruction set could be optimized for the algorithm (so fewer are required for a given function)
3. the machine architecture (including the instruction set) could be optimized for the algorithm

Hardware speed is a function of both the available technology and money. Instruction set design involves the inclusion of operations, addressing modes and branching mechanisms for implementing algorithms, and the design of a compatible processor architecture.

Architectural innovations fall into two classes; first, given the highest speed hardware and an instruction set which is compatible with the operations required in the algorithm, how can the execution rate be maximized? Second, given all the previous operations, how can data flow through the processor be maximized? The sequential fetch and execution of instructions is a feature of the von Neumann architecture. Enhancing this control flow will be discussed later. At this point we shall concentrate on data flow.

Of the many architectural innovations for increasing data flow, two form the major components of almost all systems. Processing throughput can be enhanced by replicating the processing units and thus providing a parallel set of paths or by the use of a special sequence of serial processing units which is called a pipeline. The structure and resulting performance of these mechanisms will be examined next. A scheme for classifying the gross features of the control and data flow will be discussed in Section 6.1.3.

### 6.1.2 Performance Models

From the perspective of a von Neumann machine each instruction was a new event; it was fetched, decoded, and executed. The fact that sequences of instructions were logically related in some form of task was architecturally transparent. It was the task of the programmer to code iterative loops or to use other mechanisms to enhance performance while constrained by this basic limitation. As the cost of controllers was reduced (by high speed components and microprogramming), it was possible to introduce assembly language commands which executed lengthy manipulations on designated strings of data. Even after this enhancement was in place, it became apparent at the algorithmic level that many sequences of operations have little or no precedence relationships and could be done in any order or, more importantly, together. In general we seek forms of multiplicity as a final resort in enhancing performance. We shall now introduce and address this problem.

Several performance measures have been used in the past to evaluate computer performance; two attributes of such measures seem necessary. First, they must be meaningful in the sense that they reflect a well understood criterion in a particular processing environment. Second, they must be measurable by some reasonable procedure across a range of processors. Two measures are used commonly:

MIPS—Millions of instructions per second; a measure of raw instruction execution rate without specifying the nature of the computations.

MFLOPS—Millions of floating point operations per second; a measure useful in assessing computations in floating point format.

Despite its widespread use, an MIP is perhaps the poorest definition of performance since it contains no quantifiable attributes for assessing useful processing. Usually some standard mix of instructions must be specified before any form of comparison is meaningful.

In a general purpose computer used for a variety of processing tasks it is usual to define a benchmark program which contains a representative mix of basic instructions for a class of problems. The execution time of this program serves as an overall measure of performance (for that class). On the other hand, in specifying some systems it is more useful to begin with a response time constraint, which is then translated into performance measures for each subcomponent.

The term FLOPS is widely used in signal processing applications and is a common measure of performance in comparing processors. The difference between MIPS and MFLOPS can be appreciated by considering a simple DO LOOP high level language construction:

```
DO I = 1 TO 1000000 STEP 1
   BEGIN
   Z(1) = X(I) * Y(I) + C(I);
   END;
```

Each iteration accomplishes two floating point operations, yet depending on the host computer the compiled assembly language code could occupy many bytes. The speed of execution of the two floating point operations depends therefore on the MIPS of the processor; provided that each iteration could be completed in say a microsecond, the processor would then execute at the rate of two MFLOPS. A system of a million processors could conceivably do all the iterations at once and attain a performance of two million MFLOPS.

In this section our concern is with two approaches for increasing the performance of a system: a pipeline or a parallel array of processing units. Both achieve enhanced performance and both impose data flow and control requirements which will be discussed later. To introduce the concepts, consider a situation as shown in Figure 6.5(a). Here a sequence of procedures is assumed each to process data in a time $\tau$, except for the DFT procedure which

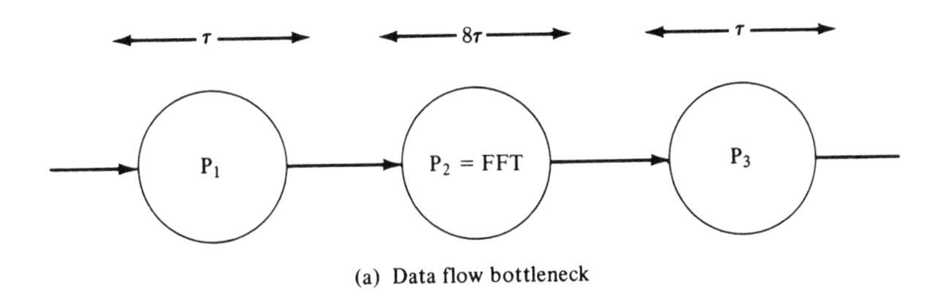


(a) Data flow bottleneck

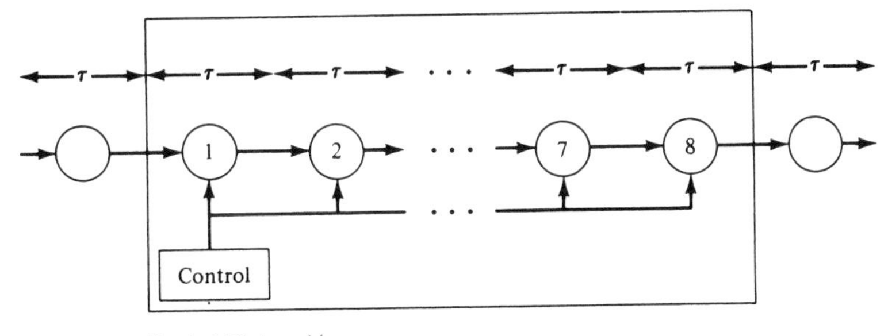


Bandwidth in = $1/\tau$
Bandwidth out = $1/\tau$ (after pipe is filled)

(b) Pipeline solution

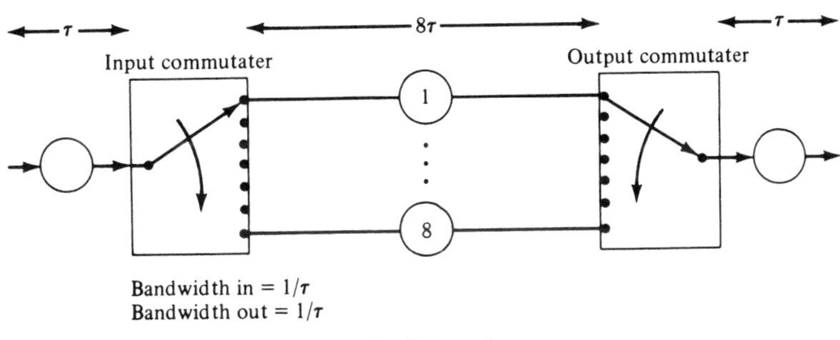


Bandwidth in = $1/\tau$
Bandwidth out = $1/\tau$

(c) Parallel solution

**Figure 6.5:** Data flow bottlenecks and solutions.

consumes 8 $\tau$. This forms a data flow bottleneck, and given that all the mechanisms for increasing throughput (i.e., for decreasing $\tau$) discussed in Section 6.1.1 have been exhausted, either a pipeline or a set of parallel paths may alleviate the constriction. In the following an overview of pipeline and array mechanisms will be discussed followed by a performance analysis.

A simple pipeline is shown in Figure 6.5(b). Here the procedure has been partitioned into a sequence of 8 processing steps each requiring $\tau$ seconds (this forms the pipeline). Input data then progresses through the pipe being successively modified until it emerges. The design must be such that a new input can be accepted every $\tau$ seconds, and also such that each unit operates independently (no inter-unit data dependencies). If this is the case the throughput matches the rest of the system, except for the transient interval for the pipe to fill.

The simplest view of a pipeline is that each stage consists of combinatorial logic driven by an input register. The output from a stage is captured by the input register of the following stage. Each stage has a delay for the initial data capture and subsequent processing. If all stages have an equal delay, then a synchronous clock can transfer results into each input register. This is the simplest control problem. If there is a large discrepancy between the various delays in each stage, then an asynchronous data transfer might be in order. It is possible to construct a pipeline in which the intermediate registers are omitted. The design of such pipes requires careful timing of data input/output.

A synchronous pipeline produces a result every clock period $\tau$, i.e. a data-flow rate of $1/\tau$ outputs per second. An $N$-stage pipeline gives an apparent $N$-fold increase in performance. If the input to the pipeline is intermittent, however, then some stages will not be processing valid data, and this must be accounted for by the control mechanism. If, on the average, only a fraction $P$ of the total stages are occupied, then the data flow falls to $P/\tau$ outputs per second.

The appropriateness of a pipeline approach clearly depends on the predictable nature of the data flow. Pipelines are often used to build mathematical processors because of the well understood nature of such operations. In general a pipeline can be considered if:

1. The procedure can be broken into a sequence of discrete steps
2. The steady state data flow matches the remainder of the system
3. Components can be found which implement the steps with the desired response

An alternate approach is to build more processing units and connect them in parallel. A system of parallel paths is shown in Figure 6.5(c). In this case the individual processes still operate with a response time of 8 $\tau$. The input commutator sequentially allocates input data which is collected 8 $\tau$ seconds later by the output commutator. Once again, after a transient interval, data flow through the system is unconstrained. The advantage compared to a pipeline is that the procedure need not be partitioned into steps. On the negative side the input/output commutation is usually difficult to implement and consumes some overhead which lowers the effective throughput.

A parallel array need not have an identical delay in each path, although this complicates the control problem. If each of $N$ units has a delay $\tau_i$, then the average delay could be used to compute data-flow. For $N$ parallel paths the response will be shown to be the same as an $N$-stage pipeline. If a proportion $P$ of units is unused then the output rate drops. The overall behavior is identical therefore with a pipeline although implementation issues are widely different.

Early implementations of these two logical concepts were in hardware; but each may be implemented in either hardware or software. Historically, implementation costs favored a pipeline; however, this has been and is changing rapidly. It is important at this point to regard them as constructs and to separate their logical attributes from current implementation mechanisms or limitations.

To model the performance of each approach in more detail it is useful to ignore initially the synchronization and other control problems. We will analyze first the parallel array and then the pipeline. The conclusion is that each has the same potential throughput and this throughput is highly dependent for efficient use on a compatible data input rate.

Generalizing the examples shown in Figure 6.5, we assume a parallel processor $Q_1$ of multiplicity $M$; and an $M$-stage pipeline processor $Q_2$. Suppose that they each perform a task $F$ in time $t_1$ and $t_2$, respectively. Figure 6.6 shows a space-time graph of the two processors in which $n$ tasks of type $F$ are processed. Suppose an elapsed time of $T_1$ and $T_2$ is required as shown.

For the parallel processor shown in Figure 6.6(a), the last $t_1$ interval will have, in general, some of the processors idle. The time for completion is therefore:

$$T_1 = \lceil n/M \rceil t_1 \quad \text{seconds}$$

1 M + 1 2M + 1
2
3
n
n − 1
M 2M n − 2
t1 2t1 kt1 T1

(a) Parallel array

(b) Pipeline

**Figure 6.6:** Task execution profiles.

where brackets indicate the smallest integer equal to or greater than $n/M$. The throughput in tasks per second is

$$T.P. = n/T_1$$

which for large n approaches $M/t_1$.

$$T_2 = \lceil n/M \rceil t_2 + [n - M(\lceil n/M \rceil - 1) - 1] t_2/M \text{ seconds}$$

where the second term is the time for the pipe to empty. For large $n$, the completion time $T_2$ approaches $nt_2/M$, and the throughput is:

$$T.P. = M/t_2$$

If $t_1 = t_2$, which is a reasonable assumption, then for a continuous flow of tasks the two mechanisms have the same performance capacity.

Identification of the intrinsic adaptability of an algorithm to either approach depends on a detailed analysis. It is often easier to detect potential pipeline operations than parallelism. In either case, however, once the hardware is in place the performance of other than an ideal sequence of tasks is of importance. This is measured by a "utilization factor" and/or a related "speedup factor."

Suppose a processor exists with a multiplicity $M$ (either kind), and a set of non-homogeneous tasks is to be executed. Suppose that the tasks are partitioned into a set $F_1, F_2, ...F_n$; and that for each task $F_i$ it is determined that the task will execute in time $T$ if a processor of multiplicity $M_i$ is available. Without loss of generality, let $T =$ unity, then if $T_i$ is the time to perform task $F_i$,

$$T_i^M = 1, \text{ if } M \geqslant M_i$$
$$= \lceil M_i/M \rceil, \text{ otherwise}$$

If a conventional processor were available, it would perform each task sequentially in time

$$T(\text{sequential}) = \sum_{i=1}^{n} T_i^1$$

The throughput gain-ratio is then

$$G.R. = \frac{\sum T_i^1}{\sum T_i^M}$$

Each task will consume different proportions of the total available processing units. A utilization ratio $\rho_M$ is the space-time profile of the tasks divided by the space-time available (space-time is the product of the number of processing elements multiplied by the processing time). Thus

$$\rho_M = \sum M_i / M \sum T_i^M$$

The gain-ratio and utilization are related since

$$\sum M_i = \sum T_i^1$$

$$\rho_M = G.R./M$$

In order to achieve the performance potential it is essential that the data flow to the processors is equal to the available processing bandwidth, and in the case of a parallel array that the tasks be distributed and the results collected. The job profile has an equally significant effect on utilization and if under-utilization becomes a cost, then jobs must be made to match capability with a greater sense of urgency. Both of the mechanisms therefore exhibit an algorithm rigidity in the sense that although a high throughput can be obtained, a varying job profile leaves the processing potential under-utilized. High performance processing systems require a large multiplicity (and hence high cost) but the cost/effective utilization of this performance narrows the range of problems to which such an architecture can be applied.

The multiplicity of parallelism gives a measure of potential speed-up. The actual speed-up depends on the architectural arrangements for supplying the required data flow. As the operations required by the processing algorithm depart from full utilization of the parallelism of the processor, the performance degrades. A measure of this degradation will now be developed.

Suppose the processor executes various operations as follows:

$P_v$ : the rate for vector operations

$P_s$ : the rate for scalar operations

and that a particular algorithm can be characterized by the ratio of scalar operations to the total operations executed:

$$f = \frac{\text{\# of scalar operations}}{\text{total operations}}$$

Thus the average time for a mix of operations is:

$$1/P = f/P_s + (1 - f)/P_v$$

or

$$P = MP_s/(1 + (M - 1)f), \text{ where } MP_s = P_v$$

This equation provides a means of estimating the processor's sensitivity to load variations contained in $f$:

e.g., for

$$f = 0 \text{ (i.e. no scalar operations)}$$

$$\text{Pmax} = MP_s$$

This rate is reduced to one-half when

$$1 + (M + 1)f = 2$$

or

$$f = 1/(M - 1)$$

which shows that for a large multiplicity even a small proportion of scalar operations reduces the throughput dramatically.

Thus in general an interesting trade-off occurs. To achieve impressive performance enhancements a large multiplicity is required; but the processor becomes more algorithm dependent in the sense that variations in operations leave large portions of the capability unused. This result is what one would expect.

***Example:***

The FFT (discussed in Section 4.1) involves executing $(N/2)\log N$ butterflies. The computational difficulty is to complete the required butterfly computations in the time allowed by the application. If each butterfly takes $T_B$ seconds and the time constraint is $T$, then

$$T_B \leqslant 2T/(N\log N)$$

If $T_B$ seconds exceeds this time-constraint, then some form of parallelism must be introduced to enhance the performance. Consider an arithmetic unit which computes a butterfly; four architectural variations can be considered:

- sequential
- pipeline with $M = \log N$ stages
- iterative parallel array with a multiplicity $M = N/2$
- fully parallel array with $M = (N/2)\log_2 N$

The table on the next page gives an estimate of the execution times. For the pipeline assume that $n >> M$ so that the time for each result is the inverse of the limiting throughput. The cost performance figure is defined as the ratio of the number of basic arithmetic units divided by the total time.

To some extent the efficiency of the architectural alternatives can be quantified by considering the time-overhead to perform an FFT in terms of the time required for a single butterfly. The overhead is a function of both the processor instructions and required data fetches, and reflects the multiplicity of this aspect of the architecture in a very pragmatic way. For example, say a single butterfly (with four multiplications and four additions) is performed in 1 microsecond; then a 1024 point transform might be expected in 5.12 milliseconds; this time is more typically 5 to 8 milliseconds, reflecting considerable overhead in other portions of the machine.

| *Architecture* | *Time T* | *Cost Performance (M/T)* |
|---|---|---|
| Sequential ($M = 1$) | $T_B(N/2)\log_2 N$ | $2/(T_B N \log_2 N)$ |
| Pipeline ($M = \log_2 N$) | $T_B(N/2)$ | $2\log_2 N/(T_B N)$ |
| Parallel-iterative ($M = N/2$) | $T_B(\log_2 N)$ | $N/(2T_B \log_2 N)$ |
| Parallel-array ($M = N/2 \log_2 N$) | $T_B$ | $N\log_2 N/(2T_B)$ |

In Section 6.2, processors using both the above mechanisms will be discussed. Before this, however, an interesting classification scheme which places data flow and control in perspective will be discussed. This scheme also forms the basis for describing the dynamic reconfiguration of architecture to achieve higher performance, to be discussed in Volume III.

### 6.1.3 A Classification Scheme

There have been many attempts to classify processor architectures. A standard classification scheme would be exceedingly useful both for discussion purposes and as a guide to processor designs. The requirements for such a scheme are at least that it be complete (i.e., include all architectures) and orthogonal (i.e., differentiate the key attributes). Unfortunately, despite the attractiveness of the concept, no such scheme exists. Of the many proposals, one [Flynn, 66] forms the basis of many others. It is neither complete nor orthogonal, yet its elegance and intrinsic simplicity are attractive and it does concentrate on data flow and control in a general way.

The basis of the scheme is that a processor processes data by a sequence of instructions regardless of the format and mechanisms whereby each arrives at the point of action. Based on the concept of a data stream and an instruction stream, four possibilities exist:

SISD—Single Instruction Single Data Stream
SIMD—Single Instruction Multiple Data Stream
MISD—Multiple Instruction Single Data Stream
MIMD—Multiple Instruction Multiple Data Stream

Examples are shown in Figure 6.7. Note that both the Babbage and von Neumann architectures are SISD, although they differ greatly in implementation. The performance of such a configuration can be thought of as unity for purposes of comparison.

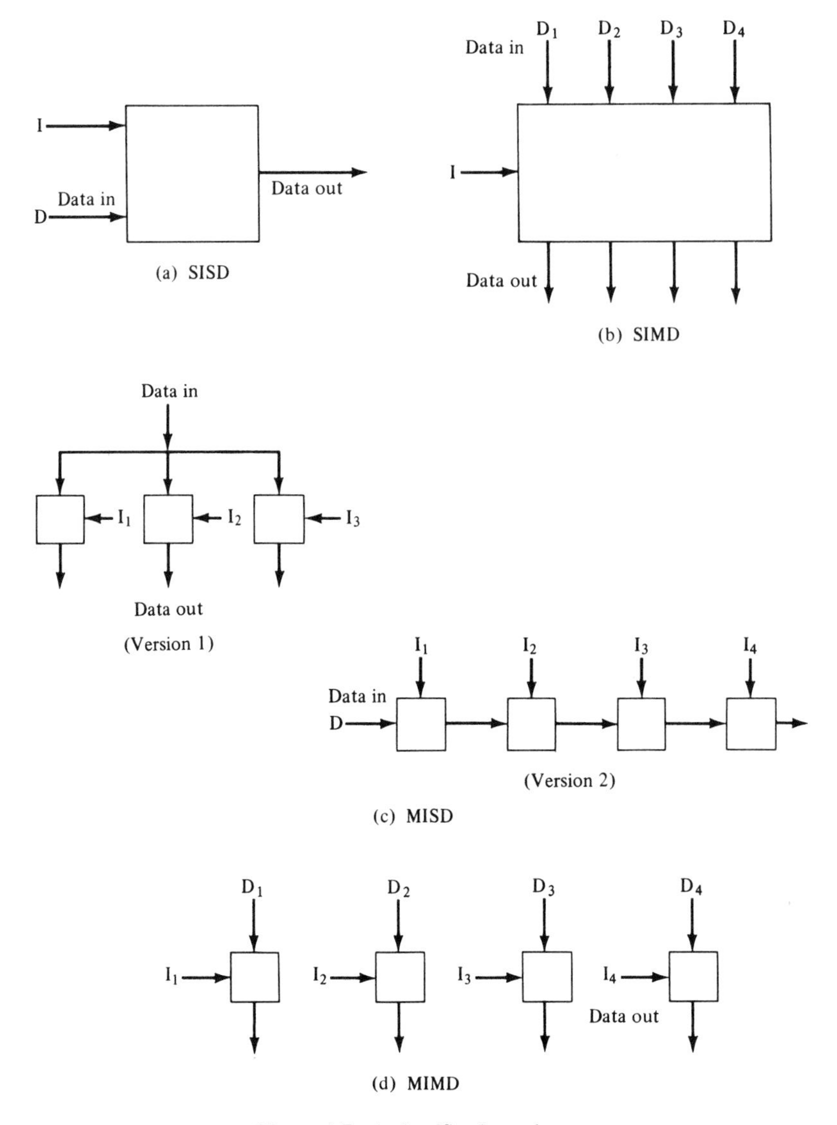


**Figure 6.7:** A classification scheme.

The SIMD architecture is an example of a parallel array in which each processing unit executes the same instruction. It can achieve an $n$-fold increase in data flow band-width for each instruction, provided that the units can be continuously utilized.

The MIMD architecture is implemented by a multiple processor system. Clearly implied is some form of cooperative network to share a computational task (completely autonomous units being of little interest). This is also an example of a parallel array in which the task assigned to each processor can be different. The performance enhancement potential is equal to the number of processors.

The MISD architecture is not widely implemented in practice and substantial disagreement exists on its exact structure. It is considered here as a pipeline in which a single data stream is modified at successive stages, and its performance enhancement potential equals the number of stages as shown in the previous section.

There is a relationship between these classifications and the structure of processing algorithms. An algorithm may contain a collection of processing tasks which could optimally be assigned to different processing configurations to achieve an overall higher performance. If components were of sufficiently low cost, a solution might be to build a conglomerate of different processing architectures and utilize the optimum one at appropriate points in the algorithm. The task assignment problem here is formidable; and as well the physical complexity and lowered reliability of such a conglomerate of components is a major limiting factor of such a scheme. This will be discussed in more detail in Volume II.

### 6.1.4 Summary

This section was a critical first step in acquiring the concepts required to achieve our eventual goal, namely processor design. The outstanding issue is that any algorithm can be viewed as a flow of data subjected to some control mechanisms. Both the data-flow and the control requirements impact processor architectures; neither can be ignored in a design methodology. The instruction set of a modern computer controls data flow and operations as part of the specifications contained within the various component fields. This is a powerful solution which, despite the performance limitations (caused by the strict sequentiality), tends to reappear in many systems as the final guarantee of proper synchronization.

Many processing requirements do not need the close coupling of data and instructions in this sense, and therefore impressive increases in performance can often be obtained with architectures that optimally combine the two requirements. Eventually we must augment and extend these concepts into a design methodology. This will be done in Volume II.

The concepts described in this section were in place by the mid-1960s. In addition, a requirement existed and the technology was also in place. The requirement was for scientific processing involving large matrix manipulations. There is an inherent parallelism (e.g., in matrix multiplication) in many of these algorithms and it was recognized that this parallelism could be reflected into the architecture of processors. The technology was emitter coupled logic (ECL), which had a cost/performance ratio which was acceptable for this application. In the next section we explore two processors developed in response to this requirement and built with this technology.

## 6.2 CLASSICAL SUPER COMPUTERS: TWO EXAMPLES

### 6.2.0 Introduction

The first generation of computers developed during the early 1940s was strongly influenced by the concepts introduced by Babbage a century earlier. von Neumann's architecture, which was to become widely used for general purpose computers, resulted from an attempt to design a special purpose machine to compute artillery trajectories. Early designers were also inhibited by mechanical components for implementation. This accounts for the current use of the word "machine" to describe devices built on silicon chips. Even the advent of vacuum tubes, which considerably increased performance, did not reduce physical size. The EDVAC, completed in 1946, weighed 30 tons and contained 18,000 vacuum tubes.

Historically, the requirement for special purpose high speed computation led to innovations in architecture, which in turn became part of general purpose machines. It is not surprising to find that the acceptable ways of increasing performance in general purpose machines were extremely costly experimental techniques a few years earlier. It was the advent of transistors and silicon technology which made meaningful design of large and fast computing systems possible. The improvements (other than technological ones) can be loosely grouped into three broad (and sometimes overlapping) categories:

1. Improvements in the basic instruction fetch and execution speeds of the basic von Neumann architectures. This was achieved by instruction prefetching, overlapping fetch and execution, memory interleaving, etc. We include here the off-loading of certain functions such as input/output onto special processors.
2. Multi-programming, which involves the overlapped execution of different programs.
3. Specialized Architectures. Problems involving certain operations, such as array manipulation, demand architectures which achieve performance by exploiting the algorithmic parallelism.

The first generation of high performance commercial machines designed for specialized tasks included the Control Data Corporation's String Array Processor—STAR-100, the Texas Instruments' Advanced Scientific Computer—ASC, and the ILLIAC-IV, designed at the University of Illinois and built by Burroughs. The performance of these machines was impressive by the standards of their day; however their specialized application areas and cost limited the number of installations (STAR-100 (4), ASC (7), ILLIAC-IV (1)). They served however as a training ground for the development of concepts and were invaluable in preparing architects to cope with the challenge of VLSI. A second generation of high performance machines emerged in 1975 with the CRAY-1. The CRAY machine was built by the Cray Research Corporation. It was an extension of the architectural concepts of the Control Data 6600 and 7600. The ILLIAC evolved into the Burroughs Scientific Processor—BSP, the STAR-100 into the Cyber 203 and the Cyber 205. These large machines are not our principal concern in this book; however, they form a useful background for illustrating concepts and performance enhancement mechanisms.

In this section we select for review two first generation super computers, the ASC and the ILLIAC-IV. These were both designed in the mid-1960s. Our overall purpose is to explore the architectural issues faced by the design team and their solutions for achieving performance. At a subliminal level a much broader range of issues will emerge, some of which will be discussed in Section 6.3.

### 6.2.1 ILLIAC-IV: An Array Processor

Design studies for the ILLIAC were begun in early 1966. At this time two complementary situations prevailed: the need for higher performance computers and the development of the required technology. Computationally massive problems were already formulated which were not computable on conventional machines. Linear programming problems required the manipulation of very large matrices. A wide range of problems in weather analysis and prediction, heat flow, acoustic propagation, etc., were modeled as sets of partial differential equations over large grids. Signal processing required the usual high speed sum of product computations. Emitter Coupled Logic (ECL) had progressed to reasonable gate density (10-100 gates/chip) with gate delays of 2 to 5 n.s. Memories of a million words with a 200 to 500 n.s. cycle time were available at reasonable cost.

Two further factors influenced the designers. First, it was felt that the various mechanisms to enhance the speed of sequential processing had been exhausted. The complexity of control for such mechanisms as multi-phased memories, pipelined arithmetic units, fetch/execute overlap, etc., had reached a point of marginal performance returns in terms of cost and complexity. Second, results were now available on the SOLOMON computer, which gained performance from a set of parallel arithmetic units under centralized control. The decision was to achieve performance through multiplicity gained in the form of parallelism.

The ILLIAC was to consist of four quadrants of 64 processors (called processing elements—PE). Each quadrant was controlled by a single control unit (CU). The quadrants could be independently reconfigured to operate independently or cooperatively on a problem. The PEs were designed to process 64 bit data, and could be internally configured to either a $2 \times 32$ or an $8 \times 8$ bit data stream.

The general structure of a quadrant is shown in Figure 6.8. The central control unit interprets instructions as in a conventional computer; however, the array of processors is used to simultaneously execute certain commands. The processors are each connected to private memory and are interconnected by a communications network. By this means, a single instruction can simultaneously operate on multiple data streams. Clearly, in such a system the organization and flow of data into the array processors are critical.

The ILLIAC-IV is an example of a parallel array discussed in Section 6.1.3. The array processors in this configuration can be controlled synchronously such that every command executes in the same time. It is therefore an example of an SIMD architecture.

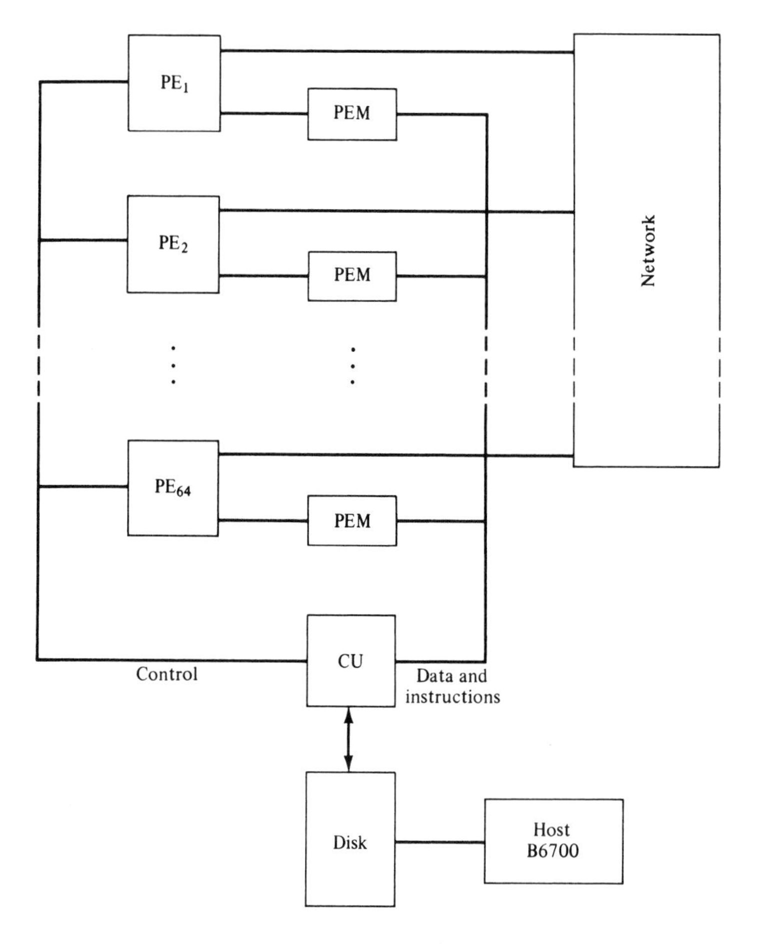


**Figure 6.8:** The ILLIAC-IV architecture (one quadrant). © 1968 IEEE Inc. Reprinted by permission, "The ILLIAC-IV Computer," G. H. Barnes, et al., *IEEE Transactions on Computers*, Aug. 1968, pp. 746-57.

To illustrate the use of an ILLIAC, consider the matrix multiplication problem for $N \times N$ matrices:

$$C = A \times B$$

where

$$c_{i,k} = \sum_{j=o}^{N-1} a_{i,j} \times b_{j,k}$$

Assuming for the moment that data is appropriately organized, this can be written in pseudo-code as:

```
FOR j:=0 to N-1 STEP 1
    BEGIN
      c[i,k] := c[i,k] + a[i,j] × b[j,k], (0 ⩽ k ⩽ N − 1)
    END
```

This is computed as follows: for $j = 0$ (i.e., first row of $A$ and first column of $B$) to $N - 1$ repetitively compute $c(i,k)$. All $N$ processors operate simultaneously and therefore a row of the $C$ matrix is computed by one iteration on $j$. The arrangement of data in each memory must accommodate these operations. Even without further reading it is apparent that the algorithm must be such as to require all 64 units to operate continuously to realize the maximum performance. Thus the ILLIAC performance is highly dependent on the algorithm among other things. A simple control mechanism has been traded for a certain rigidity in application range. This is characteristic of the SIMD concept.

In the following discussion we shall examine the ILLIAC in three parts; first, the general attributes of the array of processors, second, the CU, and third, the PEs. Our purpose is to illustrate the issues which had to be resolved to achieve performance in an SIMD architecture. The mechanisms for resolving these are changing, but the basic issues remain. Before proceeding we note that only one quadrant (64 PEs) was ever implemented.

In order to achieve reasonable performance for a varying job stream the 256 PE arrays were designed to be configurable into 2 × 128 PEs or 4 × 64 PEs. The subconfigurations were each controlled by a CU and could operate independently. This reconfigurability was an attempt to match problem requirements to hardware in order to achieve reasonable utilization as discussed in Section 6.1. It also allowed some recovery in case of failures. This will be discussed later.

The PEs in a quadrant were interconnected to their four nearest neighbors in a square grid such that processor $i$ was connected to $i - 1$, $i + 1$, $i - 8$, $i + 8$. The end connection was circular (i.e., MOD 64) with provision for alternate connection to other quadrants to maintain the pattern.

The Control Unit (CU) is shown in Figure 6.9. The CU must perform all the functions of a standard SISD computer with the additional ability to control both the quadrants and the individual processors.

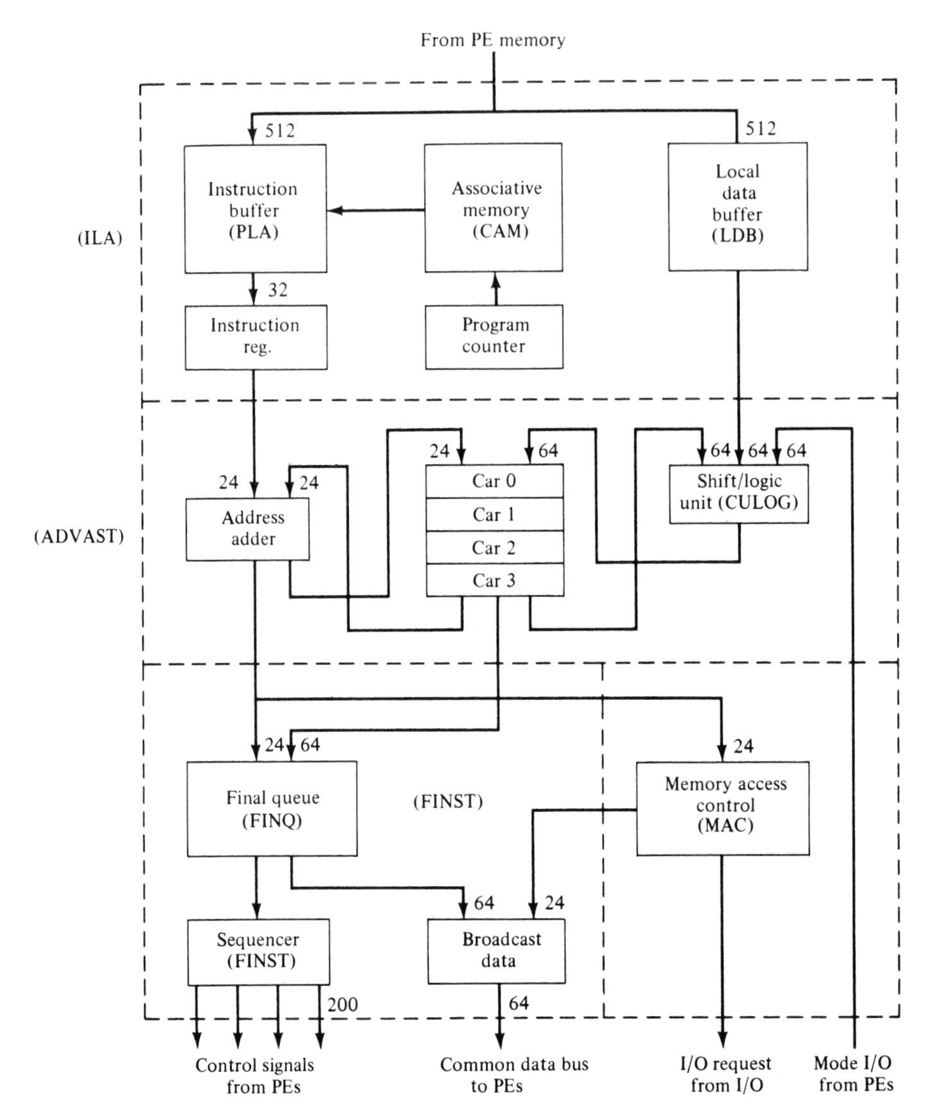


**Figure 6.9:** The ILLIAC-IV control unit. © 1968 IEEE Inc. Reprinted by permission, "The ILLIAC-IV Computer," G. H. Barnes, et al., *IEEE Transactions on Computers*, Aug. 1968, pp. 746-57.

The Control Unit is composed of five major blocks:

ILA: Instruction Look Ahead
CU instructions are fetched (from PE memories) and paged into the instruction buffer

ADVAST: Advanced Station

This unit examines each instruction and executes the non-vector operations. Vector operations are moved to the final instruction queue.

FINST: Final Station

This unit transmits control signals to the PE for execution of vector instructions.

TMU: Test and Maintenance Unit

To control the sequencing of the four quadrants.

MSU: Memory Service Unit.

Functionally the CU must

1. Fetch and decode instructions
2. Generate and transmit control pulses to the PEs
3. Compute common addresses and manipulate common data for transmission to the PEs
4. Receive and process trap signals from the PEs or the host.

These functions were executed by the components shown in Figure 6.9. The ILA operates as a cache memory, and holds 128 instructions (each 72 bits long). As the program counter requests a new instruction, an associative search is used to find the instruction and load the instruction register. Instructions are fetched in blocks of 16. The strategy adopted was to begin the fetch of a new block when 8 instructions of any block are completed. If the block was already resident, the fetch was aborted. A least-recently-used (LRU) replacement policy was chosen, thus the buffer was cyclically filled. A jump instruction to a non-resident block also initiated a fetch of the required block.

In the advanced station (ADVAST), control commands were executed. This unit consists of a set of four accumulators used for temporary data storage or as index registers. Separate ALUs were used for address computation and for data manipulation. Both the CU and the PEs contain index registers so that the final address used by a PE is

$$a_i = a + (b) + (c_i)$$

where $a$ is the base address specified in the instruction, $b$ is the CU index register (CAR) and $c_i$ the local PE index register.

Array instructions are passed from ADVAST to the final station (FINST) from which the appropriate control signals are broadcast to the PEs. Common data is also broadcast from this block, as shown.

Configuration control of the arrays was implemented with each CU containing three configuration control registers (CFCR), each of 4 bits. These registers can be set by a CU instruction or by the host computer. The first register indicates with a 1 in any of bits 0 to 3 the current array configuration. The second register specifies the instruction addressing to be used within the array, so that instructions may be fetched from any of the participating quadrants. The third register specifies the data addressing in the same way as instructions. The second and third must be consistent with the configuration code in the first (or a trap is generated).

The processing of traps and interrupts will not be considered, although they pose special problems in such a system.

The Processing Element (PE) architecture is shown in Figure 6.10. The PE uses 64-bit floating point arithmetic operations although it can be configured as 2x32 or 8x8-bit subprocessors. It contains no control logic since this is done by the CU. Two arithmetic units were used for separate manipulation of addresses and data. Functionally it executes the manipulations on all local data and computes the final address for operands.

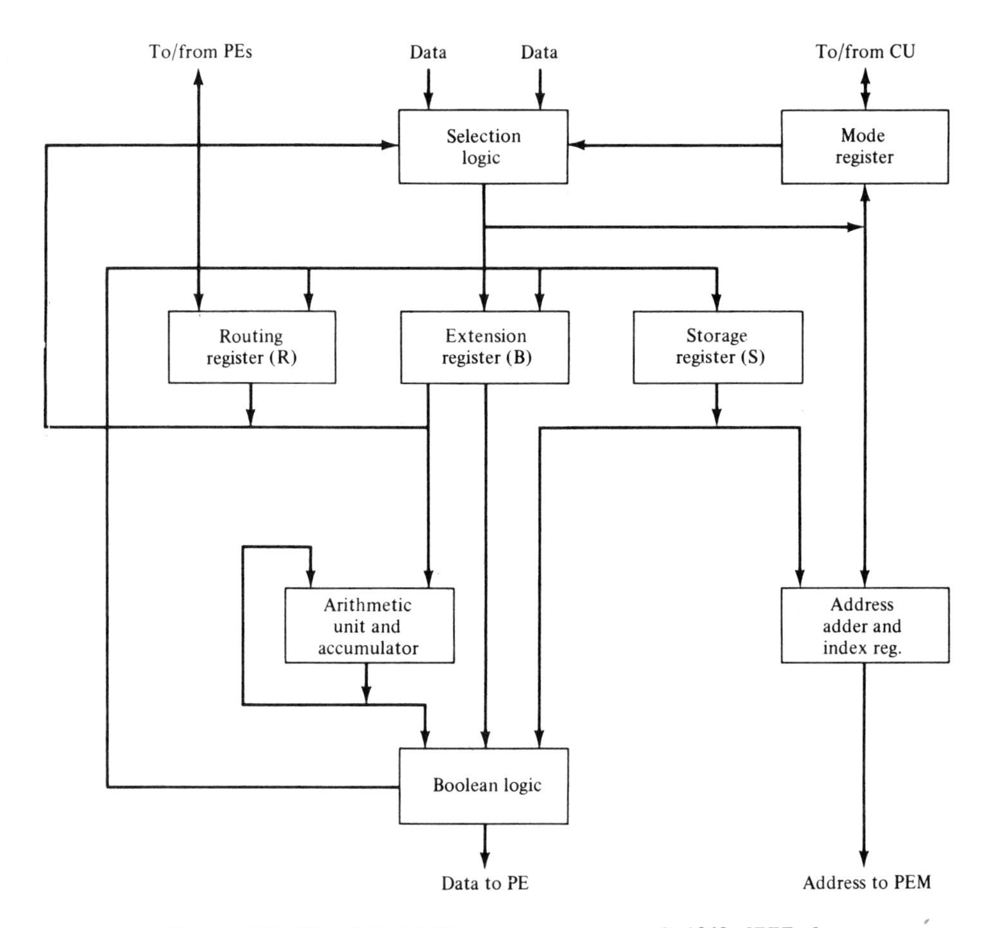


**Figure 6.10:** The ILLIAC-IV array processors. © 1968 IEEE Inc. Reprinted by permission, "The ILLIAC-IV Computer," G. H. Barnes, et al., *IEEE Transactions on Computers*, Aug. 1968, pp. 746-57.

The PE consists of four major components:

1. Four 64-bit registers
   A: Accumulator
   B: Operand
   R: Multiplicand and data routing
   S: General storage
2. Arithmetic logic unit
   This unit is an adder-multiplier with boolean functions as well as a high speed barrel-shifter
3. Memory address adder and index register (X)
4. An 8-bit mode register

The issues surrounding the logical design of both the CU and the PE are:

- Instruction set
- Mechanism for conditional branching
- Data structures
- Interprocessor Communication

In the following section, each of these issues will be developed and the requirements discussed.

**Instruction Set Design**

In general, instructions fall into two categories:

- Control Instructions executed by the control processor only
- Vector Instructions interpreted by the control processor and executed simultaneously by some or all of the array processors

To illustrate the requirements, it is more convenient to proceed in reverse order from the vector to the control instructions. The demands placed on the architecture by the instructions also motivate a discussion of branching, data structures, and interprocessor communications.

Vector instructions affect all processors which perform the designated operations simultaneously. A sample set of such instructions is shown in Figure 6.11.

A subset of the control instructions is illustrated in Figure 6.12. The control instructions are executed by the control processor only. The branching instructions provide special problems discussed next.

**Conditional Branching**

Branching forms a unique problem in SIMD computers. The problem arises from the multiplicity of accumulators and the possibility of different branching conditions in each. Thus, for instance, branch on zero, or positive, etc., will find a true condition in some processors and a false condition in others.

Data:

LDY : Load register Y, i.e., A, B, R, S, or X.
STY : Store register Y, i.e., A, B, R, S, or X.
LDCi : Load $CAR_i$, $i = 0, 1, 2, 3$.

•
•

Arithmetic:

ADB : Add bytewise.
SBB : Subtract operand from A bytewise.
ADD : Add A register and operand (64 bits).
AD[R, N, M, S] : ; Add operand to A; R-round: N-normalize; M-mantissa only; S-special treatment of signs.

Logic:

AND : AND A Register with operand
OR : OR A register with operand

**Figure 6.11:** Some ILLIAC vector instructions.

Since only one instruction stream is available, some mechanism must be in place to perform the branch computations sequentially.

A mask register can be maintained by the control processor. On receipt of a branch instruction a set of the processors which satisfy the branch condition is allowed to proceed. The remaining set is deactivated by the control mask. Those processors not satisfying the branch condition must remain inactive until those processors that took the branch complete that processing stream.

**Data Structures**

The flow of data from memory to processors was illustrated using matrix multiplication. It was implied by this example that the appropriate operands were available during the iteration. Each problem would exhibit such requirements and clearly the organization of data in the processor memories is a key issue in utilizing the ILLIAC for high speed computations.

Using the matrix multiplication example, an arrangement of data is shown in Figure 6.13. The use of index registers is assumed to enable the data elements to be sequentially read from and restored to memory. This arrange-

Index:

IX(L, E, G; I) : Set I by comparing X register and operand. (L–less than; E–equal to; G–greater than)

Mode setting comparisons:

EQB : Test A and B for byte equality.
GRB : Test B register byte greater than A.

**Figure 6.12:** Some ILLIAC control instructions.

|  |
| --- |
| A(00) |
| A(01) |
| A(02) |
| A(03) |
| B(01) |
| B(11) |
| B(21) |
| B(31) |
| C(01) |
| C(11) |
| C(21) |
| C(31) |
|  |

**Figure 6.13:** Memory organization for processor 1 (matrix multiplication).

ment is optimum for the matrix multiplication; however, if the resulting $C$ vector required further manipulation, difficulties could arise. For example, if a row of the $C$ matrix were required at another processor then the matrix would have to be rearranged or sequentially read and broadcast to the designated processor. This would be time consuming regardless of the intercommunication.

Considerable study has been published on data formats for optimizing computations on SIMD machines. These are beyond our scope; however, the issue remains for the implementation of algorithms on almost all general purpose signal processors.

### Interprocessor Communication

The computation of the FFT discussed in Chapter 3 provides an example of the requirement for interchanges between processors. In general each new problem will generate a different interconnection requirement for optimum performance.

The problem of the interconnection of multiple processors for the efficient exchange of data during the execution of an algorithm is of a universal nature and will be discussed in more detail in Volume II. The design issue is to obtain the required interprocessor bandwidth at a minimal cost. Since each problem tends to be unique, a general purpose interconnection would require each processor to be connected to all others. This complete interconnection of processors requires $N(N-1)/2$ links between $N$ processors. The cost therefore tends to increase as $N^2$ while the performance increases as $N$ (e.g., consider adding a processor to gain more throughput). For large $N$ the cost of the interconnection network begins to dominate the cost of the system. Hence a considerable study must be committed to network design if reasonable cost/performance ratios are to be obtained.

### Function of the Host

The use of a general purpose processor as a host has become a widespread solution to array processing. The ILLIAC is an attached processor, using a Burroughs B6700 as host. The host performs the following functions:

1. Overall control of the CUs and of configuration
2. Internal data transfers from secondary storage to memory, including data formating
3. Diagnostic and maintenance programming
4. External I/O processing
5. Disk file supervision
6. Program compilation, editing, debugging, etc.

### Reliability Considerations

The reliability of a large conglomerate is inversely proportionate to the number of components. The MSI circuits used at the time had about 100 gates and an estimated 30,000 parts would be required. It was proposed that for the four quadrants the reliability would be $10^5$ hours per element or a failure rate of about one every 400 hours.

In the event of system failure the host can be tested by standard diagnostics. The array memories can then be tested. The host can then test the CU which in turn can test the PEs. The number of identical elements in array systems reduces the number of spare parts required which in turn reduces the cost of inventory.

The ILLIAC forms a real example of the SIMD approach to achieving high performance. The functional partitioning of control and data manipulation into separate processors allows each to be optimized and allows many data streams to be controlled in a strictly sequential manner. A further discussion of the attributes of this architecture will be delayed until after the next section in which a distinctly different approach is studied.

## 6.2.2 TI/ASC—A Pipelined Seismic Processor

The Advanced Scientific Computer [ASC] designed by Texas Instruments is shown in Figure 6.14.

The design philosophy for the ASC was fundamentally different from the ILLIAC in several ways. First, the architecture was to be modular so that a static configuration could be implemented and optimized to a particular environment. Second, the array of pipelined arithmetic units could be viewed as a high bandwidth processing unit; and the ASC design revolved around the flow of data into and out of this unit. The key issues are the high speed flow of instructions, and of data. Significant features of the ASC are therefore the memory control and the arithmetic unit. The motivation for the design came from the requirements of seismic processing although the final design provides for a much larger application range.

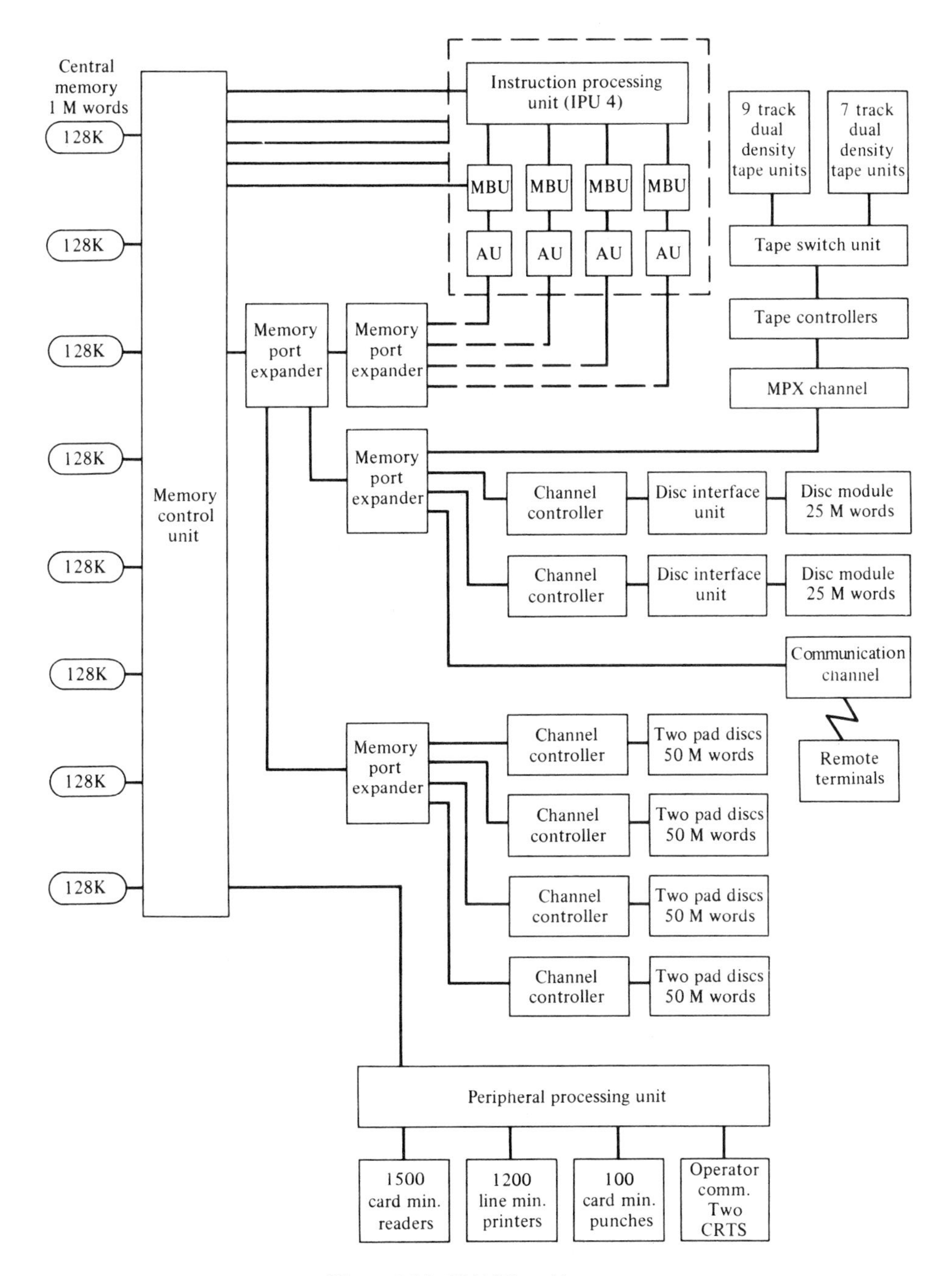


**Figure 6.14:** TI/ASC architecture.

The major components are shown in Figure 6.14: the main memory, the central processor, the peripheral processor plus mass storage, and communications facilities. The peripheral processor is in fact the host and it executes the operating system. The central processor is an attached special purpose processor designed for high performance. The main memory serves as the common communications link between all components of the system. As such it is a potential bottleneck and mechanisms for achieving a large throughput are of central importance. In the following discussion we examine the memory first followed by an examination of the central processor. Before proceeding we note that the ASC is an SISD machine in concept; however, considerable overlap in instruction execution between the processors exists. An MIMD classification might also be appropriate.

**Central Memory**

The central memory and memory control unit (MCU) are shown in Figure 6.15. The MCU provides a bi-directional 256-bit/channel data transfer network between eight processor ports and nine memory buses. The nine buses provide eight-way interleaving for the first eight buses with the ninth bus used for memory extensions. The eight processor ports provide access to the complete memory address space (24 bits—$16 \times 10^6$ words).

The central memory is divided into eight equal sized modules (to permit interleaving). The technology gives a cycle time of 160 n.s. with a read of 140 n.s., with all transfers being 256 bits. The MCU is capable of a transfer of $80 \times 10^6$ words per second per port (a total of $640 \times 10^6$ words/sec). Our primary concern is with the flow of data (and its control) between the memory and the arithmetic units. This will now be explored.

The central processor consists of three functional units: the instruction processing unit (IPU), the memory buffer unit (MBU), and the arithmetic unit (AU). The processor instruction format is 32 bits. It contains 48 internal registers (16 base and 8 index registers, 16 accumulators and 8 vector parameter registers). The vector parameter registers were used as implicit parts of the format of vector instructions.

The IPU performs the usual functions:

1. instruction fetch, decode, and execute
2. regular operand selection
3. address computations
4. generation of vector starting addresses and transmital of vector parameters to the MBU

The vector instructions include most of the standard opertions, e.g., add, subtract, multiply, divide, dot product, matrix multiplication, plus logical operations as well as such operations as merge, search, order, etc. Each vector instruction specifies a parameter file which contains pertinent information such as dimension, starting address, etc. An example of parameter files will

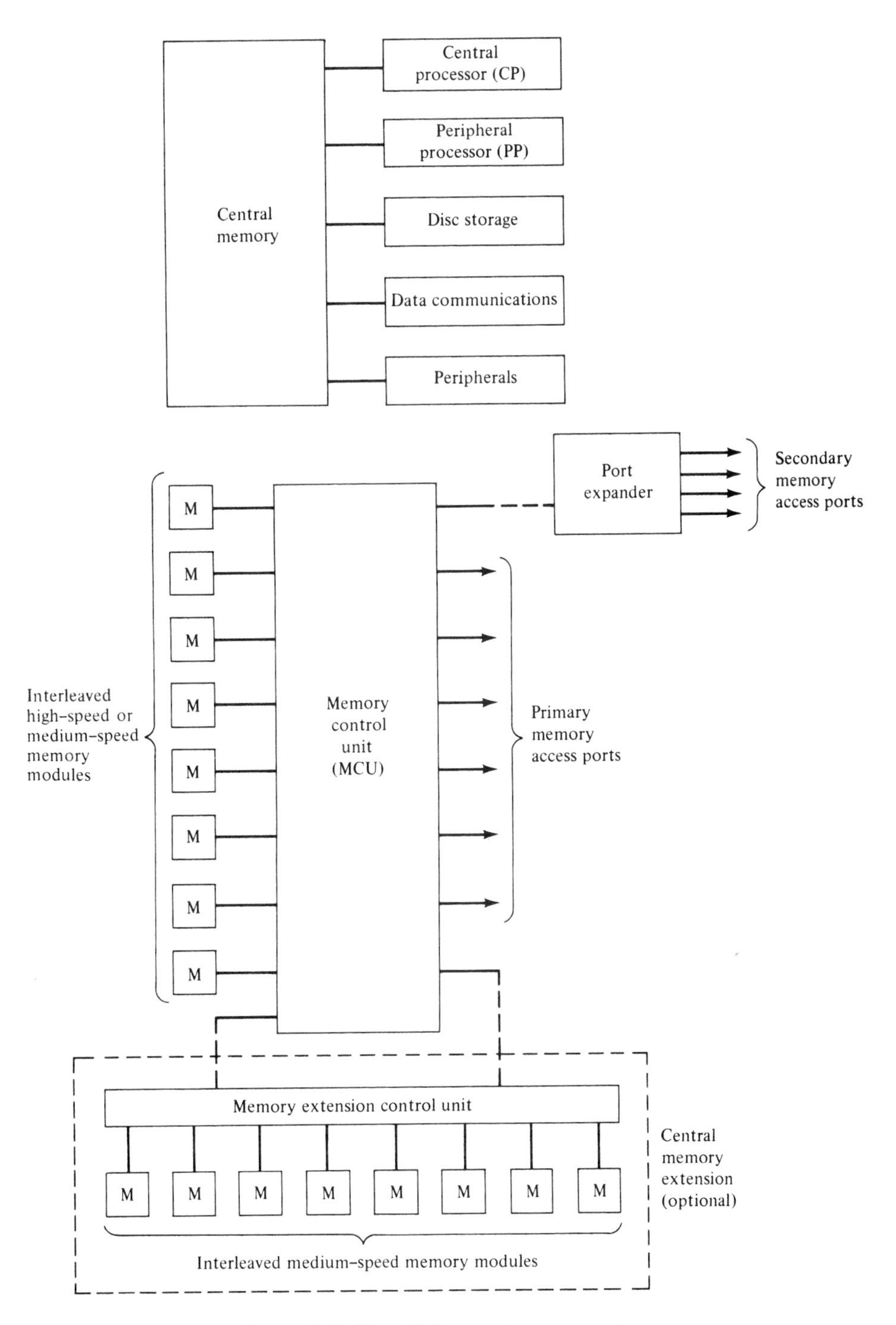


**Figure 6.15:** The ASC central memory system.

be given for the AP/130 in Chapter 7. The IPU uses two 8-word buffers for instruction prefetching (a cache similar to the ILLIAC).

The MBU receives instructions from the IPU and interfaces directly with memory to supply a continuous stream of operands to the AU. The MBU contains three sets of double buffers; two for input and one for results. The operand data fetch and the result data store are executed simultaneously. Addresses for a vector operation are compiled by the MBU from parameters specified in the vector parameter file (supplied by the IPU).

This AU offers an interesting example of a parallel array of pipelines. The ASC was designed to operate with one, two, or four pipelined arithmetic elements in parallel. The pipes were designed with eight segments to be configurable depending on the operations as shown in Figure 6.16; for example, multiplication used seven segments while addition used only six. In terms of the performance measures described in Section 6.1.3, assume the time per segment is 60 nanoseconds, therefore the pipe transit time is 480 nanoseconds.

Consider the data flow problem in order to keep the pipe full. At full utilization the input to each pipe is two 64-bit operands, each of which produces one 64-bit result. Thus every 60 nanoseconds a total transfer of 768 bits is required. The ratio of memory access time to pipe throughput is $160/60 = 2.66$, therefore $2.66 \times 768 = 2043$ bits must be transferred every memory cycle. Since each memory module has 256 bits of data, an eight-way interleave is sufficient.

The pipelined arithmetic unit and associated processors run the application programs, the rest of the system software runs on a collection of eight peripheral processors.

### 6.2.3 A Performance Comparison

The major issue faced by the architects of the ILLIAC and the ASC was how to increase the performance beyond that of a basic von Neumann machine. The first, and obvious, requirement was to completely overlap the instruction fetch cycle with the execution cycle. Beyond this the execution cycle itself presented the bottleneck for high performance processing. As we discussed in Section 6.1.2, some form of parallelism is necessary to increase the throughput of operands. The ILLIAC achieved this by a multiplicity of data streams and processing units; the ASC by an array of pipelined arithmetic units fed by a single high bandwidth data bus.

In an ILLIAC-type architecture (i.e., SIMD) the ability to utilize the multiplicity of processing elements is dependent on a problem formulation such that a central control can keep $N$ processors working on an instruction-by-instruction basis. The ASC approach appears more traditional in that it remains architecturally an SISD machine. However, in both cases the processing algorithm must support the multiplicity and the data structure must be carefully organized.

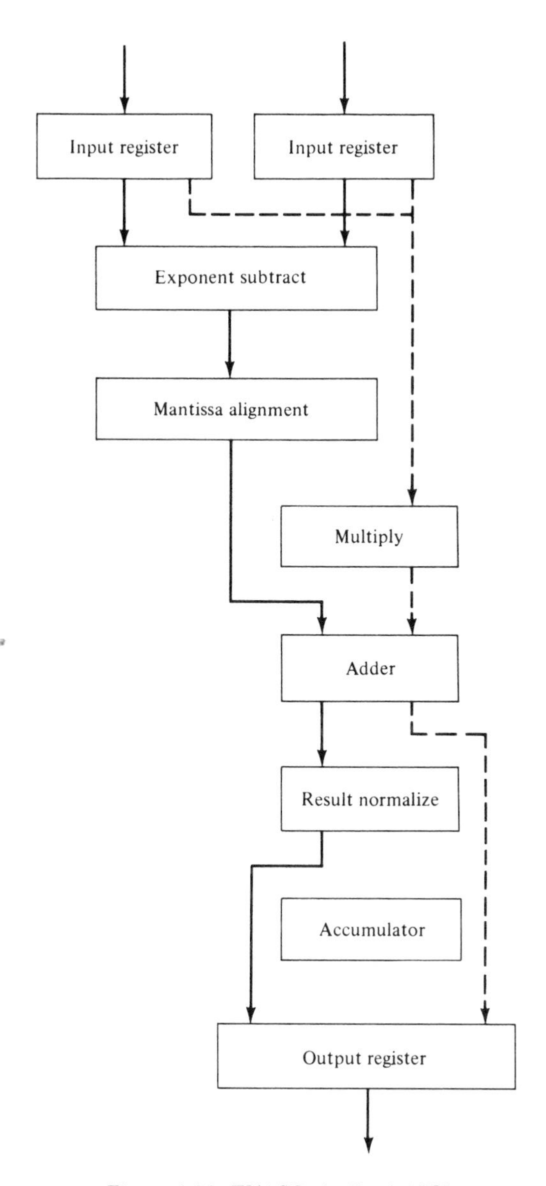


**Figure 6.16:** TI/ASC pipelined ALU.

For the ILLIAC, the multiplicity is 64 and a performance speed-up of this order would be expected. Performance is reduced to one-half for a scalar/vector operation mix of only 1/63. For the ASC the multiplicity is 4 for the 64-bit configuration. It can tolerate a 1/3 scalar/vector mix for a degradation of one-half.

### 6.2.4 Summary

The two processors discussed in this section were chosen to illustrate the two approaches to increasing performance; the ILLIAC being a parallel array and the ASC a pipeline. There is no intent to imply that these mechanisms for enhancing performance are mutually exclusive for they were used together in the ASC.

Each processor affords a profound insight into the trade-offs between control and data-flow requirements. In the ILLIAC the control processor acts as the source of synchronization and instruction execution control, which is probably the simplest way of coordinating a parallel array. The data flow on the other hand has been the major drawback of this machine. First, the data required by each array processor must be organized in the appropriate memory, and second, the requirements for data interchange must be accommodated by a connecting network. The ASC provides a high performance AU; and the architecture was optimized to provide a high bandwidth flow of data (both operands and results) between central memory and the processing units.

Any particular architectural configuration tends to be optimized for a specific set of algorithms. Strings of scalar operations which do not utilize all the processors of a parallel array are inefficient in the sense that most of the array is idle. In the ASC, the central processor controls an array of four arithmetic units, each of which is pipelined. The central processor in turn is directed by a peripheral processor. The pipelining of the AUs is in a sense transparent to the system from an architectural point of view. Their performance demands a high data-flow rate which must be accommodated. It could be argued that the central processor of the ASC is an SIMD configuration such as the SOLOMON computer which inspired the ILLIAC.

The overall problem of both architectures is the need to execute a large number of data organization operations so that vector-type manipulations can be performed at high speed. To achieve this and to maintain high performance, it was found necessary to overlap, as much as possible, routine scalar operations with the vector operations. Almost invariably this led to a host computer which supports in some fashion the routine work of data

| | *ILLIAC-IV* | *ASC* |
|---|---|---|
| Data Word Size | 64 | 16, 32, 64 |
| Instruction Size | 32 | 32 |
| Memory Size | 2K bytes/PE | 8M (64 bit) |
| Memory Cycle | 250 n.s. | |
| Multiplicity | 64 | 4 |
| Processing Speed | | |
| Add | 500 n.s. | |
| Multiply | 700 n.s. | 75 n.s. (streaming) |

**Figure 6.17:** Summary of attributes: ILLIAC and ASC.

input/output, vectorizing the data arrays, and job scheduling. These host computers were impressive machines in their own right; and the combination tended to be high in capital cost. This trend toward supporting a high performance processor with a host is now a recognized architectural mechanism. Even at the VLSI chip level it will emerge as the optimal solution as we proceed.

Programming of both systems remained a major consideration. The representation of a problem to exploit the architecture and its rendition in a high level programming language tends to demand specialists. Thus code generated for the ILLIAC is expensive. Several high level languages have been proposed (TRANQUIL and IVTRAN); however since only one operational installation exists no extensive language development seems to have occurred. The standard appears to be GLYPNIR, an extension of ALGOL. Vectorizers have been developed to examine Fortran programs in an attempt to discover parallelism. These reduce the need for specialized programmers but represent both a development and a utilization cost.

Code for the ASC was conceptually easier to generate. Higher level MACROs were available for vector operations and manipulations which were compatible with the application requirements. The architecture of the ASC was probably more transparent to the programmer than that of the ILLIAC. Unfortunately, there are only a few programmers who have had extensive experience on the ILLIAC and comparisons may not be valid for it never became a commercially successful product.

These machines, and others, were developed in response to a class of problems which are similar to those of signal processing but not identical. During the 1970s more urgency (and hence more dollars) became evident for architectures optimized for signal processing. These will be introduced in Chapter 7. Before examining these, however, we will discuss some broad issues of performance limitations based on our discussion thus far.

## 6.3 PERFORMANCE LIMITATIONS

### 6.3.0 Introduction

In the previous sections two processor implementations have been reviewed to illustrate the various mechanisms which have been used to achieve high performance. In each case the flow of data for specialized applications and the need for centralized control limited the general applicability of the resultant architectures. To some extent each machine was slightly ahead of the supporting technology.

If raw computational power were the only issue in signal processing, it is clear that an analog computer would be an optimal solution. There are, however very few groups who are anxious to commit their resources to this solution. Other factors are clearly determinant in this choice. In general the flexibility of a programmable digital processor is traded against the higher performance of an optimized special purpose processor. It is not an easy matter to decide where the compromise is to be settled.

The parallelism of the algorithm and the architecture of the processor must be compatible for a variety of reasons. These are often difficult to quantify exactly; however, a high speed array processor is obviously a poor choice for scalar operations. In order to match algorithms to architectures some classification of each is necessary. In a large problem it is usually most difficult to expose all the parallelism in the algorithm. Unfortunately, formulating the problem in the first place tends to demand an orderly sequential approach which often hides the parallelism. Thus while it is easy to see the parallelism in an array multiplication, for example, the larger structure of the algorithm may yield more substantial performance benefits. Estimates are subject to wide errors and automatic detection of parallelism is an unsolved problem.

In this section several factors which tend to limit performance will be discussed. We will draw on the previous examples to emphasize the interdependence of architecture and technology. In addition it seems appropriate to discuss the requirement for programming and the need to support complex data structures. These are really issues in the overall design which will be discussed in more detail in Volume II.

### 6.3.1 Architectures/Technology

Computing power has increased by orders of magnitude throughout the last few decades. Computer architects have indeed been given a free ride (in many respects) on the backs of innovations introduced by solid state theorists and manufacturers. Demands for performance tend, however, to always exceed even the combined efforts of technology and current architectural concepts. Architecture therefore becomes a key issue in exploiting technology. The basic mechanisms for enhancing throughput are understood; however, these must be supported by control and data-flow arrangements which are often difficult to generalize. Thus there is always an algorithmic bias in considering specific designs which almost always inhibits performance outside a narrow range of applications. Technology holds the key which should permit unlimited access not only to raw performance improvements but to innovative solutions to generalized control and data-flow architectural problems.

The major drawback of classical supercomputers was the rigidity of architectural structure. Indeed, their failure as a species is often attributed to their inability to adapt to a changing algorithmic environment. This is certainly true; however, it is not the whole truth. VLSI provides the capability to create within quite reasonable cost constraints an ILLIAC-type machine with say four times the processing units. This approach however has severe constraints for as the number of components and their interconnections increase, the system reliability decreases. No new ILLIAC-type machines have appeared.

Throughput and reliability are two requirements which impose contradictory demands on the architect. The extension of current processor architectures to super systems is intrinsically bound to create deterioration in reliability. It is necessary to evolve new concepts to permit dynamic adaptation to the structure and performance requirements of an algorithm without increasing the physical complexity of the hardware. These issues form a major portion of the motivation for examining new approaches to design and architectural concepts in Volumes II.

### 6.3.2 Programming

The requirement for generating software exists for all processors except those specialized to a single function. Even here the cost of generating the code to implement the algorithm after the processor is functional can still be a significant fraction of the total cost. For other processors the adaptation of a specific requirement to a sequence of instructions may introduce factors which limit performance. In the extreme case, for example, the ILLIAC-IV becomes more and more inefficient as longer sequences of scalar operations are introduced. While this is easily recognized, more subtle aspects of the programming may not be. In general the structure of the algorithm must be matched to that of the processor. This is somewhat more difficult to do than to prescribe.

In the programming of a general purpose computer, it is often found that certain programming structures within a high level language and executed more rapidly than others. This is a function of either the compiler or the architecture. For well supported computers, massive compilers execute algorithms to optimize at a global and a procedural level. In this environment, even poorly designed programs can be expected to execute with reasonable efficiency. Such extensive support is generally not available for signal processors because of the more restricted usage. Yet the demands for highly optimized code to achieve high performance are even more stringent.

Two further factors influence the overall programming costs. First, the nature of the problem is esoteric and the programmer must be thoroughly knowledgeable. This usually demands a team of programmers plus experts in the field, which is a costly arrangement. Second, languages, which are suitable to the problem, and which optimize the capability of the processor architecture, must be provided. To minimize the costs an approach has been to design a "vectorizer" which examines a regular high level language and optimizes its algorithmic structure to the machine. Machine independence can thus be obtained if suitable intelligence is built into the compiler. Even after compiling it is usually necessary to hand tune the code to achieve the response requirements.

In the absence of software support, most signal processing software is programmed in the assembly language of the processor. The cost of such code for algorithms with stringent performance constraints can become an order of magnitude greater than the equivalent for a more conventional architecture. The increase in cost is a function of the specialized knowledge required as well as the debugging and validation costs. The message here is that in some fashion the designer must remain cognizant of the needs of the end-user; and end-users should consider carefully what their ongoing costs of using the system could be.

A significant aspect of signal processing is the manipulation of arrays of data. Implicit in an architecture and instruction set is a limitation on the types of data structures which can be supported. This factor deserves special comment and is considered next.

### 6.3.3 Data Structures

Almost all aspects of the concept of data tend to be application dependent. We are concerned here with only two factors; the logical structure of the data and the requirement to address it for retrieval and storage. It is assumed for this discussion that data will be stored in a random access memory which is in the addressing range of a processor.

In the simplest case a sequence of data is stored in contiguous memory locations. Even in this case, the processor requires index registers and base registers to locate the start of the data and to step through the memory for storage or retrieval. In a more complex situation data may be thought of as a component of a logical structure containing several layers, in which case several index registers may be required with associated arithmetic elements to compute the required offset to a base register.

Consider, for example, a processor executing 100 MFLOPS. To maintain this throughput the system absorbs two operands and produces one result; a total flow of $300 \times 10^6$ words/second. This flow must be organized into the expected format. To sustain these data rates it has become common practice to utilize a host processor to organize the data flow while the special purpose attached processor manipulates the data. Both the ILLIAC and ASC were designed to work this way. A variety of other such processors will be discussed in Chapter 7.

### 6.3.4 Summary

This section was intended to serve two purposes. First, we hoped to introduce a thread of pragmatism into the discussion. It is easy to become entranced with the elegance of processor architectures and to forget the mundane problems of their utilization. Clearly our assessment of performance limitation must be gradually enlarged as we proceed, and our design methodology must eventually account for these limitations. Second, requirements other than programming often must be accommodated, which may directly conflict with performance requirements. For example, a design specification may require a processor to be reliable, modifiable, fault tolerant, and technology independent in addition to other possibly antagonistic attributes.

## 6.4 CHAPTER SUMMARY

This chapter represents the first step in a long sequence of steps extending through this and the next volume, relating signal processing to the design and utilization of signal processors. The most important concept in this chapter is the performance enhancement achievable with a multiplicity of processor units. The architecture of systems can be analyzed based on this multiplicity and the performance enhancement predicted. It is therefore a powerful and useful concept.

Realizations of the multiplicity concept are in the form of a pipeline or a parallel array. For each case there are two considerations. First and foremost, the data must be organized and supplied at a compatible rate. Second, deviations in the algorithm which cause departures from the full utilization of the inherent multiplicity become costly in the sense that portions of the hardware are unused. These concepts formed integral parts of the earliest super computers as illustrated by the ILLIAC-IV and the ASC.

An interesting question remains at this point: "How can we achieve high performance using parallelism without the inherent algorithmic inflexibility?" The long term answer is twofold—much cheaper hardware will help; but physical complexity multiplies rapidly as conglomerates of VLSI chips become large; more important is the concept of dynamically reconfigurable architectures which remain optimized as the algorithm is executed.

At this point we have the basic concepts of performance enhancement plus both some limited experience with processor architectures and some analytical tools for comparison. Some confidence should begin to emerge that a systematic examination of a new processor can be undertaken to determine its performance potential and its suitability to a particular application.

In the final chapter of this volume, we will examine more modern processors and look briefly at future technology capabilities and their impact on new architectures.

## 6.5 EXERCISES AND PROJECTS

### Exercises

**1.** Consider two proposals to increase the throughput potential of an SIMD architecture similar to that of the ILLIAC-IV. The first involves increasing the instruction execution speed by 3/2 and the other increasing the number of PEs from 64 to 96.

(a) Analyze the throughput enhancement potential of each scheme.

(b) List and discuss the factors which should be considered before deciding on which to adopt.

**2.** Consider the ILLIAC and the ASC as black boxes which absorb and create data; for example, a vector multiplication requires two operands and produces one result.

(a) If the two processors were to operate continuously as vector multipliers, compute the bandwidth of the data-flow.

(b) Considering the two architectures, discuss the logical complexity of data flow, i.e., the problem of arranging the data in appropriate memories for subsequent execution, and for retrieving the results to support the operation in (a).

**3.** For both the ILLIAC and ASC, the sensitivity of the architecture to changes in the job stream is of interest. From the model developed in Section 6.2.3, it is apparent that the throughput approaches the maximum value only when $f$, the fraction of scalar to vector operations, approaches zero. Find an expression to show the sensitivity of performance as a function of $f$.

**4.** Consider the ASC pipeline.

(a) Compute the performance enhancement obtained by its ability to bypass segments for addition.

(b) Suppose a mixed job stream of multiplications and additions were to be processed. Find an expression for throughput if the ratio of additions to multiplications is $f$.

(c) Consider the possibilities of enhancing performance by scheduling the use of the individual pipeline stages as jobs arrive. (A discussion of basic scheduling is contained in Volume II and in reference 5.)

**5.** Assume the ASC pipeline is to be used for a sum of product operation such as the DFT.

(a) Draw a sketch of the pipe.

(b) Compute the time for one butterfly.

(c) Compute the time for a 1024-point transform.

**6.** Consider the execution time given in the table of the example in Section 6.1.2. Assume a butterfly can be executed in 750 n.s.

(a) Compute the time for each of the alternatives (including pipe delays for filling and emptying) for a 64-point transform.

(b) Compute the gain ratio and utilization for each alternative.

(c) Consider the data-flow problems. Compute the bandwidth required to maintain data flow to (and from) each alternative.

(d) Suppose that coefficients and data are in some form of Random Access Memory (and results must be stored in a RAM). Sketch an interconnection topology for each alternative. If a single control processor is used to move operands to the FFT processor, estimate the MIPS required.

## Mini-Projects

**1.** Prepare a complete comparison of the CDC String Array Processor—STAR, and the TI-ASC ([9] contains a discussion of the STAR plus further references). The approach suggested in this chapter could be augmented by that in the reference.

**2.** Assess the performance enhancement achieved through pipeline scheduling algorithms. Compare this to static reconfiguration based on the job stream (i.e., 16-32-64 bit operations as in the ILLIAC or multiplication, addition reconfiguration as in the ASC).

**3.** The interconnection topology for an SIMD architecture is a major design issue (e.g., see [8], Chapter 8). The ILLIAC design called for each processor to be connected to its four nearest neighbors.

(a) Create a model to show the degradation if a portion $f$ of commands were to transfer data from processor $i$ to all other processors.

(b) Consider a model to predict the cost/performance ratio as the neighborhood of each processor was increased to include all remaining 63 processors.

**4.** It would seem evident that throughput could be increased indefinitely by increasing the parallelism (or pipe length).

(a) Tabulate the issues which would eventually limit this growth. How would a designer decide on the final value of multiplicity for either mechanism?

(b) Are either the pipeline or parallel array intrinsically more amenable to increased multiplicity? List and discuss the issues.

**5.** Consider redesigning the ASC central processor with gallium arsenide (GaAs) components. Obtain a set of performance parameters for these parts, then

(a) Estimate the performance of the architecture built with these components.

(b) Discuss modifications to the architecture which could simplify the CP while achieving the same performance as the original (e.g., eliminate MCU, is the pipeline necessary?, etc.).

**6.** Code the matrix multiplication given in Section 6.2.1;

(a) Use the ILLIAC instruction set given in [Barnes, 68]. From the specification compute the MIPS and FLOPS.

(b) Use the AP 130/B instruction set. Compute the MIPS and the FLOPS.

CHAPTER 7

# Signal Processors

## 7.0 INTRODUCTION

This chapter forms the conclusion of Volume I and must motivate the need for, and structure the context of, Volume II. We accomplish this by two mechanisms. First, by introducing a selection of more modern processors and, by utilizing the principles developed thus far, we provide evidence that they can be analyzed and compared in a useful way. Next, by introducing some technology trends and projections, we provide further evidence that higher performance processors are sure to evolve.

In Section 7.1 a selection of processors developed during the 1970s is presented and analyzed. These are all attached processors, relying on a host for much of the routine data flow and control. The concepts, which we hope have been developed to date, are found quite adequate both to appreciate and to understand these processors. A comparison of two processors, one developed at the beginning and one at the end of that decade provide an impressive demonstration of the impact made by advances in component technology on processor performance.

In Section 7.2 an examination of the evolutionary trends of digital component technology is presented with some projections (speculations) on future capabilities and their impact on signal processor architectures. Projections are a dubious subject for inclusion in a textbook. However, here we are concerned more with the rate of change for the 1980s and the challenges (and perhaps excitement) that this generates. The past trends of the 1970s, upon which these projections for the 1980s are based, leave little doubt that the effect of component technology evolution (in whatever form it takes) will indeed make high performance processor design both challenging and exciting in the 1980s.

## 7.1 SIGNAL PROCESSOR ARCHITECTURES

### 7.1.0 Introduction

A wide variety of architectures have been designed and implemented to deal with the high performance requirements of digital signal processing. The use of parallelism and pipelining techniques, as discussed in the previous chapter, has remained the basis for most attempts to implement high performance architectures. The variations in processor architecture tend to be with respect to the functional level at which parallelism and pipelining concepts are applied and to what degree these techniques are utilized.

In this section we shall review several signal processor architectures that have emerged over the past decade. The processors discussed in this section were designed to be attached to general purpose host computers. Thus, their architectures tend to concentrate on the efficient implementation of the high speed arithmetic operations necessary for signal processing while relying on the host machine for programming support, bulk memory, and connection to standard peripheral devices.

Each of the architectures discussed represent a variation in approach to resolving the often conflicting requirements for high volume processing and data flow, while maintaining flexible yet tenable control. The requirements as seen from the examples in Chapter 6 appear to be:

1. The AU must be designed to allow high speed computation of the basic functions.
2. Data-flow paths to and from memory must be optimized for high speed exchanges.
3. Instruction fetch cycles must be overlapped with data flow in such a way as to impose no time penalties.

A further factor which became apparent during the decade is the use of microcoding to bury the complexity of the control algorithms. There appears to be a distinct difference between an attached processor such as the ILLIAC, which had to be programmed in its own right, and a modern processor such as the Data General AP/130, in which calls for signal processing functions appear as standard assembly language commands of the host. This dramatically reduces the cost of software production.

In general we can identify several approaches to achieving performance with some form of multiplicity, e.g.,

1. Multiple general arithmetic units
2. Multiple specialized functional units
3. Multiple specialized processors
4. Multiple specialized memories and parallel buses feeding high speed pipelined arithmetic elements
5. Multiple microprogrammable modules or programmable processors

The processors examined in this section provide an example of each of these approaches. For each processor we shall look briefly at the major architectural features and then examine processor operation to see how the chosen approach to implementing high volume data flow has affected processor control.

In general we shall see that each of the designs has maintained centralized control through sequential program execution. The fundamental problem then is to utilize efficiently the potential performance enhancements of parallel or pipelined processing activities within a sequential control environment.

We also note that the functional level at which parallelism has been implemented is generally (but not always) at the level of individual arithmetic operations (i.e., add, subtract, multiply, etc.). In most cases the requirements of digital signal processing are at a larger macro level such as high speed sum of product calculations, complex and/or floating point arithmetic, second order section calculations and butterfly computations. Attempts to match processor architectures to these common signal processing macros mark a fundamental difference between the processors of this chapter and the vector processors of the previous chapters.

The processor designs discussed in this section span a time frame covering approximately the 1970s. The performance levels of the various processors vary, not only with architectural changes, but also with respect to advances in integrated circuit component technology. The impact of advancing component technology will be discussed in more detail in the final section of the chapter. However, in this section we include a brief description of a recent single chip signal processor and compare it to a processor designed almost ten years earlier. The impact of changing component technology is strikingly illustrated.

### 7.1.1 The Lincoln Laboratory Fast Digital Processor (FDP)

One of the early realizations of an attached processor aimed specifically at signal processing was the Lincoln Laboratory Fast Digital Processor (FDP). This processor was designed as a high speed arithmetic attachment to the Univac 1219 computer. The FDP was completed in 1971. It is therefore a contemporary of the ILLIAC and ASC.

The basic architecture and data-flow paths of the FDP are shown in Figure 7.1. The major components of the processor are:

1. An arithmetic section consisting of four identical arithmetic elements (AEs). Each AE contains a multiplier that operates in parallel with the adder and main arithmetic registers. The structure of an AE is shown in more detail in Figure 7.2.

2. Two data memories ($M^a$ and $M^b$) that can be addressed simultaneously. Each data memory contains 1024 words of 18 bits each.
3. A separate instruction memory, $M^c$, divided into two sections of 256 × 18-bit instruction words each.
4. A control unit that simultaneously executes a double word length instruction pair from the control memory.
5. Separate input and output buffers that connect the data memories to the Univac main memory or to external ADC and DAC units.

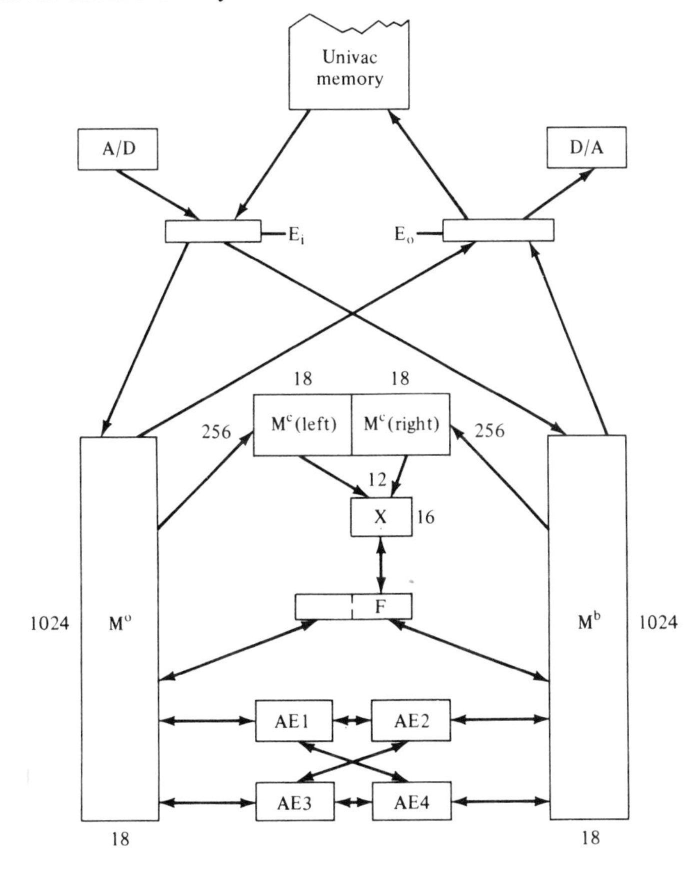


**Figure 7.1:** FDP architecture. © 1971 IEEE Inc. Reprinted with permission from *IEEE Transactions on Computers*, **C-20**, 8, Jan. 1971, pp.33-38.

Instructions are all 18 bits long except for block transfer instructions which are 36 bits long. The control memory can only be written into by block transfer from $M^a$ and $M^b$. Instructions are either AE control instructions or data-flow control instruction.

A double length instruction word allows the simultaneous execution of two 18-bit instructions on the FDP (one AE control and one data-flow control). Instructions are either AE control instructions or data-flow control instruction. The AE instructions are either arithmetic or transfer instructions of the format of Figure 7.3. Data-flow control instructions are interpreted as either data memory addressing or control instructions.

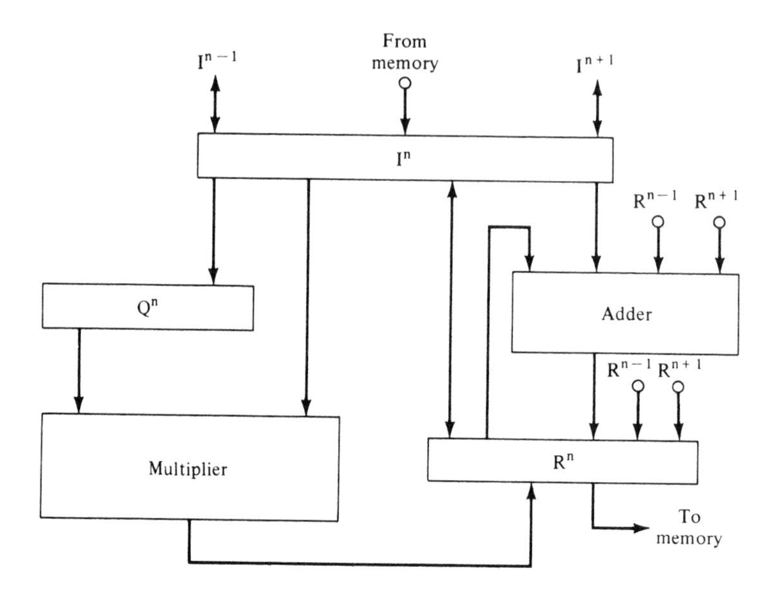


**Figure 7.2:** FDP arithmetic element. © 1971 IEEE Inc. Reprinted with permission from *IEEE Transactions on Computers*, **C-20**, 8, Jan. 1971, pp. 33-38.

The two data memories and the AEs are organized to optimize data transfer for the performance of complex arithmetic and the interconnection between AEs was optimized for the performance of recursive filtering (second order sections) and butterfly operations. The multiplicity achieved with the design permits roughly an order of magnitude speed-up compared to a standard SISD machine.

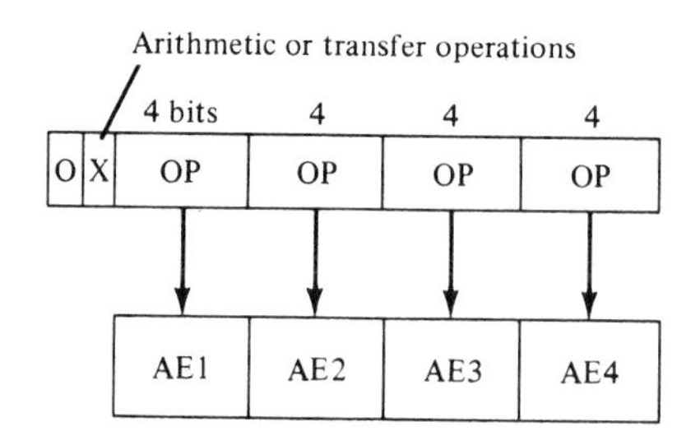


**Figure 7.3:** AE instruction format. © 1971 IEEE Inc. Reprinted with permission from *IEEE Transactions on Computers*, **C-20**, 8, Jan. 1971, pp. 33-38.

The multiplier and adders in each AE operates in parallel. In addition the simultaneous execution of two instructions adds another level of parallelism within this processor. As a result the FDP is not an easy machine to program. The programmer had to exercise great care to obtain maximum usefulness of the inherent parallelism of the system. Since all four AEs are controlled in parallel by a single instruction, which can be either of an arithmetic or transfer type, it is not possible to transfer data between two AEs while others are doing arithmetic operations. This restricts their flexibility.

The requirement for 3 instruction cycles to complete a multiply operation also complicates the control of the AEs. In this respect the control of the AEs is similar to the ILLIAC. However, in the FDP each AE may be given a different instruction rather than have all of them execute the same instruction. This gives the AE array an MIMD type of control character.

The transfer of data into and out of the FDP is restricted in that all data transfers and program transfers are done through the input and output buffers. These constitute a data-flow bottleneck. Since all data transfers are controlled by the execution of instructions from the program memory, the dual instruction execution system is necessary to allow simultaneous data flow and the execution of arithmetic operations in the processor. This is a clear example of how the concept of simple von Neumann sequential control is complicated by the necessity of controlling parallel activities. It also illustrates the tradeoff necessary in operational flexibility of the various parallel activities when they must be controlled by a single sequence of instructions.

The FDP provided a training environment and was followed by the LSP/2. It was becoming more urgent for architects to achieve high performance while maintaining logical simplicity for software generation.

### 7.1.2 The LSP/2

The LSP/2 was designed at Lincoln Laboratories based on experience gained with the FDP. It is an example of a programmable signal processor realization based on multiple functional units. Figure 7.4 illustrates the general architecture of this processor. The main architectural features are 64 dual-copy 32-bit general registers, three 32-bit parallel data buses, and a set of dedicated functional units. The dedicated functional units include an index arithmetic unit (XAU), an arithmetic logic unit (ALU), a multiplier unit, a division unit, a shift and normalization unit, a 4K × 32 bit data memory, a programmed I/O port and DMA unit. A separate 4K × 32 bit program memory is provided which can be loaded from the data memory by block transfer.

The LSP/2 instruction set is divided into four classes: arithmetic instructions, constant handling, control and memory addressing. This processor also depended on execution of a single sequence of instructions for control. The separation of the program memory from data memory allows execution of instructions to be done in parallel with the fetching and decoding of the next

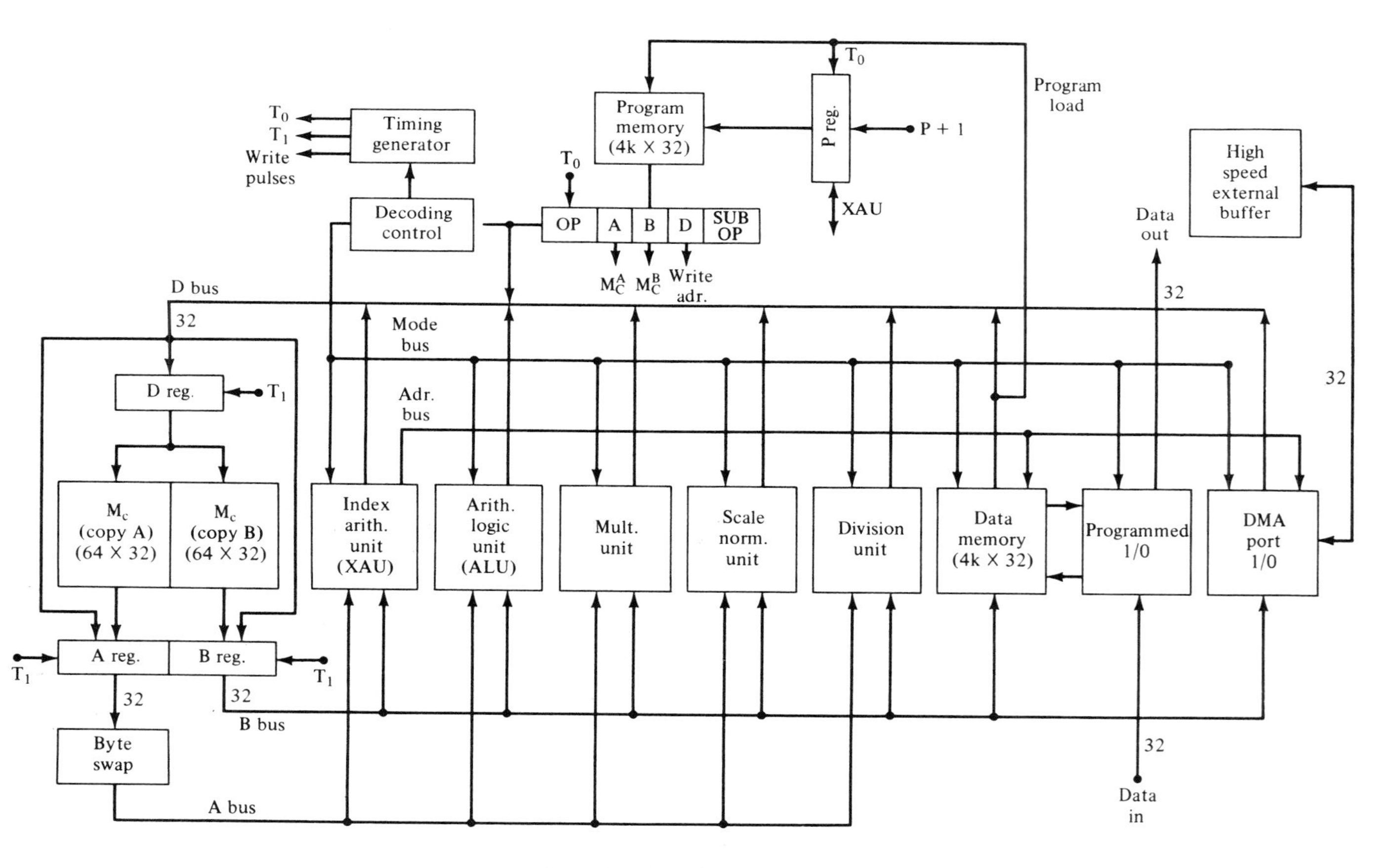


Figure 7.4: LSP/2 architecture.

instruction. The dual copies of the 64 general registers and the 3 parallel buses allow a simultaneous access of two operands for an operation and the capture of the results from the last operation in every instruction cycle. Since the various functional modules are essentially connected in parallel between the three buses and are effectively self-contained, they can run in parallel. However, the common source and destination buses and registers limit the initiation of an operation in any one instruction cycle to a single functional module and the capture of a single result. In this manner the sequential control is mapped to the parallel functional modules. Hence the various modules will not be kept busy and the full utilization of the potential parallelism of these multiple functional modules is limited by the sequential control mechanism.

Essentially, the parallelisms realized in this processor are threefold: the program memory access, the general register reading and writing, and the operation of functional units. The overlapped timing of these operations is illustrated in Figure 7.5. The $T_0$ to $T_0$ timing cycle of the processor is 70 nanoseconds for most operations. However, to accommodate the variation in execution time for some functional units such as multiply, which is greater than 70 nanoseconds, the timing is adaptive, thus causing a further complication in system control. While the LSP/2 architecture provides functional parallelism to a considerable extent, the basic arithmetic operations available are still at the add/subtract/multiply/divide level. To implement digital signal processing functions the machine has to be programmed to carry out the basis operations in terms of standard arithmetic operations and data transfers.

The measure of the LSP/2 design is the FDP and the results are encouraging. The LSP/2 could perform a 16 × 16 bit complex multiply in 300

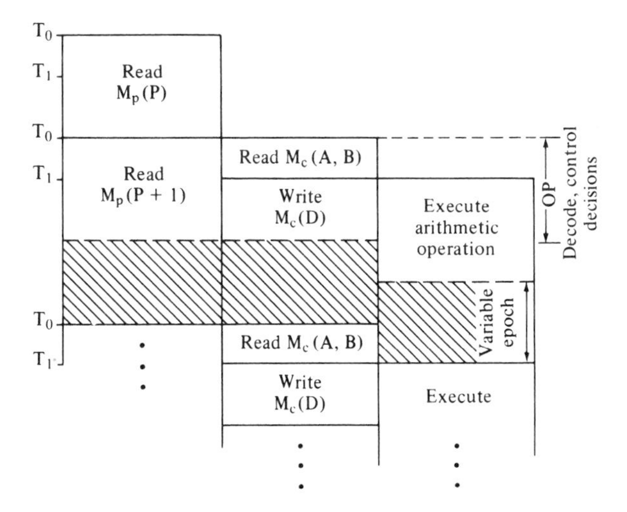


**Figure 7.5:** LSP/2 overlapped adaptive timing.

nanoseconds (FDP-600 nanoseconds). A complex butterfly in 340 nanoseconds (FDP-1360 nanoseconds). The LSP/2 achieved about four times the performance of the FDP with about one-third as many integrated circuits. This trend increases dramatically throughout the 1970s. Performance enhancement is achieved by a combination of higher speed, denser technology, parallelism, and larger functional modules.

### 7.1.3 The SPS 41/81

The functional partitioning of the LSP/2 architecture of the previous subsection provided parallelism at the level of basic arithmetic operations and data transfer under a single sequential control stream. Another approach to partitioning processor functions is to separate major activities such as I/O, internal data handling, and actual arithmetic processing into synchronized but separately controlled activities. This leads to an architecture of multiple specialized processors working in a tightly synchronized interaction.

A detailed discussion of the control complexities of concurrently operating processors in an MIMD architecture requires the acquisition of several new concepts to be presented in Volume II. We note at this time, however, that by effecting this control at the microcode level the assembly level programmer is provided with a virtual machine of higher functional performance and lower apparent programming complexity. This represents an important concept in processor useability. As we shall see in the next two subsections, the use of microcoding techniques has become extremely important in the design and implementation of modern signal processing systems.

The SPS 41/81 architecture exhibits an extension of the attached processor concept of separating the main arithmetic processing from the support function processing. In the SPS case a further partitioning has been effected to separate the control of I/O operations, address generation and control of the main arithmetic operation sequencing and the actual function units that carry out the primary computation. By going to this higher level of functional parallelism the SPS processors have achieved reasonable performance with slower, less expensive TTL component technology.

### 7.1.4 Sperry Programmable Acoustic Receiver (SPAR)

The Sperry Programmable Acoustic Receiver (SPAR) is an example of combining the high level functional partitions of the SPS-41 and 81 with the dedicated function unit concepts of the LSP-2. In the SPAR processor a set of microprogrammable functional modules are interconnected by a set of buses to an address processor. Each microprogrammed module is programmed to carry out a specific set of signal processing tasks. The address processor controls all main memory addressing and overall system control. The microprogrammed modules access common memory to obtain blocks of

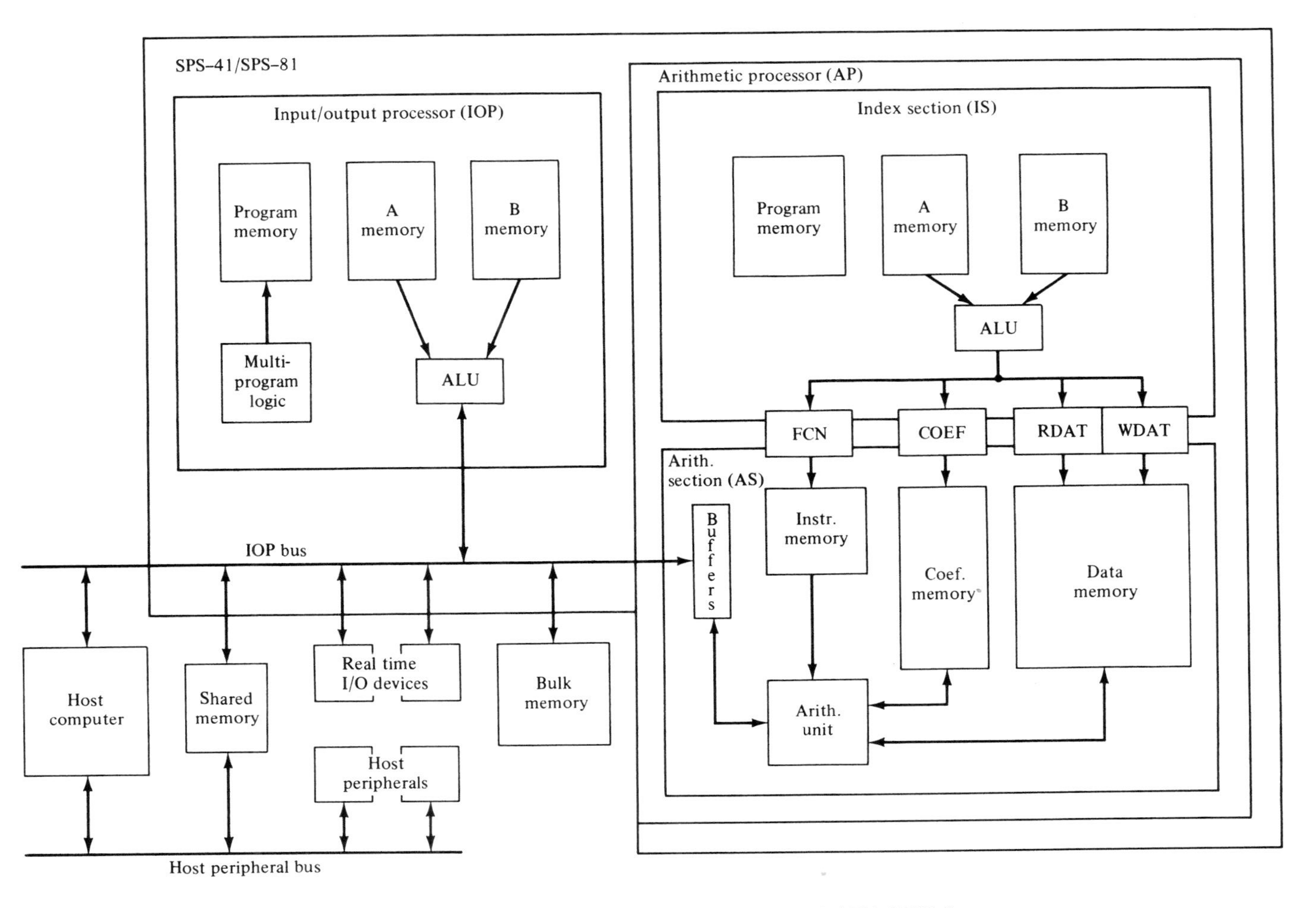


**Figure 7.6:** SPS-41/SPS-81 processor configuration. © 1974 IEEE Inc. Reprinted with permission from *IEEE EDSCON*, **74**, Oct. 4-9, 1974, pp. 674-78.

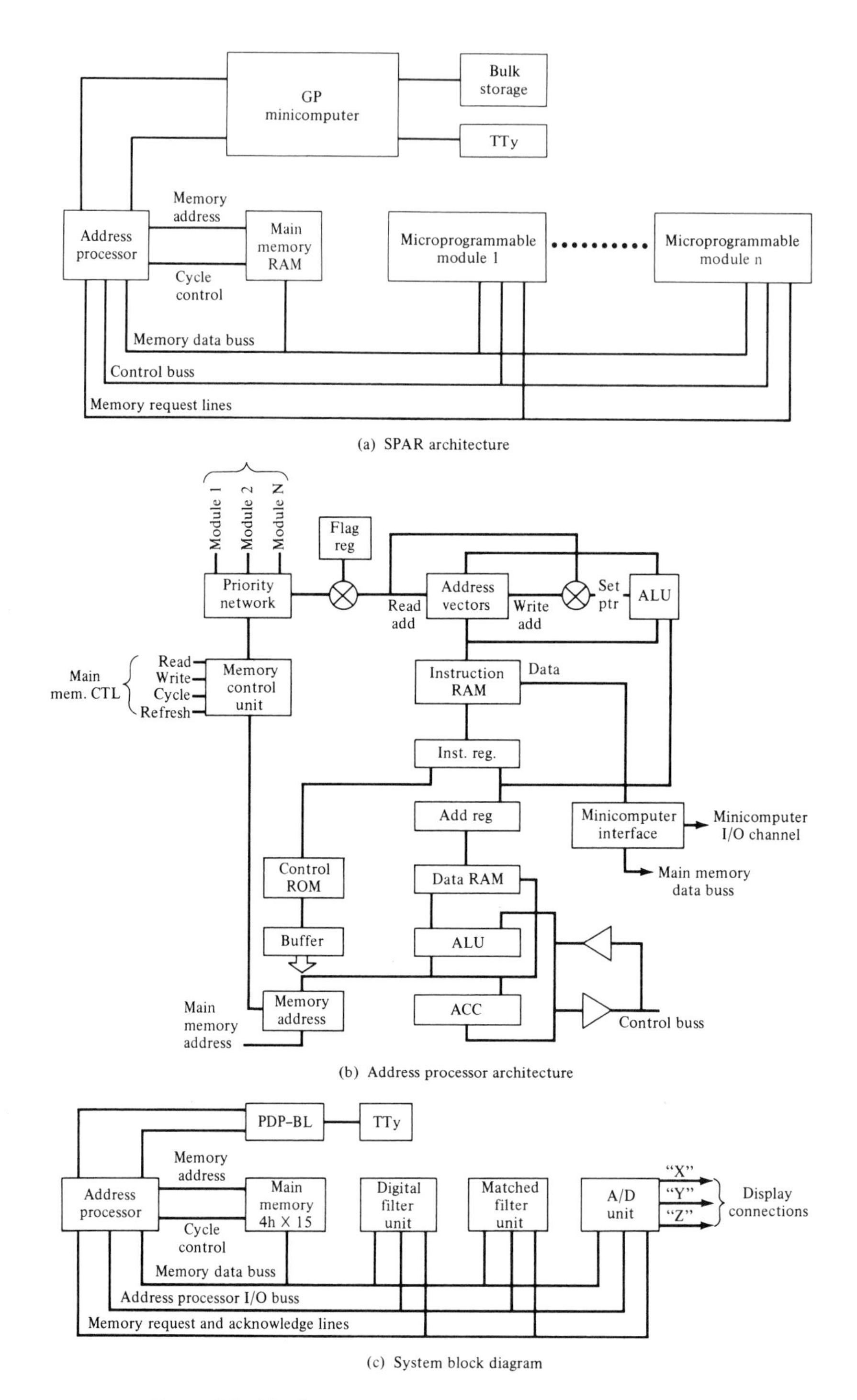


(a) SPAR architecture

(b) Address processor architecture

(c) System block diagram

**Figure 7.7:** The Sperry programmable acoustic receiver.

data and then run autonomously in a concurrent manner. The major limitation of this architecture is main memory access. This system is an example of the classic memory access bottleneck of a common bus interconnection topology for multiple processors.

Again we must defer a more detailed discussion of the multiprocessor control problem to Volume II. However, the SPAR serves to point out the central limitations of such architectures, namely the data transfer requirements associated with multiple processing elements manipulating a common data structure. We note that a common sequential control mechanism has to be maintained in the SPAR system to coordinate the various functional units. In Volume II we shall examine the problems of data flow and control synchronization as the overall system control is allowed to overlap and become distributed throughout the functional units.

### 7.1.5 The Data General AP/130 Array Processor

A more recent signal processor is the Data General AP/130 array processor. This processor typifies the architecture of current low cost array processors designed as an attachment to a general purpose minicomputer.

The AP/130 was designed to augment the throughput of the Data General S/130 minicomputer. The design goal was to permit extensions to the standard instruction set of the S/130 with an array processing instruction set (APIS). This APIS can be interspersed with the standard instructions in a transparent manner. The APIS is shown in Figure 7.9. These large macro instructions permit rapid execution of array manipulations as well as standard computations (e.g., a 1024-point complex FFT in 8.75 milliseconds; a 1024-point real array multiply in 0.921 milliseconds).

The architecture of the AP/130 is shown in Figure 7.8. The major functional units of the processor are a pipelined floating point adder and a pipelined floating point multiplier, multiple parallel data buses and multiple specialized memories and register files, including sine and cosine lookup tables. The pipeline arithmetic units produced normalized results at full utilization at a rate of 400 nanoseconds for the multiplier and 200 nanoseconds for the adder. The sine/cosine table provides a resolution of 1024 points on the unit circle. The internal bus is 64 bits wide and operates at 5 MHZ (40 M bytes/sec).

#### Data Structures

The processor is designed so that the APIS augments the standard instruction set of the S/130 CPU. The architecture is simple and the execution throughput depends to a large extent on the data structures and the manipulative capability of the component instructions.

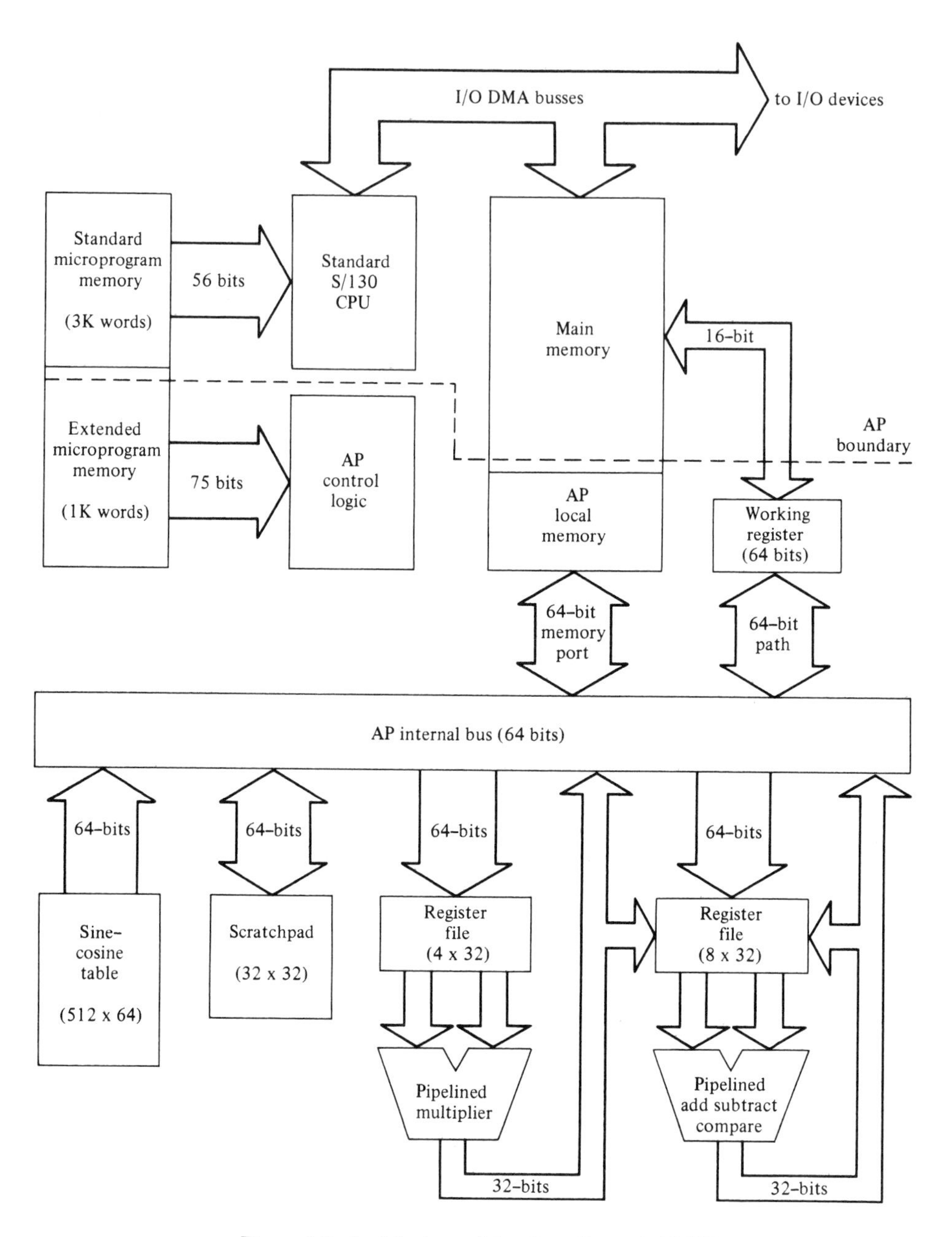


**Figure 7.8:** Architecture of the Data General AP/130.

Scalar Instructions

| | |
|---|---|
| ARS | Add real scalar to array |
| SRC | Subtract real scalar from array |
| MRS | Multiply real scalar by array |
| MCS | Multiply complex scalar by array |
| SPS | Signed product of real scalar and array |

Array Instructions

| | |
|---|---|
| ARA | Add real arrays |
| SRA | Subtract real arrays |
| MRS | Multiply real arrays |
| MCA | Multiply complex arrays |
| NRA | Negate real array |
| SMA | Square magnitudes of complex array |
| SPR | Signed product of real arrays |
| SER | Sum of elements of real array |
| PER | Product of elements of real array |
| PEC | Product of elements of complex array |
| IPR | Inner product of real arrays |
| IPC | Inner product of complex arrays |
| EPR | Evaluate polynomial in real array |
| EPC | Evaluate polynomial in complex array |
| MXR | Maximum element of real array |
| MNR | Minimum element of real array |
| CMA | Compare arrays |

Signal Processing Instructions

| | |
|---|---|
| CONR | Convolution of real arrays |
| CONRZ | Convolution of real arrays with zero initial conditions |
| CONC | Convolution of complex arrays |
| CONCZ | Convolution of complex arrays with zero initial conditions |
| CORR | Correlation of real arrays |
| CORC | Correlation of complex arrays |
| RFR | Recursive filter of real data |
| RFC | Recursive filter of complex data |
| INR | Integrate real array |

Fast Fourier Transform Instructions

| | |
|---|---|
| FFTC | Fast Fourier transform of complex array |
| FFTR | Fast Fourier transform of real array |
| BRC | Bit-reverse indices of complex array |
| SCB | Store complex array with bit-reversed indices |

Miscellaneous Instructions

| | |
|---|---|
| FLL | Float and load (convert integer to real) |
| FXS | Fix and store (convert real to integer) |
| LSR | Load scratchpad registers |
| SSR | Store scratchpad registers |
| SRW | Store real scalar form working register |
| SCW | Store complex scalar form working register |
| LDR | Load real array (from main memory) |
| LDC | Load complex array (from main memory) |
| STR | Store real array (to main memory) |
| STC | Store complex array (to main memory) |
| CRE | Create a real array |
| MOD | Modify a real array |

**Figure 7.9:** The array processing instruction set of the AP/130.

The following data formats are supported:

Integers: 16 bit, two's complement
Real Scalars: 32 bit, normalized floating point
Complex Scalars: ordered pairs of real scalars
Structured Array: an ordered sequence of scalars

It is the array processing instructions supported by the underlying hardware which provide the potential performance enhancement. The array instructions must manipulate one or more arrays and return results to a third array. Each array instruction has an associated parameter block. This block consists of up to 16 words and contains all relevant control information such as array lengths and starting addresses. In general each word has a specific function and is ignored by an instruction which does not use that function. This permits several instructions to use the same parameter block when the data are compatible.

The instructions for manipulating arrays cover all but one of the basis functions proposed in Chapter 5:

1. Difference Equation

$$Y(n) = \sum_{k=0}^{N-1} a_k x(n-k) - \sum_{k=1}^{M-1} b_k y(n-k)$$

AP/130—Recursive Filter for Complex Data = RFC (using local notation)

$$Q_n = \sum_{k=0}^{N_p-1} A_k x_{n-k} - \sum_{j=1}^{N_Z-1} B_j Q_{n-j}$$

$N_p$ = number of poles (1 or 2)

$N_z$ = number of zeros (0, 1, 2)

A similar command exists for real data—RFR.

2. Correlation Calculation

$$X_k = \sum_{n=0}^{N-1} x(n) e^{-j(2\pi/N)nk}$$

AP/130—Convolution of Real or Complex Arrays
CONR Convolve Real
CONC Convolve Complex

3. Complex Frequency

$$y(n) = x(n) e^{j2\pi nk}$$
$$= x(n) Z_n$$

AP/130—Evaluate Polynomial (EPC, EPR)

$Z$ = any complex scalar

4. Discrete Fourier Transform

$$X_k = \sum_{n=0}^{N-1} x(n) e^{-j(2\pi/N)nk}$$

AP/130—FFTR, FFTC—FFT of real or complex array

5. Power Spectral Density or Magnitude Squared

$$|X_k|^2 = X_k X_k^*$$

AP/130—SMA—Square magnitudes of complex array.

6. Matrix Arithmetic and Matrix Manipulation: Matrix arithmetic and manipulation operations can be readily programmed using the many array handling instructions of the AP/130.
7. Logarithms and Exponentiation: There are none in the AP/130.

The instruction set design of the AP/130 does not require a detailed knowledge of signal processing in order to code an algorithm. In this respect the processor as an augmenter to the supporting minicomputer offers considerable potential savings in software costs. For example, the FFT is coded as follows:

```
; Forward Real FFT

    ELEF 2, PBLK     ;  Load Acc 2 with
                     ;  Parameter block address
    FFTC             ;  FFT of complex array
    BRC              ;  Bit reverse the indices
                     ;  of the complex array
    FFTR             ;  FFT of real array

; Inverse Real FFT

    ELEF 2, PBLK 1   ;  Set up two parameter blocks
    FFTR             ;  FFT of real array
    ELEF 2, PBLK 2   ;  get address of second block
    FFTC             ;  FFT of complex array
    BRC              ;  bit reverse the results
```

These large macrocommands bury the internal complexity of the machine and provide a reasonable approach to writing processing algorithms.

The hidden advantage here is that the AP/130 architecture and implementation could be changed without destroying the usefulness of existing software.

### 7.1.6 A Comparison of the FDP and the S2811

In our brief review of signal processor architectures we have been mainly concerned with exposing the various levels and approaches to parallelism and pipelining that have been employed to enhance processor throughput. Clearly one aspect of performance improvements has been the continued advances being made in component technology. As we shall discuss in the final section of the chapter, there has been an exponential growth in component capabilities over the past decade and industry projections indicate a continuation of this trend. To illustrate the impact of component technology advances on processor design we shall briefly compare two attached processors whose implementation were separated by a period of less than ten years: the FDP and the single chip S2811 signal processing peripheral manufactured by American Microsystems Inc.

The FDP was constructed of approximately 10,000 SSI integrated circuit packages of the fastest available technology of the time, Emitter Coupled Logic (ECL). The basic architecture of the FDP was presented in Section

7.1.3. The S2811 is a single 28 pin dual in line package fabricated from v-mos technology. A block diagram of the S2811 internal architecture is shown in Figure 7.9. Figure 7.10 shows the instruction format and addressing modes of the S2811.

Architecturally there are several similarities between the two processors. Both processors employ two independently addressable data memories and a separate program memory. The arithmetic section of both processors employs separate adders and multipliers that execute in parallel.

With respect to the arithmetic sections of the two processors, we note a similarity in the structure of the S2811 and that of the individual AEs of the FDP. The FDP contains 4 identical AEs, each with an adder and registers and a separate combinational multiplier that executes in parallel with the adder. The S2811 also employs a separate adder and multiplier that execute concurrently. Indeed, the multiplier on the S2811 chip executes in a single 300 nanoseconds instruction cycle while the FDP multipliers requires 450 nanoseconds or 3 instruction cycles.

Another similarity is the dual instruction execution format. The FDP executes a double word length instruction and the S2811 has an instruction set consisting of two sets of operational codes, one for data flow and one for arithmetic control, that are executed simultaneously.

In terms of processing performance the S2811 with a single arithmetic section compared to the four AEs of the FDP, provides approximately one-fourth the throughput of the FDP for standard signal processing operations such as fast Fourier transform computations. However when one recalls that a reduction in component count of four orders of magnitude has been achieved in the S2811, this performance comparison appears dramatic.

This brief comparison of the FDP and the S2811 clearly illustrates the impact that advances being made in component technology are having on our perspectives and our expectations about processor performance. The next section of the chapter deals with the evolution of integrated circuit technology and reinforces the impact on future processor architectures.

### 7.1.7 Summary

Our primary concern in this section has been to reinforce the concepts and techniques developed in Chapter 6; that is, the performance potential of a processor can be determined in a systematic way by examining the multiplicity of processing units. At this level also, the applicability of the processor to a particular task stream can be computed. Supporting the basic architecture there must be mechanisms for supporting the data flow and for programming specific operations. Thus in an expanding sequence any processor can be examined and evaluated in terms specific to an application.

It has become very clear that architects must be concerned with software and programming when considering a processor. Designs which require the

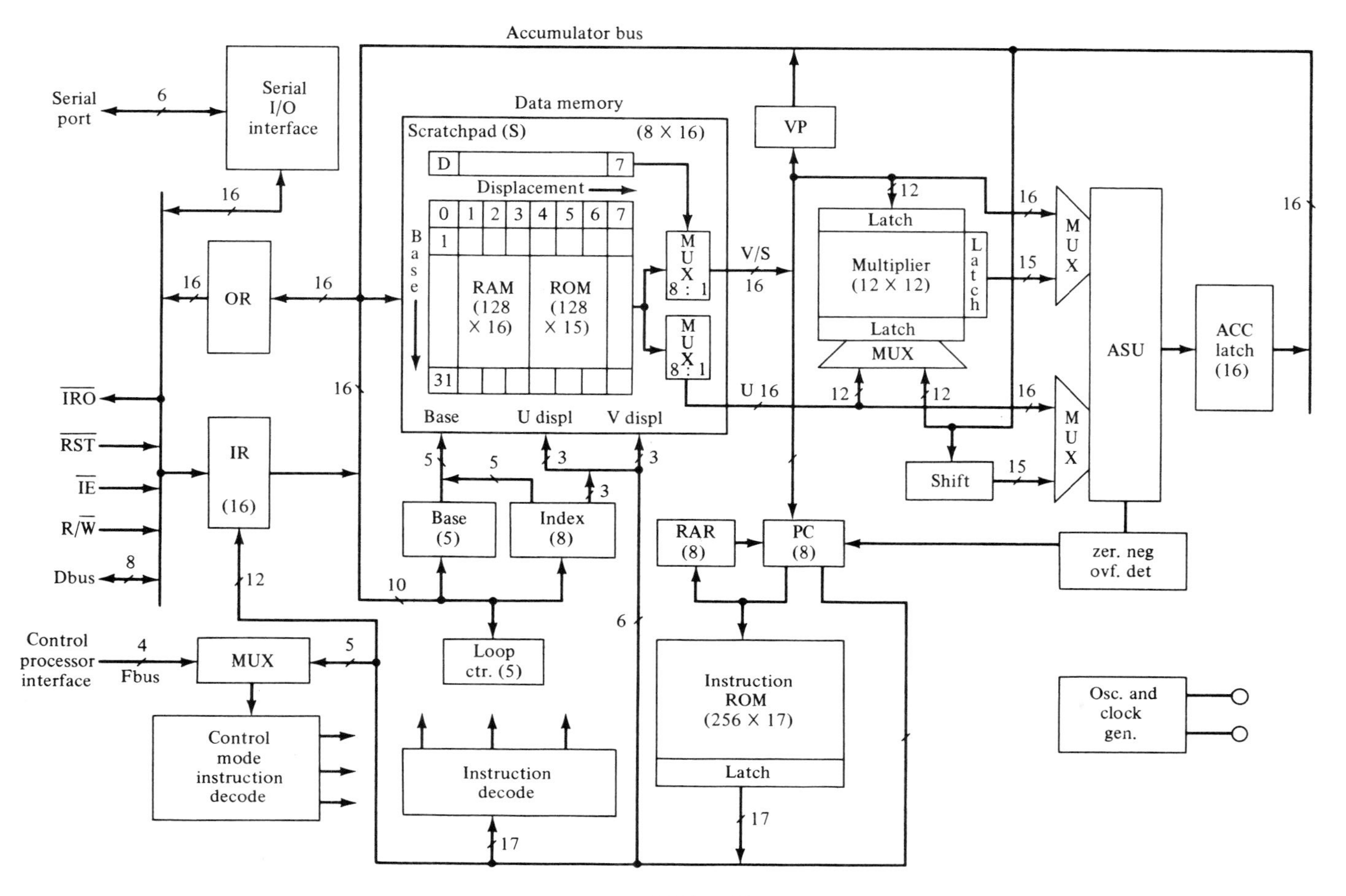


**Figure 7.10:** S2811 architecture.

| | $I_{16}$ – $I_{12}$ | $I_{11}$ – $I_8$ | $I_7$ – $I_0$ |
|---|---|---|---|
| SPP instruction format | OP2 | OP1 | Operand |

SPP addressing modes ← 17 bits →

| | 5 bits | 4 bits | 3 bits | 3 bits | 1 bit | 1 bit |
|---|---|---|---|---|---|---|
| Offset addressing (UV/US) | OP2 | OP1 | $O_1$ | $O_2$ | 0 = US<br>1 = UV | O |
| Direct addressing (D) | OP2 | OP1 | Address (OH) | | | 1 |
| Direct transfer (DT) | OP2 | OP1 | Transfer address (HH) | | | |
| Literal (L) | OP2 | Data word (HHH) | | | | |

| Addressing mode | Effective address U | Effective address V/S | Multiplier operands |
|---|---|---|---|
| UV | $(BAS) + O_1$ | $V = (BAS) + O_2$ | $P = U \cdot V$ |
| US | $(BAS) + O_1$ | $S = O_2$ | $P = U \cdot S$ |
| D | — | OH | $P = A \cdot V$ |

Note: O indicates an octal digit (3 bits) and H indicates a hexadecimal digit (4 bits)

**Figure 7.11:** S2811 instruction format and addressing modes.

designer's knowledge of the hardware for effective program generation are doomed to a restricted application (i.e., market oblivion). The trend which is evident is toward attached processors with a host requesting computational services from a specialized processor, the computational services being microprogrammed and callable on demand. Transparency can be achieved in this fashion and standard compiler operations used to create operating software. This approach is a retreat to the logical simplicity of sequential control in the von Neumann sense and depends on high speed components to reduce execution time rather than parallelism in the SIMD or MIMD sense. The rapid increase in VLSI performance may reinforce this trend for some time to come.

We note, however, that architectures such as the SPS 41/81 and the SPAR system have moved toward multiprocessor solutions with parallelism at a high level.

As we shall see in the next section, component technology is accelerating in two primary areas: increased speed and increased functional complexity of single components. While the speed increases have enabled simple architecture such as the AP/130 to realize impressive performance at low cost under strictly sequential control, the ability for continued speed increases will ultimately be limited by signal propagation time on the chip itself (i.e., the speed of light forms the ultimate speed limit). On the other hand, the S2811 example indicates that future systems may well be composed of collections of components that are each complete individual processors with architectures optimized to specific types of processing tasks.

Major issues in the design and implementation of such systems include consideration of system partitioning the interconnection, synchronization, scheduling and coordinated control of many functional elements to realize overall processing algorithms. These issues will be considered in detail in Volume II.

## 7.2 INTEGRATED CIRCUIT TECHNOLOGY

### 7.2.0 Introduction

In this section we deal with integrated circuit technology and the limitations it imposes on components and on the realization of systems. We are concerned both with what is available and what is probable in the next several years. In the background is the thesis that, as the needs of signal processing are articulated, the semi-conductor industry is now capable of responding with components. At a more basic level the availability of high performance components is creating dynamic alterations of all aspects of signal processing. These components are affecting not only considerations at the implementation level, but are reverberating into algorithm design and even representations.

There is an inherent trade-off between algorithmic complexity and their supportative control and data structures. The decrease in execution time which elegant algorithms yield is usually paid for in more complex control and/or data structures. Booth's algorithm for multiplying binary numbers, for example, provided a substantial average decrease in execution time at the expense of additional control circuitry to detect strings of 1's and 0's in the multiplier. Similarly the fast Fourier transform increased the real time computational ability of general purpose computers at the cost of rather complex data flow and control. The advent of VLSI and the realization of high speed, low cost components have drastically altered our ability to create hardware for these problems. At the system level the shift is clearly toward simpler algorithms with associated simplicity in data flow and control. This shift is made possible by the migration of algorithm complexity into silicon as component speeds increase and component architectures are optimized to execute signal processing algorithms.

Two factors dominate any discussion of the semi-conductor industry: first, component complexity and capabilities are increasing exponentially, and second costs are decreasing almost as fast. This has made possible the utilization of these components in an ever expanding range of application areas. On the other hand, that critical point in a system design where the components must be specified becomes agonizing. It is clear that any design will be technologically obsolete almost immediately. It appears that designers must live in this environment for the foreseeable future. The major problem in terms of any discussion on design philosophies and methodologies is that the original technological constraints which bias design choices may suddenly alter drastically or even vanish; however, the effects which have permeated throughout the design may be difficult to trace.

A second factor is the growing collaboration of systems architects and component designers. The semi-conductor industry grew over several decades with a capability to increase continuously the gate density and speed of chips. To some extent each manufacturer grew and sold what they were able to. It was not clear in the beginning, for example, what market a microprocessor would fill. The results of this innovation have surprised even those who first created it. Capability during the 1980s has reached the point where the major question is "what to build"—not "how to build it." Thus, perceived demand will strongly influence component availability. Signal processing demands are now being accommodated and an increasing emphasis is being placed on components for this market. This suggests that preconceived notions on target components and resulting architectures should be viewed with caution; which is of course the underlying philosophy of the whole book.

### 7.2.1 IC Technologies

The technologies currently available to fabricate commercially viable components can be partitioned into two categories: MOS (Metal Oxide Semiconductor) and bipolar. Within each category a variety of implementation mechanisms offer a differing set of characteristics. In terms of overall component characteristics MOS and bipolar components can be differentiated with respect to device density, clock speeds (gate delays) and power consumption. In general, MOS circuits are characterized by a high device density per chip at low to moderate clock speeds, and moderate to low power consumption. Bipolar circuits offer high clock speeds at the expense of higher power consumption and lower device density.

Bipolar technologies provided the earliest integrated circuits. The higher speed (and power comsumption) of bipolar devices results from the ability to accurately control diffusion depths of doping impurities and thus define narrow base regions in the junction transistors. However, due to the three dimensional nature of the junction transistors of bipolar integrated circuits, they do not scale down as easily as the essentially planar field effect transistors (FETs) of MOS technologies. Thus it is in the area of MOS technology where the most spectacular increases in gate density are occurring as minimal resolvable lithographic feature sizes decrease.

Several fabrication technologies are currently used for bipolar integrated circuits. Among the more common are Transistor-Transistor Logic (TTL), Schottky and low power Schottky Transistor-Transistor Logic (STTL and LSTTL), Emitter Coupled Logic (ECL), which is currently the fastest bipolar logic, and Integrated Injection Logic ($I^2L$). New technologies and fabrication techniques continue to be developed, such as Integrated Schottky Logic (ISL) and Polysilicon Self-Aligning (PSA).

MOS has demonstrated a remarkable tenacity and capability for enhanced performance over the years. Despite periodic predictions of a limit being reached, innovations continue to increase performance with respect to speed, and device density and power consumption. Within the MOS family, variations exist based on fabrication techniques, e.g., N-MOS, V-MOS, and C-MOS, although a search through the literature will reveal H-MOS, D-MOS, and more.

P-MOS was the first successful MOS technology and the slowest. However, the faster N-MOS dominates the market at the beginning of the 1980's. Fabrication is well understood and yields are satisfactory. V-MOS permits a higher component density; however, manufacturing is more complex. C-MOS also suffers from a complex manufacturing process. However its unique features are exceedingly low power required during idle periods and the ability to support on the same chip both analog and digital circuits.

New technologies may also become feasible over the next decade, e.g., C-MOS on sapphire, silicon on sapphire (SOS), galium arsenide (GaAs), and the Josephson junction. It is worth reiterating, however, that commercial viability implies, as a minimum, that the component can be manufactured with a high enough yield to be sold at a profit into some market. It is perhaps significant that none (with the exception of P-MOS) of these technologies existed in 1970 and only a few in 1975. Projections to 1990 may therefore reflect an incredible myopia when viewed in retrospect on that date.

### 7.2.2 Evolution and Projections

The increases in integrated circuit capability and complexity have evolved over the past decades as if in response to an organic growth law. The consistency of this exponential growth has been such that quite reasonable projections over the next decade seem possible.

To examine the evolution of integrated circuit technology consider the following component parameters:

- device densities
- gate delay
- speed-power products
- lithographic resolution

The device density of an integrated circuit depends on two factors: chip area and minimum lithographic resolution. Figures 7.12, 7.13, and 7.14 due to [Keyes, 79] illustrate the trends and relationship of chip size and minimum lithographic resolution to device density for both memory and logic circuits from 1960 projected through to the year 2000. Figure 7.15 (also due to [Keyes, 79]) indicates actual levels of integration for MOS devices from 1970 projected through to 2000. Note that while the projected rate of increase begins to level off considerably beyond 1990, the past trends con-

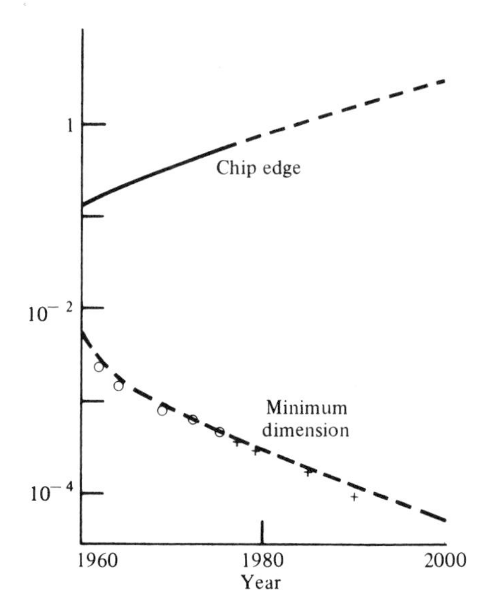


**Figure 7.12:** Lithographic and chip edge projection.

tinue essentially unchanged until the late 1980's. For comparing bipolar and MOS device densities, a rough rule of thumb is that N-MOS provides up to an order of magnitude increase in density over TTL at essentially equivalent speeds. It is interesting to note that the general distinction between MOS and bipolar technologies with respect to device density, speed, and power consumption is not entirely clear. As MOS technologies are scaled to smaller dimensions, they begin to challenge bipolar speed.

The speed-power product provides a measure combining gate delay (which relates directly to clock speed) and power consumption. The exponential decrease in speed-power products is also expected to continue into the late 1980s.

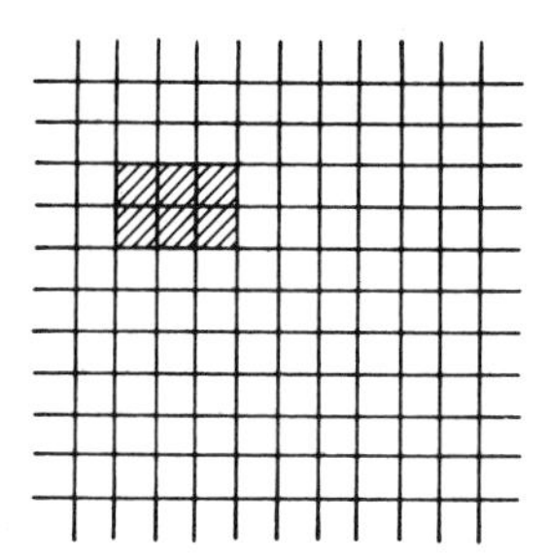

**Figure 7.13:** A resolvable element (a square with side equal to the minimum lithographic dimension). © 1979 IEEE Inc. Reprinted with permission from *IEEE Transactions on Electron Devices*, **ED-26**, 4, April 1979, p. 272.

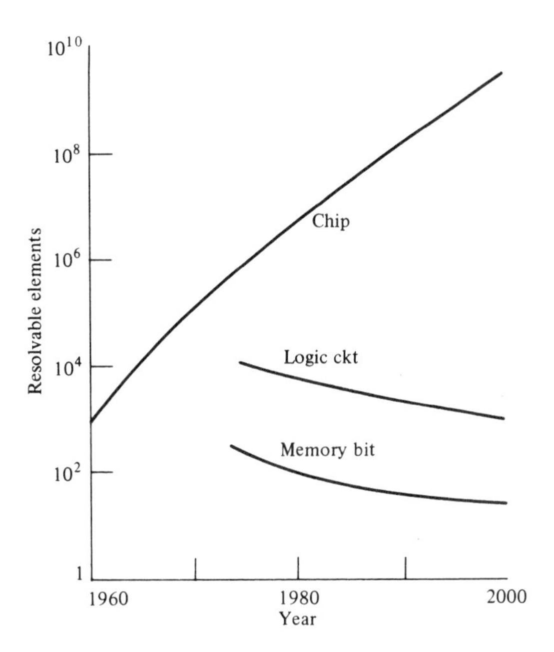


**Figure 7.14:** Resolvable elements on a chip of maximum dimension.

A minimum of 1 micron is commonly considered the useful limit of optical lithographic techniques for production IC fabrication. Figure 7.12 projects this limit by 1990. However, electron beam and X-ray lithography techniques promise to enable minimum feature sizes to drop possibly as low as 0.1 mi-

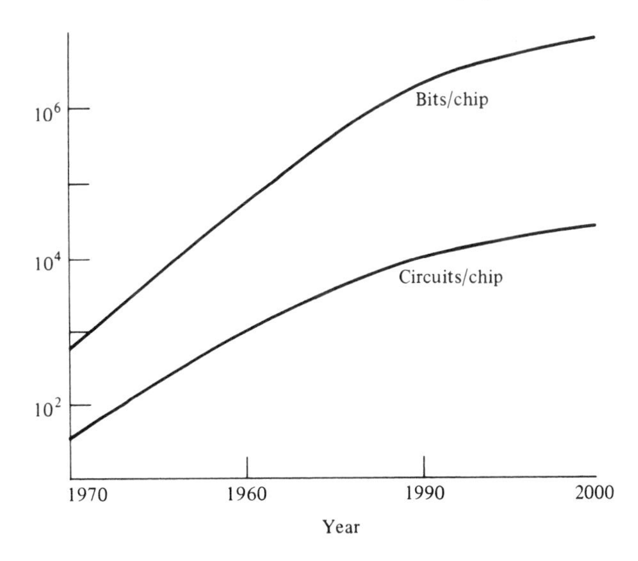


**Figure 7.15:** Projections from main memory and logic chips.

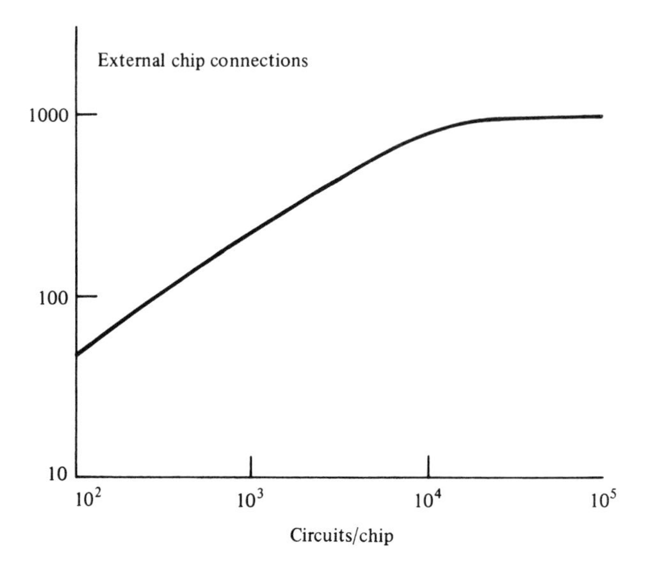


**Figure 7.16:** Signal connections to a logic chip. © 1979 IEEE Inc. Reprinted with permission from *IEEE Transactions on Electron Devices*, **ED-26**, 4, April 1979, p. 274.

cron by 1990. The ability to achieve such reductions in minimum feature size implies staggering increases in component capabilities of the next decade.

Throughout the literature a constant opinion prevails that no major scientific breakthrough is required to maintain the rate of growth. In order to place these projections in some comparative framework, consider the influence of two attributes:

- feature size (depends on lithography)
- throughput (function of gates and gate delays)

If $N$ represents a shrinkage factor for minimum lithographic feature size, i.e.,

$$N = \text{Present capability/future capability}$$

then the projections suggest an improvement from 3 to 0.2 microns ($N = 15$) by 1990. Various chip attributes are related to $N$ as follows:

Current reduction $1/N$ : 1/15
Voltage Supply $1/N$ : 1/15
Gate Delay $1/N$ : $(1/15) \times 1.5\text{ns} = 0.1\text{ns}$
Power Dissipation $1/N^2$ : 1/225
Speed Power
  Product $1/N^2$ : $(1/225) \times 2 = .06\ PJ$
Device Density $N^2$ : $225 \times 200 = 45,000/\text{mm}^2$

Processing
Throughput/
Unit Area $N^3$ : $3375 \times 10^9 = 3.375 \times 10^{12}$

Consider the last parameter, processing throughput per unit area ($TP$), defined as

$$TP = (\text{Gate density}) \times (\text{clock rate})$$
$$= \text{gates/mm}^2 \times \text{Fmax}$$
$$= \text{Gate-Hertz/mm}^2$$

This throughput figure of merit (attributed to the United States Military development program for Very High Speed Integrated Circuits (VHSIC)) provides a measure of processing throughput that is independent of fabrication technology.

Consider a contemporary microprocessor such as the S2811 with 30,500 devices on a 26 $\text{mm}^2$ chip operating at 20 MHz. Assuming four devices per gate, we have

$$TP(S2811) - 293 \text{ gates/mm}^2 \times 20 \text{ MHz}$$
$$= 5.86 \times 10^9 \text{ gate-HZ/mm}^2$$

Since $TP$ is proportional to $N^3$, a shrinkage factor of $N = 15$ indicates a potential component by 1990 with a processing throughput per unit area of approximately $2 \times 10^{13}$ gate-Hz/mm$^2$. This figure corresponds to a component with a gate density of 100,000 gates/mm$^2$ operating at 200 MHz, or possibly 200,000 gates/mm$^2$ operating at 100 MHz. A single component with a 1 square centimeter area gives a 2 million gate circuit operating at 1 giga Hertz. This is clearly an imposing processor in its own right.

Many such projections can be made. They indicate a staggering increase in processing power per component. Similar projections indicate the feasibility of at least a 4-megabit random access memory chip with an access time of less than 100 nanoseconds by 1990.

Consider, however, several factors involved in the production and use of such chips. First consider the requirements to test a 4-megabit RAM chip. Assume a fully automatic tester which cycles through read, write, read modify write for each cell in total time of 600 n.s. This requires 2.5 seconds to test the chip. To complete a 5-test sequence on, say, a month's production of a million chips would take $12.5 \times 10^6$ seconds. Depending on an estimate of available person-seconds in a month (let's say $2.5 \times 10^6$) it would take five testers running full-time. At perhaps one-half million dollars each, this represents a high investment in its own right just for electrical testing. The implications for more complex chips is equally staggering.

The testing problem may be resolved by allocating some position of the chip to self-test circuitry. Here one can contemplate a new discipline of silicon deviant psychology—the study of chips who claim they are sane but are not.

Next, consider the number of interconnections required as circuit complexity rises. Providing bonding pads for hundreds of pin connections could easily use up 50 to 60 percent of total chip area. Further, the chip area needed just for wire runs may use similar percentages of the remaining chip area. As a result, the chip area remaining for actually creating active devices may not increase substantially from that available today. Such factors could severely inhibit the realization of our simple predictions.

Finally manufacturing facilities in terms of simple requirements such as raw material purity and clean rooms become more significant. Insuring freedom from environmental contaminents of submicron dimensions is a requirement which is rather capital intensive. Such considerations support the conclusion that participation in VLSI manufacturing will become increasingly the domain of large companies with huge capital resources.

The projections discussed in this section are based on extensions to capabilities which have had almost purely exponential growth over the last few decades. Behind this growth has been two requirements: massive research and a growing market demand. Thus a new component on entry to the market had a large development cost which was amortized over the first few years' sales. Price decreases, as this cost was dissipated, or in response to competition, were dramatic and concurrent increases in volume usually occurred. The intrinsic feature of the market is an insatiable demand for higher performance components. The cost/performance ratio of current components has, however, created application areas of considerable volume in which further increases in performance are not required, for example, appliance controllers, calculators, games, mechanical switch replacements, automotive, etc. It is proposed here that these represent a huge market which most manufacturers cannot ignore, and thus they represent a major inhibiting factor to the continued growth of high performance components.

Exponential growth environments in nature tend to have inhibitors which eventually act to bend the projections onto some asymptote. There are a host of such inhibitors becoming visible in the semi conductor industry, although their effects are not always quantifiable. A detailed discussion is given by [Keyes, 79].

Certain physical phenomena are already recognized as potential inhibitors. Current flow in fine conducting portions of a chip may introduce physical migration of atoms, which in turn may alter the characteristic of the chip. Dielectric breakdown due to high field intensities in narrow regions may occur. These phenomena could demand a material other than silicon or some as yet unknown alloy to maintain the projected figures. This would represent potentially a huge investment which could act as a delaying mechanism.

Other factors are intuitively obvious from considering a large complex chip. Ohmic resistance of fine wires of even small lengths becomes significant. The complexity of interconnecting the gates internally and to the external pins will obviously increase. Indeed the mask layouts and sheer magnitude of designing a huge chip will consume many man-years of effort. Clearly the need for massive computer aided design (CAD) aids is indicated. The total power dissipation of several watts becomes increasingly significant and may force larger package sizes or even liquid cooling.

Perhaps the key attribute to track is the lithography resolution capability with time. If unforeseen difficulties delay progress toward a submicron line, then there will be a corresponding delay in attaining all the other attributes.

The reduction of feature size on a chip and the larger physical area begins to introduce significant problems which may present some asymptotic limit on the use of silicon. The amortization of manufacturing costs must be continuously considered. And if the manufacturing costs of increased miniaturization exceeds some limit, it is evident that such devices will not be pursued.

It has been shown here that the semiconductor industry has exhibited a remarkably consistent exponential growth characteristic for more than a decade. Projection of this exponential growth environment leads to speculation on dramatic increases in component capabilities over the next decade; yet lurking problems may inhibit or delay these projections. The message is clear. Both general and special purpose signal processors with innovative architectures and greatly enhanced performance are natural results of these component advances. Thus designers must track technology with care and become increasingly knowledgeable in multi-processor architectures.

### 7.2.3 The Impact on Architecture

In the previous section we have provided the reader with some intuitive insight into the justification for statements that claim the inevitable appearance of exceedingly complex high performance processors as single components in the near future. If the reader can accept such a statement at this time, then it follows directly that creating signal processing systems that utilize such VLSI components is clearly a multiple processor design problem. It is this conclusion that motivates Volume II, which deals with the system design concepts.

To reinforce this conclusion and bring some realism to the somewhat abstract nature of the discussion of component technology thus far, this section is devoted to several new chips which graphically illustrate the capabilities as discussed in the preceding section and which, as well, indicate the increased availability of components for signal processing.

**Memories**

It is perhaps axiomatic that the first manifestation of increased chip capacity is a higher volume memory. Memories are reasonably regular in structure and are absorbed by the market in increasing numbers. The availability of large high speed memories is of particular significance to signal processing. Here the manipulation of large arrays of sampled data in real time demands high performance memory.

In addition to the appearance of higher density (currently up to 64K bits) and higher speed (as low as 5 nanoseconds) memory components there are also available registers and multiplexers which make possible complex shifting and other bit manipulation functions. Also there now exist hardware structures which implement a variety of queueing disciplines (e.g., LIFO and FIFO). The immediate impact on processor architectures is the removal of the cost/bit memory constraints which has traditionally acted to inhibit the implementation of architectures dependent on high performance memories.

**Signal Processors**

In Section 7.1.6 the AMI S2811 signal processing chip was introduced. This chip represents only one of several LSI components aimed specifically at signal processing applications. These components range from high speed multiplier-accumulator chips to complete single chip processors for either general purpose signal processing or for special dedicated functions. Figure 7.17 lists several recently announced components for DSP. Several of these will be discussed in more detail in the design examples contained in Volume II.

At present single chip signal processors are primarily aimed at audio communication applications, whereas higher speed but less sophisticated components such as bit-sliced parts are utilized in higher performance applications.

**User Definable Chips**

In order to create the hardware required to implement a high performance system two extreme possibilities exist: first, to use existing components and second, to fabricate special purpose components. The latter choice involves such expense that it is currently a non-choice for most applications. The high performance bit-sliced microprocessors provide a programmable capability which extended the affordability of some systems. Two further developments are apparent which reduce the cost of creating special components: the gate array and the macrocell.

S2811
NEC PD 7720
NEC DSP chip set
TRW MAC
INTEL 2920
MMI (high speed multiplier)
Motorola Macro cells and Gate Arrays
Multiport memories
High speed memories
AMD 2900 bit slice family
Motorola 10800 ECL bit slice family

**Figure 7.17:** LSI components for DSP.

These devices consist of an array of basic elements which can be custom configured and interconnected by additional metal layers to perform a wide variety of functions. A prototype configuration can be specified and with extensive computer aided design assistance the required metal layer configuration can be automatically masked for production.

The basic component in a gate array is a four emitter input transistor, one output transistor (plus some clamping devices and a few resistors). As well, drivers for off-chip communications are provided. These provide a basic four input logic gate which in combinations can execute combinatorial functions or form registers of differing complexities.

The macro cell is an extension of the gate array. It is composed of larger functional blocks (rather than individual gates). In the current Motorola device, the basic component, called a major cell, is a D flip-flop. Typically 48 exist on a chip. Interface and output cells provide communication with the off-chip logic. These also can be configured in a variety of ways. By using high speed technology (ECL in the case of the Motorola device) the response of these devices can be extremely high despite the rigidity in layout. An 8-bit multiplier (giving a 16-bit product) for example has been designed with a response of 23 n.s.

The availability of the capability suggested here will significantly alter the ability of manufacturers to create signal processing equipment even for low volume applications. It remains to be seen how the fabrication costs of a chip responds to market demand.

### 7.2.4 Summary

The purpose of this section was to plot the growth of technology and to discuss its impact on signal processors.

Two aspects of the presentation are noteworthy. First the projections on capabilities are contingent on some high cost development work. A complex interaction of inhibiting factors may delay the attainment of some of the proposed goals. It should be easy to plot over the next decade the actual product announcements and to determine which factors are causing delays if any. Perhaps the key factor is the feature size as defined by the lithography capabilities. The dependence of all other factors on this capability was illustrated. It seems necessary for all designers to track and project lithography advances on an ongoing basis.

The second factor is the growth of components aimed specifically at signal processing. New signal processing components will tend to have characteristics which reflect the fabrication constraints imposed by packaging as well as the performance obtained from high density chips.

It will be necessary in the future to use processors designed in this environment and, at a more basic level, to design new processors which exploit the available components.

## 7.3 CHAPTER SUMMARY

The primary objective of this chapter has been to defend a proposition that high performance digital signal processors will be composed of a distributed array of high performance components. This was accomplished in two ways. The massive changes in progress were illustrated in two ways. In Section 7.1 a review of signal processors over a decade was used to demonstrate that functions are already appearing on programmable chips. These new chips represent the components used in the new era of processor design. To reinforce this, a review and projection of technology was presented in Section 7.2 which shows continued exponential improvements in all attributes which enhance performance. At a technical level the implications of a change in one attribute should be understood by any practitioner in this field.

We leave begging at this point the key question: How are such systems designed? It is the purpose of Volume II to address this. At the highest level this requires a design methodology which will create a processor from a set of specifications. This methodology must have many attributes, but as a minimum it must accommodate the distributed function environment introduced by technology.

Volume II undertakes three main goals: First to introduce the essence of design methodologies; second to provide a descriptive mechanism for both hardware and software, and, finally, to reinforce the new concepts by many examples.

## 7.4 EXERCISES AND PROJECTS

### Exercises

**1.** Consider the AP/130 operating in such a way as to occupy both the multiplier and adder at full utilization. If all data moves over the common bus, compute the total data bandwidth required to support the required flow (this problem is similar to problem exercise 6.2a).

**2.** Using the array instructions of the AP/130, program an $n \times n$ matrix multiplication. From your program estimate the multiplication time as a function of the component instructions. Assume a set of support instruction to set up this program (or obtain a programmer's manual).

**3.** Consider the problem of matrix inversion.

(a) Discuss an optimal inversion algorithm.

(b) Program your choice and estimate the execution time.

**4.** A recent measurement of a new microprocessor claimed 15000 gates and a 25-$MM^2$ chip operating with a clock speed of 200 n.s. Compute the gate-Hertz/$MM^2$.

Suppose that in a redesign it is hoped to double the gate density.

(a) What improvement on lithography is necessary?

(b) What is the new clock rate?

(c) What is the change in speed-power product?

(d) Compute the new gate-Hertz/$MM^2$.

## Mini-Projects

**5.** Obtain the programmer's reference manual for the S2811.

(a) Program a butterfly operation for an FFT.

(b) Estimate the time required to do a 1024-point FFT.

(c) How much of the program is devoted to data addressing and manipulation?

(d) Estimate the efficiency in terms of $(N/2 \log N) \times$ the butterfly time and the final time.

# Bibliography

This bibliography is intended as a guide to further study and to more detailed references. It is organized to reflect the sections of each chapter in order. The references are stated in two formats: for texts, [number, location] where "number" refers to the reference and "location" if present is normally a chapter, e.g., C-4; a section, S-7; or pages, e.g., pp. 14-17. Thus a complete reference might be [7, C-6, S-2, pp. 11-22] or perhaps just [8]; for papers or books from the supplementary reading list appear as [Thur, 78], where the first entry is the first four letters of the author's name and the two numbers refer to the date of publication.

## Part A: Signal Processing

The reader is reminded that a general overview of the early classical texts is contained in the introduction to Part A. These texts, [3], [6], [9], [10], [16], with the addition of [2], [7], [8] and [17], essentially cover, in considerable detail, all the major concepts discussed in Part A. For detailed discussions of DSP topics, a review of the IEEE Acoustics, Speech and Signal Processing Society Transactions and Annual Conference Proceedings should provide a virtual lifetime of study material.

### Chapter 1: Digital Signals

1.1 Basic Concepts
[6,C-1] [10, C-2] [16, C-2, C-3] [11, C-2]
Sampling Theorem
[16, C-2] [6, C-7] [10, C-2] [3, C-2] [Papo, 62, p. 50]

Analog to Digital Conversion
[16, C-19] [6, C-9] [ 10, C-5] [3, C-2] [Zuch, 70]
Digital to Analog Conversion
[10, C-5] [16, C-3]
1.2 Transform Analysis
Z-Transforms
[3, C-2] [6, C-2] [10, C-2] [16, C-4]
Hilbert Transforms
[6, C-7] [16, C-8] [10, C-2] [8, C-10]
Chirp Z-Transform
[6, C-6] [10, C-6] [3, p. 213]
1.3 A Statistical Signal Model
[8]: an excellent text on random variables and stochastic processes covering a range of topics from elementary probability theory to detailed problem analysis in the application of complex stochastic processes.
Discrete Random Signals
[6, C-8] [16, C-7]

For further reading and references on statistical signal analysis see [Davenport and Root, 1958], [Box and Jenkins, 1970], [Jenkins and Watts, 1968], [Whalen, 1975], [Wozencraft and Jacobs, 1965].

**Chapter 2: Linear Systems and Digital Filters**

2.1 Discrete Time Linear Systems
[10, C-2] [3, C-2] [6, C-1] [16, C-5] [2]
Response to Random Signals
[16, C-2] [6, C-8]
Two Dimensional Signal Processing
[6, C-1] [10, C-7]
2.2 Time Domain Representations of Linear Systems
Time Series Analysis
[Boje, 70] [Bhat, 72] [Robi, 67] [Olen, 72] [Rosi, 78] [Olen, 78]
State Space Representations
[16, C-6] [6, C-4] [2]
2.3 Digital Filters
[9] [3] [6] [10] [16] [2] [Ackr, 79] [Cadz, 73] [Liu, 75] [Terr, 80]
[Bozi, 79] [Homm, 77]
Frequency Domain Representation
[3, C-3] [6, C-1] [9, C-1] [10, C-3, C-4] [16, C-5]
Network Representation
[6, C-4] [16, C-8] [10, C-1]

State Space Representation
[16, C-6] [6, C-4] [2]

2.4 Digital Filter Design Techniques and Implementation Issues
Design of IIR Filters
[6, C-5] [10, C-4] [9, C-2] [3, C-3] [16, C-8]
Design of FIR Filters
[6, C-5] [10, C-3] [9, C-2] [3, C-3] [16, C-8]
Finite Word-Length Effects
[6, C-9] [10, C-5] [3, C-4] [16, C-9] [9, C-6]

**Chapter 3: Detection and Estimation**

3.1 Basic Concepts of Detection Theory
[17] [20]

3.2 Basic Concepts of Estimation Theory
[17] [20]

3.3 Linear Minimum Mean Square Error Estimation
[16, C-12, C-13, C-14] [17] [20] [8] [7, C-7]
Hilbert Spaces [16, C-13]
Linear Parameter Estimation [16, C-14] [17]
Kalman Filtering [Kalm, 60] [KaBu, 61] [16, C-14]
[17, C-4] [Gelb, 74]
Smoothing, Filtering, and Prediction
[8, C-11] [7, C-7]

3.4 Spectral Estimation
Conventional Spectral Analysis
[16, C-11] [6, C-11] [3, C-6] [10, C-6] [8] [7, C-6]
Autoregressive Spectral Analysis
[16, C-11] [Burg, 67]

An excellent coverage on the comparison of conventional and autoregressive spectral analysis techniques is given by [Mark, 1976].

**Chapter 4: Digital Signal Processing Algorithms and Techniques**

4.1 The Fast Fourier Transform
[9, C-3] [10, C-5, C-10] [6, C-6] [3, C-6, C-7] [16, C-10]

4.2 Generalized Linear Filtering
[3, C-8] [6, C-1]

4.3 Data Compression
[7]

## Chapter 5: From Processing to Processor

5.1 Applications of Digital Signal Processing
[7]

# PART B: Signal Processors

## Chapter 6:

- The history of the development of computing machinery is facinating reading and is well documented: [4] [ROSE, 69], [Reid, 79], [Will, 78], [Penn, 78], [5, C-1].
- The origins of the von Neumann architecture were published in a report, co-authored by A.W. Burks and H.H. Goldstine, for the U.S. Army Ordinance Department in 1946: ("Preliminary Discussion of the Logical Design of an Electronic Computing Instrument"). It has been widely reprinted, e.g., [1, pp 92-119], [13, pp 221-259], [14].
- Architectural classification schemes are discussed in [Flyn, 66], [Hand, 77]. An overview is contained in [15].
- The ILLIAC-IV is described in many books. [5] contains a brief discussion; [12, C-8] contains more material with emphasis on interconnection topologies; [15] is devoted to SIMD machines including the ILLIAC-IV and others. The original paper was by [Barn, 68].
- The TI-ASC is described by [Wats, 72], and by [Thur, 79] which also contains a wide ranging discussion of high performance processors.
- Performance measures for computers are discussed in [12, C-11].
- A tutorial introduction to pipelines and parallel arrays is given in [5, C-3], [12, C-9]. A comparison of processors using these techniques is available in [Hufn, 79].

## Chapter 7:

- Technology forecasts are widely available. Our references are limited to 1979. These include projections over the 1980's: [Keye, 79], [Alla, 79], [Moor, 79], [Mack, 78].
- A collection of reprints which contain a wealth of processor architectures, comparisons, and tutorial material is contained in [Sala, 77].
- The FDP and LSP2 are discussed in [10, C-11]; of particular interest is a discussion of the FFT using the FDP. This reference also contains some useful overviews on how to approach the design problem in general.

# References

## TEXTS

[1] BELL, C.G., A. NEWELL, *Computer Structures, Readings and Examples*, McGraw-Hill, New York, 1971.

[2] DERUSSO, P.M., R.J. ROY, C.M. CLOSE, *State Variables for Engineers*, John Wiley, New York, 1965.

[3] GOLD, B., C.M. RADER, *Digital Processing of Signals*, McGraw-Hill, New York, 1969.

[4] GOLDSTINE, H.H., *The Computer from Pascal to von Neumann*, Princeton University Press, Princeton, N.J., 1972.

[5] HAYES, J.P., *Computer Architecture and Organization*, McGraw-Hill, New York, 1978.

[6] OPPENHEIM, A.V., P.W. SCHAFFER, *Digital Signal Processing*, Prentice-Hall, Englewood Cliffs, N.J., 1975.

[7] OPPENHEIM, A.V. *Applications of Digital Signal Processing*, Prentice-Hall, Englewood Cliffs, N.J., 1978.

[8] PAPOULIS, A., *Probability, Random Variables and Stochastic Processes*, McGraw-Hill, New York, 1965.

[9] PELED, A., B. LIU, *Digital Signal Processing: Theory, Design and Implementation*, John Wiley, New York, 1976.

[10] RABINER, L.R., J.B. GOLD, *Theory and Applications of Digital Signal Processing*, Prentice-Hall, Englewood Cliffs, N.J., 1975.

[11] RABINER, L.R., R.W. SCHAFER, *Digital Processing of Speech Signals*, Prentice-Hall, Englewood Cliffs, N.J., 1978.

[12] STONE, H.S. (editor), *Introduction to Computer Architecture*, SRA Inc., Chicago, Ill., 1975.

[13] SWARTZLANDER, E.E. (editor), *Computer Design Development, Principal Papers*, Hayden, Rochelle Park, N.J., 1976.

[14] TAUB, A.H. (editor), *J. von Neumann Collected Works, Vol. 5: Design of Computers, Theory of Automata and Numerical Analysis*, Pergamon Press, New York, 1963.

[15] THURBER, K.J., *Large Scale Computer Architecture*, Hayden, Rochelle Park, N.J., 1976.

[16] TRETTER, S.A., *Introduction to Discrete Time Signal Processing*, John Wiley, New York, 1976.

[17] VanTREES, H.L., *Detection, Estimation, and Modulation Theory, Part I*, John Wiley, New York, 1968.

[18] VanTREES, H.L., *Detection, Estimation and Modulation Theory, Part II*, John Wiley, New York, 1971.

[19] VanTREES, H.L., *Detection, Estimation and Modulation Theory, Part III*, John Wiley, New York, 1971.

[20] WHALEN, A.D., *Detection of Signals in Noise*, Academic Press, New York, 1975.

## SUPPLEMENTARY BOOKS AND TEXTS

The following set of books were not explicitly referenced; however they form valuable references and also indicate the rapid growth in available material over the last decade.

### Pre-1970

BLACKMAN, R.B., J.W. TUKEY, *The Measurement of Power Spectra from the Point of View of Communications Engineering*, Dover Publications, New York, 1958.

BLACKMAN, R.B., *Linear Data Smoothing and Prediction in Theory and Practice*, Addison-Wesley, Reading, Mass., 1965.

COOLEY, J.W., J.W. TUKEY, *An Algorithm for the Machine Calculation of Complex Fourier Series, Mathematics of Computation, Vol. 19*, April 1965, pp. 297-301.

DAVENPORT, W.B., W. ROOT, *An Introduction to the Theory of Random Signals and Noise*, McGraw-Hill, New York, 1958.

DERUSSO, P.M., R.J. ROY, C.M. CLOSE, *State Variables for Engineers*, John Wiley, New York, 1965.

FORSYTHE, G.E., C.B. MALER, *Computer Solution of Linear Algebraic Systems*, Prentice-Hall, Englewood Cliffs, N.J., 1967.

GOLD, B., C.M. RADAR, *Digital Processing of Signals*, McGraw-Hill, New York, 1969.

HORTON, C.W., Sr., *Signal Processing of Underwater Acoustic Waves*, U.S. Government Printing Office, Washington, D.C., 1969.

JURY, E.I., *Sampled-Data Control Systems*, John Wiley, New York, 1958.

JURY, E.I., *Theory and Application of the Z-Transform Method*, John Wiley, New York, 1964.

KALMAN, R.E., "A New Approach to Linear Filtering and Prediction Problems," *Trans. of the ASME, Journal of Basic Engineering*, Vol. 82D, March 1960, pp. 34-45.

KALMAN, R.E., R. BUCY, "New Results in Linear Filtering and Prediction Theory," *Trans. of the ASME, Journal of Basic Engineering*, Vol. 83D, March 1961, pp. 95-108.

JENKINS, G.M., D.G. WATTS, *Spectral Analysis and Its Applications*, Holden Day, San Francisco, 1968.

KUO, F.F., J.F. KAISER (editors), *System Analysis by Digital Computer*, John Wiley, New York, 1966.

MONROE, A.J., *Digital Processes for Sampled Data Systems*, John Wiley, New York, 1962.

MORRISON, N., *Introduction to Sequential Smoothing and Prediction*, McGraw-Hill, New York, 1969.

PAPOULIS, A., *Probability, Random Variables and Stochastic Processes*, McGraw-Hill, New York, 1965.

PAPOULIS, A., *The Fourier Integral and Its Applications*, McGraw-Hill, New York, 1962.

RAGAZZINI, J.R., G.F. FRANKLIN, *Sampled Data Control Systems*, McGraw-Hill, New York, 1958.

ROBINSON, E.A., *Multichannel Time Series Analysis with Digital Computer Programs*, Holden Day, San Francisco, 1967.

WOSENCRAFT, J.M., I.M. JACOBS, *Principles of Communication Engineering*, John Wiley, New York, 1965.

## 1970

ANDREWS, H.C., *Computer Techniques in Image Processing*, Academic Press, New York, 1970.

BOX, G.E., G. JENKINS, *Time Series Analysis, Forecasting and Control*, Holden Day, San Francisco, 1970.

## 1972

ANDREWS, H.C., *Introduction to Mathematical Techniques in Pattern Recognition*, John Wiley, New York, 1972.

BHAT, U.N., *Elements of Applied Stochastic Processes*, John Wiley, New York, 1972.

FLANAGAN, J.L., *Speech Analysis, Synthesis and Perception*, Springer-Verlag, New York, 1972.

OTNES, R.K., L. ENOCHSON, *Digital Time Series Analysis*, John Wiley, New York, 1972.

RABINER, L.R., C.N. RADER, *Digital Signal Processing*, John Wiley, New York, 1972.

## 1973

AKROYD, M.H., *Digital Filters*, Butterworth, London, 1973.

BEAUCHAMP, K.G., *Signal Processing Using Analog and Digital Techniques*, Allen and Unwin, Ltd., London, 1973.

CADZOW, J.A., *Discrete-Time Systems*, Prentice-Hall, Englewood Cliffs, N.J., 1973.

FLANAGAN, J.L., L.R. RABINER (editors), *Speech Synthesis*, Academic Press, New York, 1973.

JURY, E.I., *Theory and Applications of the Z-Transform Method*, Krieger, New York, 1973.

## 1974

BRIGHAM, E.O., *The Fast Fourier Transform*, Prentice-Hall, Englewood Cliffs, N.J., 1974.

STEIGLITZ, K., *An Introduction to Discrete Systems*, John Wiley, New York, 1974.

## 1975

AHMED, N., K.R. RAO, *Orthogonal Transforms for Digital Signal Processing*, Springer-Verlag, New York, 1975.

BOGNER, R.E., A.G. CONSTANTINIDES (editors), *Introduction to Digital Filtering*, John Wiley, New York, 1975.

CHILDERS, D., A. DURLING, *Digital Filtering and Signal Processing*, West Publishing, St. Paul, Minn., 1975.

LIU, B. (editor), *Digital Filters and the Fast Fourier Transform*, Dowden, Hutchinson and Ross, Stroudsburg, Pa., 1975.

OPPENHEIM, A.V., R.W. SCHAFFER, *Digital Signal Processing*, Prentice-Hall, Englewood Cliffs, N.J., 1975.

RABINER, L.R., B. GOLD, *Theory and Application of Digital Signal Processing*, Prentice-Hall, Englewood Cliffs, N.J., 1975.

REDDY, D.R. (editor), *Speech Recognition*, Academic Press, New York, 1975.

SCHWARTZ, M., L. SHAW, *Signal Processing: Discrete Spectral Analysis, Detection and Estimation*, McGraw-Hill, New York, 1975.

STANLEY, W.D., *Digital Signal Processing*, Reston, Reston, Va., 1975.

STEARNS, S.D., *Digital Signal Analysis*, Hayden, Rochelle Park, N.J., 1975.

**1976**

CLAERBOUT, J.F., *Fundamentals of Geophysical Data Processing,* McGraw-Hill, New York, 1976.

FU, K.S. (editor), *Digital Pattern Recognition*, Springer-Verlag, New York, 1976.

JOHNSON, D.E., *Introduction to Filter Theory*, Prentice-Hall, Englewood Cliffs, N.J., 1976.

PELED, A., B. LIU, *Digital Signal Processing: Theory, Design and Implementation*, John Wiley, New York, 1976.

TRETTER, S.A., *Introduction to Discrete-Time Signal Processing*, John Wiley, New York, 1976.

MARKEL, J.D., and A.H. GRAY, Jr. (editors), *Linear Prediction of Speech*, Springer-Verlag, New York, 1976.

TAYLOR, F., S.L. SMITH, *Digital Signal Processing in FORTRAN*, D.C. Heath & Co., Lexington, Mass., 1976.

**1977**

AGGARWAL, J.K., et al. (editors), *Computer Methods in Image Analysis*, IEEE Press, New York, 1977.

ANDREWS, H.C., B.R. HUNT, *Digital Image Restoration*, Prentice-Hall, Englewood Cliffs, N.J., 1977.

HAMMING, R.W., *Digital Filters*, Prentice-Hall, Englewood Cliffs, N.J., 1977.

ROBERTS, R., *Signal Processing Techniques*, Interstate Electronics Corporation, Anaheim, Ca., 1977.

SALAZER, A.C. (editor), *Digital Signal Computers and Processors*, IEEE Press, New York, 1977, and John Wiley, New York, 1977.

SMITH, J.M., *Mathematical Modeling and Digital Simulation for Engineers and Scientists*, John Wiley, New York, 1977.

## 1978

CAPELLINI, V., A.G. CONSTANTINIDES, P. EMILIANI, *Digital Filters and Their Applications*, Academic Press, London, 1978.

CHILDERS, D.G. (editor), *Modern Spectrum Analysis*, IEEE Press, New York, 1978.

KUCK D., *The Structure of Computer and Computations*, John Wiley, New York, 1978.

OPPENHEIM, A.V. (editor), *Applications of Digital Signal Processing*, Prentice-Hall, Englewood Cliffs, N.J., 1978.

OTNES, R.K., L. ENOCHSON, *Applied Time Series Analysis*, John Wiley, New York, 1978.

ROBINSON, E.A., M. SILVIA, *Digital Signal Processing and Time Series Analysis*, Holden Day, San Francisco, 1978.

## 1979

AGGARWAL, J.K., *Digital Signal Processing*, Western Periodicals Co., North Hollywood, Ca., 1979.

ANTONIOU, A., Digital Filters: *Analysis and Design*, McGraw-Hill, New York, 1979.

BEAUCHAMP, K.G., C.K. YUEN, *Digital Methods for Signal Analysis*, Allen & Unwin, Winchester, Mass., 1970.

BOZIC, S.M., *Digital and Kalman Filtering*, John Wiley, New York, 1979.

CASTLEMAN, K.R., *Digital Image Processing*, Prentice-Hall, Englewood Cliffs, N.J., 1979.

DIXON, N.R., T.B. MARTIN, *Automatic Speech and Speaker Analysis*, IEEE Press, New York, 1979.

EVANS, C., *The Micro Millennium*, Viking Press, New York, 1979.

HUANG, T.S. (editor), *Picture Processing and Digital Filtering*, Springer-Verlag, New York, 1979.

McCLELLAN, J.H., C.M. RADER, *Number Theory in Digital Signal Processing*, Prentice-Hall, Englewood Cliffs, N.J., 1979.

SCHAFER, R.W., J.D. MARKEL (editors), *Speech Analysis*, IEEE Press, New York, 1979.

SRINATH, M.D., P.K. RAJASEKARAN, *An Introduction to Statistical Signal Processing with Applications*, John Wiley, New York, 1979.

TRIBOLET, J.M., *Seismic Applications of Homomorphic Signal Processing*, Prentice-Hall, Englewood Cliffs, N.J., 1979.

WILLSKY, A.S., *Digital Signal Processing and Control and Estimation Theory*, MIT Press, Cambridge, Mass., 1979.

## 1980

BENDAT, J.S., A.G. PIERSOL, *Engineering Applications of Correlation and Spectral Analysis*, John Wiley, New York, 1980.

LEA, W.A. (editor), *Trends in Speech Recognition*, Prentice-Hall, Englewood Cliffs, N.J., 1980.

TERRELL, T.J., *An Introduction to Digital Filters*, John Wiley, New York, 1980.

## PAPERS

ALLAN, R. "VLSI: Scoping Its Future," *IEEE Spectrum*, April 1979, pp. 30-37.

BARNES, et al, "The ILLIAC-IV Computer," *IEEE Trans. on Computers*, Aug. 1968, C-17, pp. 746-757.

BERNHARDT, P.A., "Digital Processing of Ionospheric Electron Content Data," *IEEE Transactions on Acoustics, Speech and Signal Processing*, Vol. 27, No. 6, December 1979, pp. 705-712.

BURG, J.P., *Maximum Entropy Spectral Analysis*, presented at 37th Meeting of the Society of Exploration Geophysicists, Oklahoma City, Okla., 1967.

CAPECE, R.P., "Tackling the Very Large Scale Problems of VLSI: A Special Report," *Electronics*, Nov. 23, 1979, pp. 111-125.

COOLEY, J.W., J.W. TUKEY, "An Algorithm for the Machine Calculation of Complex Fourier Series," *Mathematics of Computation*, Vol. 19, April 1965, pp. 297-302.

DURNIAK, A., "VLSI Shakes the Foundations of Computer Architecture," *Electronics*, May 24, 1979, pp. 111-133.

FAGGIN, F., "How VLSI Impacts Computer Architecture," *IEEE Spectrum*, March 1978, pp. 54-56.

FLYNN, M.J., "Very High Speed Computing Systems," *Proc. of the IEEE*, Vol. 54, Dec. 1966, pp. 1901-1909.

GELB, A. (editor), *Applied Optimal Estimation*, MIT Press, Cambridge, Mass.

GOLD, B., et al, "The FDP, A Fast Programmable Signal Processor," *IEEE Trans. Comput.*, Vol. C-20, pp. 33-38.

GOLDSTINE, H.H., *The Computer from Pascal to von Neumann*, Princeton University Press, Princeton, N.J., 1972.

HANDLER, N., *The Impact of Classification Schemes on Computer Architecture*, Parallel Processing Symposium, Aug. 1977.

HUFNAGEL, S.P., "Comparison of Selected Array Processor Architectures," *Computer Design*, March 1979, pp. 151-158.

IEEE Digital Signal Processing, *Proc. IEEE* (Special Issue), Vol. 63, April 1975.

KEYES, R.W., "The Evolution of Digital Electronics Towards VLSI," *IEEE Transactions on Electron Devices*, Vol. ED-26, No. 4, April 1979.

KLASS, P.J., "GE Emphasizing Complex Microcircuits," *Aviation Week and Space Technology*, March 26, 1979, pp. 62-67.

KOZDROWICKI, E.W., D.J. THEIS, "Second Generation of Vector Super Computers," *Computer*, Vol. 13, No. 11, Nov. 1980, pp. 71-83.

LIPOVSKI, G.J., K.L. DOTY, "Developments and Directions in Computer Architecture," *Computer*, Aug. 1978, pp. 54-67.

LYMAN, J., "Lithography Chases the Incredible Shrinking Line," *Electronics*, April 12, 1979, pp. 105-116.

MacKINTOSH, I., "Large Scale Integration: Intercontinental Aspects," *IEEE Spectrum*, June 1978, pp. 51-56.

MOORE, G., "VLSI: Some Fundamental Challenges," *IEEE Spectrum*, April 1979, pp. 30-37.

PENNIMAN, C.F., "Philadelphia's 179 Year Old Android," *Byte*, August 1978, pp. 90-94.

POMERANZ, J., R. Nijhuis, C. Vicary, "Customized Metal Layers Vary Standard Gate-Array Chip," *Electronics*, March 15, 1979, pp. 105-108.

PRIOSTE, J., R. RAO, W.R. Blood, Jr., "Function Array Eases Custom ECL Design," *Electronics*, Feb. 15, 1979, pp. 113-117.

REID-GREEN, K.S., "A Short History of Computing," *Byte, Comput. Surv.*, Vol. 1, March 1969, pp. 7-36.

ROSEN, S., "Electronic Computers: A Historical Survey," *Byte, Comput. Surv.*, Vol. 1, March 1969, pp. 7-36.

STOCKMAN, T., "The Application of Generalized Linearity to Automatic Gain Control," *IEEE Trans. Audio Electroacoustics*, Vol. AU-16, June 1968, pp. 267-279.

TEXAS INSTRUMENTS, INC., *The ASC System Central Processor*, Publ. H. 1005P, Austin, Tx. 1973.

THURBER K.J., "A Parallel Processor Architecture—Part 1: General Purpose Systems, and Part 2: Special Purpose Systems," *Computer Design*, Jan. 1979, pp. 89-97, Feb. 1979, pp. 104-144.

WATSON, W.J., "The TI ASC—A Highly Modular and Flexible Super Computer Architecture," *Proc. AFIPS*, 1972 FJCC, Vol. 41, pp. 221-228, AFIPS Press, Montvale, N.J., 1972.

WEINSTEIN, C. Round-off Noise in Floating Point Fast Fourier Transform Computation, *IEEE Trans. Audio Electroacoust.*, Vol. AU-17, Sept. 1967, pp. 209-215.

WELCH, P.D., A Fixed-Point Fast Fourier Transform Error Analysis, *IEEE Trans. Audio Electroacoust.*, Vol. AU-17, June 1969, pp. 151-157.

WIDROW, B. et al., "Stationary and Nonstationary Learning Characteristics of the LMS Adaptive Filter," *Proc. IEEE*, Vol. 64, No. 8, Aug. 1976.

WILLIAMS, J.M., "Antique Mechanical Computers, Part I: Early Automata, Part II: 18th and 19th Century Mechanical Marvels," *Byte*, July 1978, pp. 48-58.

ZUCH, E.L., "Interpretation of Data Converter Accuracy Specifications," *Computer Design*, September 1970, pp. 113-121.

# Index